Reading Randomness I

Reading Randomness I: New Horizons for Understanding Fractal and Fractional Calculus shows readers how to understand and analyze seemingly "noisy" and irregular data. It offers practical methods for comparing and labeling fluctuations, ensuring reproducible results across laboratories, industries, and research fields.

The book begins with a narrative tour of how people have understood chance, from ancient divination and games to modern probability, information theory, and computing. In seven chapters, the authors provide step-by-step instructions for extracting reliable patterns from raw measurements. These instructions include removing trends, checking whether a signal has long-term memory, and identifying stable patterns hidden inside apparent randomness. The book also presents new tools for analyzing "trendless" sequences, extending Fourier-style analysis to complicated, multi-period data, and measuring correlations in a way that distinguishes between contributions from the system itself and the environment. Using a reader-friendly approach, the authors explain how "memory" kernels capture slow, history-dependent behavior. Throughout the book, the authors emphasize independent checks, surrogate tests, and instrument-path corrections to ensure that conclusions can be reliably and safely transferred across place and time.

With case studies ranging from transcendental numbers and electrical circuits to earthquake records and precious metal price data, this valuable guidebook will appeal to students, researchers, and professionals working with complex data in science, engineering, and finance.

Raoul R. Nigmatullin is Professor at the Radioelectronics and Informative Measurements Technics Department, Kazan National Research Technical University, named after A.N. Tupolev. His research interests include dielectric spectroscopy in electrochemistry and new treatment methods in radiometry.

YangQuan Chen, a Clarivate Highly Cited Researcher, is Professor at the School of Engineering, University of California, Merced. His research interests include smart control engineering via digital twins and applied fractional calculus in STEM.

Fractional Order Thinking in Exploring the Frontiers of STEM

Fractional Order Thinking (FOT) is about solving today's complex problems in the physical, social, and life sciences using the tools of fractional order calculus (FOC). Very soon after Mandelbrot introduced the fractal paradigm into the scientific lexicon, it was shown that the integer order calculus (IOC) could not describe the dynamics of fractal processes. A new kind of calculus was required to construct the equations of motion for fractal dynamic processes, which turned out to be fractional order Hamiltonian equations (FOHEs). The FOHEs are one tool in the FOC toolbox which concerns how to apply the operators of differentiation and integration of non-integer orders. Rejecting the fractional calculus is equivalent to saying there are no numbers between neighboring integers.

In this book series, we explore the core motivation of fractional calculus by first showing the "core motivation" of IOC invented by Newton and Leibnitz, the fundamental ideas of which can be traced back to the time of Heraclitus of Ephesus. Our ultimate message is that the "IOC" is driven by "the desire and the need" for the "quantification of changes" based on energy gradients in complex dynamic networks, whereas "FOC" is driven by "the desire and the need for understanding complexity" based on information gradients in those same networks. In science, technology, engineering, and mathematics (STEM), metaphorically speaking, there is plenty of room between the integers to enable the archiving of better-than-best modeling, control performance, robustness, resilience, and even intelligence.

There is an increasing interest in fractional order dynamic systems (FODS) and controls in the recent research literature, not only because of their novelty but also due to their practical applications. But to accomplish all of this requires a new way of thinking, the Fractional Order Thinking (FOT) referred to above, which in turn must be preceded by a new STEM curriculum. This series will offer a unique platform to demonstrate such additional benefits in our improved understanding of complexity and stochastic dynamics via FOT.

This exciting book series features:

- A forum demonstrating the good consequences of using fractional order thinking (FOT).
- Fractional calculus concepts are presented in the context of complex systems characterization while assuming minimal background in math and physics.
- A broad audience including professionals across many fields, the general public, and courses in colleges and even high schools.
- Wide STEM topics ranging from batteries to medicine, with a writing style easy to follow.
- Short and inexpensive books of 120–150 pages main text that can be written and read in a reasonable amount of time.

Please contact the series editors, Bruce J. West (North Carolina State University, USA) and YangQuan Chen (University of California Merced), and Taylor & Francis Publisher Lian Sun (Lian.Sun@informa.com), if you have an idea for a book for the series.

Titles in the series currently include:

Fractional Calculus for Skeptics I
The Fractal Paradigm
Bruce J. West and YangQuan Chen

On the Fractal Language of Medicine
Bruce J. West and W. Alan C. Mutch

Fractional Order Intelligent Modeling for Lithium-Ion Batteries
Theory and Practice
YaNan Wang and YangQuan Chen

Fractional Random Vibrations I
Theories
Ming Li

Fractional Random Vibrations II
Applications
Ming Li

Fractional Calculus for Skeptics II
Quantifying Roughness
Bruce J. West and YangQuan Chen

Reading Randomness I
New Horizons for Understanding Fractal and Fractional Calculus
Raoul R. Nigmatullin and YangQuan Chen

Reading Randomness I

New Horizons for Understanding Fractal and Fractional Calculus

Raoul R. Nigmatullin and YangQuan Chen

CRC Press
Taylor & Francis Group
Boca Raton London New York

CRC Press is an imprint of the
Taylor & Francis Group, an **informa** business
A CHAPMAN & HALL BOOK

Designed cover image: Shutterstock (No. 2320497767)

First edition published 2027
by CRC Press
2385 NW Executive Center Drive, Suite 320, Boca Raton FL 33431

and by CRC Press
4 Park Square, Milton Park, Abingdon, Oxon, OX14 4RN

CRC Press is an imprint of Taylor & Francis Group, LLC

ISBN: 978-1-041-34814-6 (hbk)
ISBN: 978-1-041-34827-6 (pbk)
ISBN: 978-1-003-79469-1 (ebk)

DOI: 10.1201/9781003794691

Typeset in Latin Modern font
by KnowledgeWorks Global Ltd.

Publisher's note: This book has been prepared from camera-ready copy provided by the authors.

To our mentors and families

Contents

Preface

Randomness is not noise—it is structure waiting to be read.
Raoul R. Nigmatullin

God made the integers, all else is the work of man.
Leopold Kronecker (1823–1891)

Reading the unseen: the science behind randomness and complexity.
YangQuan Chen

Modern science was built on the premise that if we observe carefully and model faithfully, the world will reveal its rules. Randomness met in Nature represents itself the "50 shades of gray" metaphor. Yet much of what matters arrives as fluctuations: jitter in a sensor, tremors of the earth, voltage flickers in an electrochemical cell, or the erratic rhythms of a living organism. Are these fluctuations merely nuisance "noise," or are they bearing intelligible information about mechanisms and memory? This book takes a clear stand: by linking fractal geometry to fractional calculus and by practicing *fluctuation metrology*, we can *read* randomness and turn apparent disorder into insight.

This volume is our invitation to that practice. It offers a grammar for randomness, a set of methods that distill robust signatures from messy time series, and a set of case studies that demonstrate how those signatures travel across domains. We aim for a book that is rigorous enough for researchers yet navigable by motivated newcomers from engineering, physics, materials science, biosciences, and data science.

In late 2021, while the world was still adapting to the constraints of a pandemic, one of us (YQC) invited the other (RRN) to co-offer a short course titled "New Signal Processing," hosted by colleagues at Xi'an Technological University. The course brought together more than 50 registered students and a much wider online audience. The energy of those exchanges—and, frankly, the urgency of better tools for complex data—convinced us to consolidate scattered methods into a coherent, application-driven text. *Reading Randomness I: New Horizons for Understanding Fractal and Fractional Calculus* grew from that seed. Our goal is practical: provide workflows that yield comparable, explainable labels for fluctuations across very different systems, and connect those labels to compact models with memory.

We use "reading randomness" deliberately. Reading emphasizes *extracting robust signatures*—scaling exponents and Hurst indices; nonlocal memory kernels; sparse oscillatory atoms; and compact correlation measures—that summarize the essential behavior of a signal. Reading does not preclude explanation; instead, it builds the disciplined summaries that good explanations require. In practice, this means moving from raw records to curated signals; from exploratory diagnostics to model families; from integer-order baselines to fractional kernels; and from oversmoothed spectra to stable, sparse decompositions. Throughout, we stress cross-validation, surrogate tests, and apparatus-function correction—because a signal is only as good as the pathway that generated it.

Reading Randomness I is organized as follows:

- *Chapter 1* lays the conceptual foundation: working notions of randomness from probability, information, and algorithmic complexity; a brief cross-cultural history of chance; and modern views emphasizing long memory, self-similarity, and fractional operators. It closes with a practical workflow for going from raw time series to interpretable structure, and it distinguishes "reading" from "explaining".

- *Chapter 2* introduces **Discrete Geometrical Invariants (DGI)** as a tool for comparing trendless sequences (TLS). The 3D-DGI method is developed and illustrated with three applications that foreshadow the book's cross-domain reach: distinguishing transcendental numbers, characterizing earthquake signals, and comparing trendless sequences from electric circuits. We also outline the CAPoNeF (Comparative Analysis of Positive and Negative Fluctuations) alternative.

- *Chapter 3* presents **NOCFASS**—Non-Orthogonal Combined Fourier Analysis of Smoothed Signals—which extends classical Fourier analysis to multi-periodic, long-memory data. Here, we emphasize stable atom extraction using band selection and smoothing, and we test the method on mimic data, economic time series (e.g., precious-metal prices), sequences derived from irrational numbers, and geophysical records.

- *Chapter 4* develops **generalized correlations** that complement spectral tools. We separate internal from external correlations, derive an algorithm for estimating them robustly, discuss the relative character of correlation, and demonstrate verification on representative mimic and real datasets.

- *Chapter 5* moves to **generalized methods using fractional calculi**. We introduce a reader-friendly "fuzzy" calculus, show how memory kernels emerge from averaging on fractal-like structures, and verify the key expressions on mimic and real data. We highlight the role of Prabhakar/Mittag–Leffler families for modeling anomalous relaxation and point to their practical implementation.

- *Chapter 6* revisits dynamics from a **hereditary** point of view. We ask when—and how—Newtonian-style balance laws with memory can be derived, show partial cases and generalizations, and connect the theory to generalized Hurst exponents and recurrence solutions.

- *Chapter 7* collects **conclusions and future opportunities**, including open problems, challenges to the community, and ideas for hybrid integer–fractional modeling, data-driven kernel discovery, and benchmark datasets.

It is interesting to point out that the readers will encounter three recurring pillars in *Reading Randomness I*:

1. *Fractals → Fractional operators.* Fractal geometry suggests averaging on nonlocal sets, leading naturally to fractional integrals/derivatives whose kernels encode memory. This operator view helps unify disparate observations: stretched-exponential relaxations, anomalous diffusion, and long-range dependence.

2. *Fluctuation metrology and sparse spectra.* Many complex signals mix long-memory backgrounds with quasi-periodic components. By pre-smoothing and seeking sparse atoms ("Prony-like" exponentials/sinusoids), we obtain stable *amplitude–frequency responses* that act as fingerprints. In some contexts, a reliable Prony spectrum serves as a compact surrogate for a full probability distribution.

3. *Generalized correlations.* Correlations, carefully defined and robustly estimated, provide complementary summaries that survive across domains and sampling regimes. They help distinguish intrinsic structure from apparatus-induced artifacts.

These pillars are tied together by a pragmatic workflow: instrument awareness and curation → exploratory reading → baseline modeling → fractional kernels and sparse atoms → validation and uncertainty. We repeatedly return to this workflow in the case studies.

If you work with signals from complex systems—in laboratories, in the field, or in silico—we wrote this for you. The text balances intuition and mathematics with deliberately minimal prerequisites. Chapters can be read independently, but we recommend starting with the conceptual map in Chapter 1, then sampling the method Chapters (2–6) according to your needs. Instructors will find natural units for short courses or graduate modules; practitioners can reuse the step-by-step recipes and checklists.

We do not claim that fractional models or Prony spectra explain everything. They do not. Our aim is more modest and, we hope, more durable: provide *readings* that are stable, comparable, and mechanistically suggestive, across electrochemistry and materials, biosignals and plants, earthquakes and infrastructure, communications and finance. Explanations should be tested against such readings, not the other way around. Where appropriate, we point to alternative baselines (e.g., ARFIMA, DFA/MFDFA, matrix-pencil methods) and emphasize model checks and surrogate-data tests.

Acknowledgments. This book reflects decades of conversations with students, colleagues, and collaborators across continents and disciplines. We are grateful to the organizers and participants of the 2021 short course that catalyzed this project; to our home institutions for intellectual shelter; and to the CRC Press book series FOT4STEM [1] community for championing cross-disciplinary thinking about memory and complexity. Most of all, we thank our families and mentors for their steady support while we wrote amid uncertain times. In particular, in the writing of Sec. 3.6 we received essential help from the geophysicists Dr. Ksenia S. Nepeina and Pavel A. Kaznacheev. Thanks go to Dr. Irina Chelnokova from the A.E. Arbuzov Institute of Organic and Physical Chemistry (Kazan, Russia) for offering electrochemical data used for analysis presented in Secs. 4.3–4.7.

YQ wishes to thank the funding support from the Center for Methane Emission Research and Innovation (CMERI) through the Climate Action Seed Funds grant (2023–2026) at the University of California, Merced. During preparation of this book we used `www.Overleaf.com` with LaTeX files converted from MS Word documents via `www.docx2latex.com`. During this conversion process, the following students were very helpful: Drs. Furkan Guc and Justus Nwoke as well as PhD candidate Mr. Osama Abdel Aal. Thank you for your hard and careful TeXnician work in 2022.

The authors would like to thank Ms. Xiaoyin Feng, Editorial Assistant, Routledge & CRC Press, Taylor & Francis Group, for her professional coordination of the book project; Ms. Lian Sun, Publisher, Head of China Books Publishing and International Cooperation Taylor & Francis Advanced Learning for her professional vision and effort in establishing this FOT4STEM book series to which this book belongs.

DECLARATION OF AI ASSISTANCE

During the preparation of this work we made judicious use of large language models (ChatGPT 5.3 and Gemini 3 Pro) for drafting, editing, and organization. We reviewed and revised all content and bear full responsibility for its accuracy and originality.

Kazan, Tatarstan, Russia
Merced, California, USA

Raoul R. Nigmatullin
YangQuan Chen
Feb. 2026

[1] `https://mechatronics.ucmerced.edu/fot4stem`

Acknowledgments

With this book, our intent is to provide a comprehensive summary of our research efforts during the past years. For that reason, this text contains materials from papers and articles that were previously published. We are thankful and would like to acknowledge the copyright permissions from Elsevier who have released our work on that topic:

©Elsevier. R. R. Nigmatullin. Recognition of nonextensive statistical distributions by the Eigen-coordinates method. Physica A: Statistical Mechanics and its Applications. 285(3-4):547–565, 2000.

©Elsevier. Raoul R Nigmatullin and Alain Le Mehaute. Is there geometrical/physical meaning of the fractional integral with complex exponent? Journal of Non-Crystalline Solids, 351(33-36):2888–2899, 2005.

©Elsevier. Raoul R Nigmatullin. The statistics of the fractional moments: Is there any chance to "read quantitatively" any randomness? Signal Processing, 86(10):2529–2547, 2006.

©Elsevier. Nigmatullin R, Toboev V, Lino P, and Maione G. Reduced fractal model for quantitative analysis of averaged micro motions in mesoscale: Characterization of blow-like signals. Chaos, Solitons & Fractals, 76:166–181, 2015.

©Elsevier. Nigmatullin Raoul, Maione Guido, Lino Paolo, Saponaro Fabrizio, and Zhang Wei. The general theory of the quasi-reproducible experiments: How to describe the measured data of complex systems? Communications in Nonlinear Science and Numerical Simulation, 42:324–341, 2017.

©Elsevier. Raoul R. Nigmatullin and Ildar A. Gubaidullin. NAFASS: Fluctuation spectroscopy and the Prony spectrum for description of multi-frequency signals in complex systems. Communications in Nonlinear Science and Numerical Simulation, 56:252–269, 2018.

List of Abbreviations

Abbreviation	Definition
ACF	autocorrelation function
ADC	analog digital converter
AFR	amplitude-frequency response
AI	Analog Input
ARFIMA	autoregressive fractional integral moving average
ARIMA	autoregressive integral moving average
CAPoNeF	Comparative Analysis of Positive and Negative Fluctuations
CEEMD	complete ensemble empirical mode decomposition
CF	correlation factor
CIs	confidence intervals
CWP	correlation weakening principle
DFA	detrend fluctuation analysis
DFT	discrete Fourier transformation
DGI	Discrete Geometrical Invariants
DNS	direct numerical simulation
DRT	Diffusion-Reaction-Transport
ECG	Electrocardiogram
EEG	electroencephalogram
EIS	Electrochemical Impedance Spectroscopy
EQ	earthquake
ES	Expected Shortfall
ETAS	Epidemic Type Aftershock Sequence
EVT	Extreme value theory
F-analysis	Fourier analysis
fBm	fractional Brownian motion
FD	fractal geometry
FDT	Fluctuation-Dissipation Theorem
FFT	fast Fourier transform
fGn	fractional Gaussian noise
FLLS	functional linear least squares
FOT4STEM	Fractional Order Thinking in Exploring the Frontiers of STEM
GARCH	Generalized Autoregressive Conditional Heteroskedasticity

GC	gravity center
GEV	Generalized Extreme Value
GGGM	Generalized Geometric Group Mean
GGM	Generalized Geometric Mean
GMV	generalized mean value
GPCF	Generalized Pearson Correlation Function
GPD	Generalized Pareto Distribution
GPFCM	Generalized Pearson Correlation Function for the Complex Moments
GR law	Gutenberg–Richter law
GTQRE	General Theory of the Quasi-Reproducible Experiments
HF	high frequency
HN	Havriliak–Negami relaxation
IDFT	inverse discrete Fourier transformation
KK test	Kramers-Kronig test
KS entropy	Kolmogorov-Sinai entropy
LF	low-frequency
LLSM	linear least squares method
LRD	long-range dependence
LTS	long-time series
MDL	minimum description length
MF	multifractal
MFDFA	multifractal detrend fluctuation analysis
MI	mutual information
MRW	Multifractal Random Walks
MSE	mean squared error
MUX	multiplexer
NAFASS	Non-Orthogonal Amplitude-Frequency Analysis of the Smoothed Signals
NIHS	Non-Integer Harmonic Series
NIST	National Institute of Standards and Technology
NOCFASS	Non-Orthogonal Combined Fourier Analysis of Smoothed Signals
PACF	partial autocorrelation function
PDF	probability density function
PGIA	Programmable Gain Instrumentation Amplifier
POT	Peak Over Threshold
PRNG	Pseudo-Random Number Generation
PSD	power spectrum density
R/S	range/scale
R-L integral	Riemann-Liouville integral
RMS	root mean squares
RRN	Raoul R. Nigmatullin
RSA	Real-time Spectrum Analyzers
RTIP	reduction to three incident points
SDA	standardized dispersion analysis
SDR	Software Defined Radio
SFM	Statistics of the Fractional Moments
SGS	seismic gradient system
SNR	signal-to-noise ratio

SRA	sequence of the ranged amplitudes
STEM	science technology engineering mathematics
SV	Stochastic Volatility
TLS	trendless sequence
UCI	University of California, Irvine
UQ	uncertainty quantification
VAGs	voltammograms
VaR	Value at Risk
VI	Volterra Integral

Authors

Raoul R. Nigmatullin earned a PhD (Hons.) from the Theoretical Physics department, Kazan State University in 1970 and 1974, respectively. He was an assistant and then assistant to the professor with the Theoretical Physics department. In 1982, he won a competition and visited the Dielectric Laboratory of Professor A. K. Jonscher, Chelsea College, London University. His doctorate theses was in fractal physics in 1992. Since 2014, he has been with the Radioelectronics and Informative Measurements Technics Department, Kazan National Research Technical University, named after A.N. Tupolev. He has authored more than 250 publications. His research interests include dielectric spectroscopy, in electrochemistry and in application of new treatment methods in radiometry, and different regions of physics and chemistry. His Hirsch index is almost 39 and he has received more than 5970 citations from his colleagues working abroad in different regions of technics, physics, and chemistry. (`https://scholar.google.com/citations?hl=en&user=qxyA4zMAAAAJ`)

YangQuan Chen earned his PhD from Nanyang Technological University, Singapore in 1998. He was a faculty member of then Electrical Engineering department at Utah State University (USU) from 2000–2012. He joined the School of Engineering, University of California, Merced (UCM) in 2012 teaching Mechatronics, Engineering Service Learning, and Unmanned Aerial Systems for undergraduates; Fractional Order Mechanics, Nonlinear Controls, and Advanced Controls: Optimality and Robustness for graduates. His research interests include mechatronics for sustainability, cognitive process control, small multi-UAV-based cooperative multispectral "personal remote sensing", applied fractional calculus in controls, modeling and complex signal processing; distributed measurement and control of distributed parameter systems with mobile actuator and sensor networks. He is listed in the Highly Cited Researchers by Clarivate Analytics from 2018–2021. He received Research of the Year awards from USU (2012) and UCM (2020). In 2018, for his cumulative work in drones, Dr. Chen won the Senate Distinguished Scholarly Public Service award that recognizes a faculty member who has energetically and creatively applied his or her professional expertise and scholarship to benefit the local, regional, national, or international community. He received the Control Engineering Practice Best Journal Paper award at the IFAC World Congress 2011 in Milan, Italy. He is the proud recipient of a "Counselor Award" from Utah State University Student Branch of IEEE for *explaining human relationship using control theory.* (`https://scholar.google.com/citations?hl=en&user=RDEIRbcAAAAJ`)

CHAPTER 1

An Introduction to Reading Randomness

This chapter builds the conceptual foundation for *reading randomness* as advocated by Prof. Raoul R. Nigmatullin. Aimed at interdisciplinary readers with minimal mathematics, we (i) survey the historical and cross-cultural origins of chance and randomness, (ii) present modern notions of randomness across mathematics, physics, and computer science, and (iii) summarize practical methods that "read" structure from noisy data, emphasizing fractal/fractional models, fluctuation metrology, and Prony-spectrum techniques such as NAFASS. Throughout, we highlight how long memory, power laws, and nonlocal kernels turn apparent noise into interpretable structure.

1.1 WHAT DO WE MEAN BY "RANDOMNESS"?

1.1.1 Working notions

At an intuitive level, a *random* outcome is unpredictable in detail. In modern probability, randomness is modeled by a probability space (Kolmogorov's axioms) [1]; in information theory, a source is "random" insofar as it produces highly *uncertain* (high-entropy) messages [2]. In algorithmic information theory, an individual sequence is "random" if it is incompressible (high Kolmogorov complexity) [3, 4]. In physics, randomness spans quantum indeterminacy and classical *deterministic chaos*—systems obey deterministic laws yet exhibit sensitive dependence rendering long-term prediction practically impossible [5].

1.1.2 Randomness vs. regularity: three lenses

- **Entropy and unpredictability** (Shannon): higher entropy implies less predictable symbol outcomes [2].
- **Complexity and compressibility** (Kolmogorov/Chaitin): incompressible sequences are random in the strongest sense [3, 4].

DOI: 10.1201/9781003794691-1

- **Correlation and memory**: empirical "noise" can display long-range dependence (LRD) quantified by the Hurst exponent H or via ARFIMA models [6, 7, 8, 9].

1.2 ORIGINS: A CROSS-CULTURAL HISTORY OF CHANCE

1.2.1 Antiquity and non-Western traditions

Ancient societies read chance through divination and games: casting lots, dice, oracles (e.g., the Chinese *I Ching*). Rather than quantifying probabilities, these cultures often interpreted randomness as fate. For the longue-durée view, see [10] and [11].

1.2.2 Classical to enlightenment developments

Renaissance gamblers (Cardano) and the Pascal–Fermat correspondence (1654) reframed games of chance in combinatorial terms; Huygens (1657) wrote the first probability text. The 18th century brought Bernoulli's *Ars Conjectandi* (1713), de Moivre's *Doctrine of Chances* (1718), and Laplace's analytical theory (1812) [12, 13, 14]. A major second turning point came with [15], who introduced Brownian-like price diffusion, anticipating stochastic calculus.

1.2.3 Axioms, noise, and information

Kolmogorov's measure-theoretic axioms (1933; Eng. 1956) gave probability a rigorous foundation [1]. Wiener and successors formalized Brownian motion; Shannon (1948) tied randomness to information content via entropy [2]. Mandelbrot & van Ness (1968) introduced fractional Brownian motion, explaining $1/f^{\beta}$-type spectra and long memory in natural data [9].

1.3 MODERN NOTIONS OF RANDOMNESS

1.3.1 True randomness, chaos, and pseudo-randomness

True randomness can be modeled as incompressibility or (in physics) quantum indeterminacy. **Deterministic chaos** generates unpredictable trajectories from deterministic laws (e.g., the Lorenz system) [5]. **Pseudo-randomness** arises from algorithmic generators: unpredictable to an observer lacking the seed, yet reproducible.

1.3.2 Long memory and fractal scaling

Many "random" signals display self-similarity: fluctuations scale statistically with time window. The Hurst exponent H and detrended fluctuation analysis (DFA) quantify such behavior [6, 16]. Fractional Gaussian/fractional Brownian models provide Gaussian long-memory baselines [9]. Non-Gaussian LRD is captured by ARFIMA models and related spectra [7, 8].

1.3.3 Fractional calculus and memory kernels

Fractional integrals/derivatives introduce power-law memory via convolution kernels. In linear viscoelasticity and complex media, Caputo, Riemann–Liouville, and Prabhakar operators are standard [17, 18, 19, 20]. Solutions often involve the Mittag–Leffler and Prabhakar functions, leading to stretched-exponential relaxation and anomalous diffusion.

Minimal math (reader-friendly): a (Caputo) fractional derivative of order $0 < \alpha < 1$ acts like a "long-memory average"

$$^{C}D_t^{\alpha} x(t) = \frac{1}{\Gamma(1-\alpha)} \int_0^t \frac{\dot{x}(\tau)}{(t-\tau)^{\alpha}} \, d\tau,$$

so the current rate depends on a weighted history; power-law weights encode slow memory decay.

1.4 HOW TO "READ" RANDOMNESS IN PRACTICE

1.4.1 From raw time series to interpretable structure

A general workflow:

1. **Acquisition & curation:** sampling, detrending, artifact handling; align instrument (apparatus) effects.
2. **Exploratory reading:** histograms, power spectra, DFA, Hurst H, stationarity checks.
3. **Modeling:** choose a baseline (white/pink noise, ARMA/ARFIMA, fbm) and test for long memory, heavy tails, or quasi-periodicity.
4. **Fractional kernels:** fit memory kernels (Mittag–Leffler/Prabhakar) in relaxation/correlation functions; compare with integer-order models.
5. **Spectral atoms & Prony spectra:** represent signals via a sparse sum of (damped) sinusoids or exponentials; cross-validate stability.
6. **Uncertainty & robustness:** bootstrap surrogates; sensitivity of labels/parameters to windowing and noise.

1.4.2 NAFASS and fluctuation metrology (Nigmatullin's approach)

Nigmatullin's NAFASS (Non-Orthogonal Amplitude–Frequency Analysis of Smoothed Signals) constructs an *amplitude–frequency response* (AFR) that fingerprints a random sequence by a non-orthogonal set of spectral atoms [21]. Key pillars:

- **Pre-smoothing $\Rightarrow$ stable atoms:** mitigate instability of classical Prony by smoothing before atom extraction.

- **Selective band reading:** emphasize informative bands tied to physical mechanisms.
- **"Universal" labels:** ordered amplitudes of fluctuations follow regular laws enabling cross-dataset comparison and detection of hidden periodicities [22].
- **Prony spectrum as a surrogate PDF:** in some settings, the distribution can be replaced by a reliable Prony spectrum for compact parameterization [23].

Why it works: many complex systems yield signals with mixed quasi-periodic structure and long memory. NAFASS isolates sparse oscillatory content while fractional kernels summarize the background memory.

1.4.3 Complementary toolkits (short guide)

- **DFA/MFDFA** for monofractal/multifractal scaling in nonstationary data [16].
- **ARMA/ARFIMA** for short/long memory with hypothesis tests on differencing order d [7, 8].
- **Fractal/fractional baselines** (fbm/fGn; Caputo/Prabhakar) for anomalous relaxation and transport [9, 17, 19].
- **Sparse spectral methods** (Prony, matrix pencil) for damped oscillations; see classical origins in [24].

1.5 READING RANDOMNESS VS. EXPLAINING IT

Reading emphasizes robust, reproducible extraction of *signatures* (labels, exponents, kernels, Prony spectra) that summarize fluctuations across domains (materials, biosignals, finance). **Explaining** aims to map these signatures to mechanistic models. Both are necessary: the former for diagnostics and cross-domain transfer; the latter for scientific understanding and control.

1.6 COMMON PITFALLS AND GOOD PRACTICES

- *Mistaking trends for memory:* always deseasonalize/detrend; compare DFA with ARFIMA fits [16, 7, 8].
- *Overfitting spectra:* validate Prony atoms with out-of-sample segments and noise surrogates.
- *Ignoring apparatus functions:* correct the instrument path to approach an "ideal experiment" before reading structure (core in Nigmatullin's workflow).
- *Equating randomness with white noise:* many real processes are colored and/or long-memory [9].

1.7 SUMMARY AND OUTLOOK

Randomness has traveled from omen-reading and games to axioms, entropy, and algorithmic complexity. For complex time series, the frontier blends *fractal/fractional* memory with *sparse spectral* structure. Nigmatullin's NAFASS and fluctuation-noise spectroscopy provide an engineering-grade reading of randomness, complementing fractional-kernel modeling. Subsequent chapters develop the mathematics (fractional operators and kernels), the metrology (NAFASS/Prony workflows), and cross-domain case studies.

Further reading: [10, 12, 13, 14, 17, 20].

1.8 SYNONYMS OF *RANDOMNESS*: A CROSS-CONTEXT TABLE WITH METHODS FOR "READING" IT

How to use this table. Each row lists a *context* and common *synonyms/near-terms* people use in that context for "randomness," plus brief nuance and the typical analysis methods (*reading* tools) suited to that usage. Citations anchor terminology or methods to canonical sources.

TABLE 1.1 Cross-context synonyms/near-terms for "randomness" and methods for *reading* them.

Context/Domain	Subdomain	Synonyms & Near-Terms	Nuance/ Connotation	Typical Reading Methods
Probability & Statistics [25, 26, 27]	Foundations	chance, alea, indeterminacy, unpredictability, variability	Underlying uncertainty in outcomes; measure via probability axioms	probability laws, Bayesian inference, likelihoods
	Estimation	error, residual, noise term, disturbance, sampling variation	Deviations from model fit; finite-sample effects	estimators, CIs, hypothesis tests
	Design/ Inference	randomness of assignment, randomization	Tool for causal identification	randomized experiments, randomization tests
Information Theory [2, 28, 29]	Entropy	uncertainty, surprise, unpredictability, information content	Average surprise per symbol	entropy/rate, mutual information, entropy rate
	Algorithmic	incompressibility, Kolmogorov complexity	Individual-sequence "randomness"	compression tests, MDL, complexity proxies

Continued on next page

Context/Domain	**Subdomain**	**Synonyms & Near-Terms**	**Nuance / Connotation**	**Typical Reading Methods**
	Cryptography	nondeterminism, unpredictability (adversarial), pseudorandomness	Statistical vs computational notions	NIST SP 800-22 batteries [30]
Signal Processing &	Spectral	noise, colored noise ($1/f^{\beta}$), jitter, drift	Frequency-structured variability	PSD, periodogram, multitaper, Allan variance [31]
Time Series [32, 33, 34]	Temporal	fluctuations, irregularity, volatility (generic)	Time-domain erraticness	ACF/PACF, DFA, ARIMA/ GARCH
	Residualization	innovations, shocks, white noise	Model-orthogonal randomness	innovations tests, whiteness diagnostics
Dynamical Systems &	Chaos	sensitive dependence, aperiodicity, unpredictability	Deterministic yet effectively random	Lyapunov exponents, correlation dimension
Chaos [5, 35]	Noise-driven	stochastic forcing, perturbations	External randomness on dynamics	stochastic differential equations
	Complexity	irregular transients, strange attractors	Geometry of "random-looking" sets	recurrence plots, surrogate tests
Statistical Mechanics	Microscopic	disorder, molecular chaos	Many-body origin of randomness	ensembles, ergodicity tests
Thermodynamics [36]	Macroscopic	fluctuations, noise, thermal agitation	Observable variability around equilibria	fluctuation/ dissipation, spectra
	Disordered media	heterogeneity, quenched disorder	Structural origins of irregularity	random fields, percolation
Turbulence &	Cascades	intermittency, bursts, multifractality	Scale-dependent irregularity	structure functions, wavelet leaders
Multifractals [37, 38]	Scaling	roughness, self-similarity	Power-law variability	$\zeta(q)$ exponents, H
	Surrogates	universal multifractals	Parametric variability classes	MRW/MF calibrations
Extreme-Value/	Tails	extremeness, rarity, outliers	Distribution tails dominate risk	GEV/GPD, POT, return levels

Continued on next page

Context/Domain	Subdomain	Synonyms & Near-Terms	Nuance / Connotation	Typical Reading Methods
Risk [39, 40, 41]	Temporal	clustering, burstiness	Extremes cluster in time	extremal index, declustering
	Finance	volatility, jumps, tail risk	Heavy-tailed returns & volatility clustering	GARCH/ MRW, VaR/ES, EVT
Biosignals & Physiology [42]	HRV	variability, irregularity, nonstationarity	Health-linked fluctuations	DFA/MFDFA, MSE, POT
	EEG/MEG	$1/f$ background, oscillatory bursts	Scale-free + transient activity	PSD/leaders, Prony modes
	Movement	sway variability, gait fluctuations	Control-loop noise & memory	SDA/DFA, EVT
Electrochemistry/ Materials [43, 44]	Relaxation	dispersion, non-Debye behavior	Broad time constants	Cole–Cole/HN, Prabhakar kernels
	Porous media	heterogeneity, tortuosity	Geometry-driven variability	transmission lines, DRT
	Measurement	apparatus noise, fixture effects	Instrument-induced variability	KK tests, positive-real checks
Hydrology/ Geoscience/Climate	Hydroclimate	persistence, Hurst effect, intermittency	Long memory + bursts	DFA/leaders, EVT (floods)
	Seismicity	triggering, aftershock decay	Clustered irregularity	ETAS, Omori/GR laws
	Wind/Marine	gustiness, wave irregularity	Hazard-focused variability	POT/GEV, spectra
Networks/ Computing [45]	Traffic	self-similarity, long-range dependence	Aggregated source variability	Hurst estimation, wavelet spectra
	Systems	jitter, latency variation, drops	Performance-level randomness	queueing tail analysis, EVT
	Security	entropy, randomness quality	Generator/PRNG assessment	statistical batteries [30]
Philosophy & Epistemology [26, 46]	Categories	aleatoric vs epistemic uncertainty	Intrinsic vs knowledge-based	Bayesian updating, sensitivity
	Common speech	luck, chance, arbitrariness, caprice	Colloquial ambiguity	clarify with formal labels

Notes. Many terms are *near-synonyms*: they emphasize particular mechanisms (e.g., *intermittency* for bursty cascades; *volatility* for scale-dependent variance; *noise* for unmodeled variation). The right way to *read* them is to pick methods matched to the connotation and to validate with independent lenses (scaling, oscillatory modes, and tails).

1.9 NOTIONS OF *RANDOMNESS*: CROSS-CONTEXT TABLE WITH DATASETS AND FAILURE MODES

With the added columns for *canonical test datasets* and *failure modes & caveats*, one might use it as a design-time checklist.

TABLE 1.2 Notions of "randomness" with reading methods, canonical datasets, and failure modes (apparatus-aware).

Context	Notion (term)	Definition/ criterion (sketch)	Reading methods/ labels	Canonical test datasets	Failure modes & caveats
Probability foundations [25, 26, 27]	Chance / alea	Measure-theoretic probability; Bayesian degrees of belief	Likelihoods, posteriors, posterior predictive checks	UCI small tabular sets [47] (toy)	Overconfident priors; ignoring identifiability; p-hacking
Algorithmic / complexity [3, 4, 48]	Incompressibility	Passes all effective tests; $K(x) \approx \|x\|$	Compression tests, universal batteries	PRNG bitstreams; NIST examples [30]	Finite-sample overfitting to test suite; conflating statistical with algorithmic
Information theory [2, 28]	Entropy / entropy rate	Average surprise; entropy rate for processes	Entropy estimators; mutual information	Text corpora; sensor symbol streams	Bias in plug-in estimators; nonstationarity inflates MI
Time series / spectra [32, 33, 34]	White vs colored noise	Flat PSD vs $1/f^{\beta}$	PSD, multitaper; DFA for H	Synthetic ARFIMA; metrology traces [31]	Aliasing; window leakage; apparatus $A(\omega)$ confounds β
Martingale / innovations	Unpredictability given past	$\mathbb{E}[X_t \| \mathcal{F}_{t-1}] = 0$	Ljung–Box, runs tests	Financial returns [49]	Microstructure effects; calendar seasonality
Chaos / dynamical [35, 5]	Sensitive dependence	$\lambda_{\max} > 0$, KS entropy > 0	Lyapunov exponents; surrogate tests	Lorenz, logistic maps	Noise contamination; short data; embedding errors
Ergodic / mixing [50]	Mixing rates	Correlations $\to 0$	Auto/cross-correlation decay	Synthetic mixing maps	Regime switches mimic mixing; nonstationary drift
Statistical mechanics [36]	Thermal fluctuations	Ensemble variability	FDT, spectra	Physics lab datasets	Non-equilibrium conditions; drift
Turbulence / multifractals [37, 38]	Intermittency, cascades	Concave $\zeta(q)$; $c_2 < 0$	Structure functions; leaders; H, c_2	Synthetic cascades; DNS fields	Narrow scale range; misconstrued curvature from noise

Continued on next page

Context	Notion (term)	Definition/ criterion (sketch)	Reading methods/ labels	Canonical test datasets	Failure modes & caveats
Extremes / risk [40, 41, 39]	Tail randomness	GEV/GPD limits; extremal index θ	POT/GEV; tail index ξ; θ	River/gauge highs; market tails [49]	Threshold mischoice; un-declustered clusters; parameter drift
Finance / economet-rics [49, 34]	Volatility clustering	Time-varying conditional variance	GARCH/SV; MRW λ^2	Daily returns (indices)	Leverage/ microstructure bias; regime changes
Hydrology / climate	Hurst effect / persistence	$0.5 < H < 1$ long memory	DFA; Whittle; leaders	Gauge/radar rainfall; streamflows	Seasonal mixing; gridding biases; censoring at low flows
Biosignals [42, 51]	Physiologic variability	Fractal scaling with crossovers	DFA/MFDFA; MSE; Prony RSA	PhysioNet ECG/EEG [51]	Motion artifacts; filtering biases; nonstationary states
Networks / computing [45]	LRD traffic	Aggregated ON/OFF $\Rightarrow H > 0.5$	Wavelet spectra; R/S; Hill	Packet/flow traces	Sampling policy changes; diurnal cycles; privacy aggregation artifacts
Metrology / instrumen-tation [31, 43, 44]	Measurement noise; apparatus function	Shot/thermal/$1/f$; causal $A(\omega)$	Allan dev.; KK/positive-real	Frequency standards; EIS sweeps	Nonlinearity; time variance; failed KK; calibration drift
Philosophy / uncertainty [26, 46]	Aleatoric vs epistemic	Intrinsic vs knowledge-based	Prior/posterior partitions	Toy Bayesian models	Confounding aleatory with model misspecification
Seismology [52]	Triggering & clustering	Omori decay; ETAS branching	ETAS; b, p; θ	Regional catalogs	Magnitude incompleteness; clock sync; site effects
Electrochemi / materials [43, 44]	Non-Debye relaxation	Broad τ spectra	DRT; Prabhakar fits	Calibrated EIS cells	Ignoring apparatus; violating linearity; poor grid/dwell

Use this like a checklist. Pick the notion aligned to your hypothesis and instrument; run the matching methods; validate with a second lens (scale/memory vs oscillation vs tails); and consult the failure-mode column before drawing conclusions.

Finally, we remark that it is Prof. Raoul R. Nigmatullin who inspired the "reading randomness" paradigm with decades of methodological innovation that bridge fractal geometry, fractional kinetics, fractional calculus, and in particular, fluctuation metrology. The content of Tables 1.1 and 1.2 demonstrates the variety of definitions of "randomness" that is needed to be read quantitatively.

BIBLIOGRAPHY

[1] A. N. Kolmogorov, *Foundations of the Theory of Probability.* New York: Chelsea Publishing Company, second english edition ed., 1956.

[2] C. E. Shannon, "A mathematical theory of communication," *Bell System Technical Journal,* vol. 27, no. 3,4, pp. 379–423, 623–656, 1948. Also published in BSTJ Oct 1948; Part II DOI: 10.1002/j.1538-7305.1948.tb00917.x.

[3] M. Li and P. M. B. Vitányi, *An Introduction to Kolmogorov Complexity and Its Applications.* New York: Springer, 3rd ed., 2008.

[4] P. Martin-Löf, "The definition of random sequences," *Information and Control,* vol. 9, no. 6, pp. 602–619, 1966.

[5] E. N. Lorenz, "Deterministic nonperiodic flow," *Journal of the Atmospheric Sciences,* vol. 20, no. 2, pp. 130–141, 1963.

[6] H. E. Hurst, "Long-term storage capacity of reservoirs," *Transactions of the American Society of Civil Engineers,* vol. 116, pp. 770–799, 1951.

[7] C. W. J. Granger and R. Joyeux, "An introduction to long-memory time series models and fractional differencing," *Journal of Time Series Analysis,* vol. 1, no. 1, pp. 15–29, 1980.

[8] J. R. M. Hosking, "Fractional differencing," *Biometrika,* vol. 68, no. 1, pp. 165–176, 1981.

[9] B. B. Mandelbrot and J. W. Van Ness, "Fractional Brownian motions, fractional noises and applications," *SIAM Review,* vol. 10, no. 4, pp. 422–437, 1968.

[10] F. N. David, *Games, Gods and Gambling: The Origins and History of Probability and Statistical Ideas from the Earliest Times to the Newtonian Era.* New York: Hafner, 1962.

[11] J. Needham, *Science and Civilisation in China, Vol. 3: Mathematics and the Sciences of the Heavens and the Earth.* Cambridge: Cambridge University Press, 1959.

[12] I. Hacking, *The Emergence of Probability: A Philosophical Study of Early Ideas about Probability, Induction and Statistical Inference.* Cambridge: Cambridge University Press, 1975.

[13] S. M. Stigler, *The History of Statistics: The Measurement of Uncertainty before 1900.* Cambridge, MA: Harvard University Press, 1986.

[14] L. Daston, *Classical Probability in the Enlightenment.* Princeton, NJ: Princeton University Press, 1988.

[15] L. Bachelier, "Théorie de la spéculation," *Annales scientifiques de l'École Normale Supérieure (Série 3),* vol. 17, pp. 21–86, 1900.

[16] C.-K. Peng, S. V. Buldyrev, S. Havlin, M. Simons, H. E. Stanley, and A. L. Goldberger, "Mosaic organization of dna nucleotides," *Physical Review E*, vol. 49, no. 2, pp. 1685–1689, 1994.

[17] I. Podlubny, *Fractional Differential Equations.* Academic Press, 1999.

[18] M. Caputo, "Linear models of dissipation whose q is almost frequency independent," *Geophysical Journal of the Royal Astronomical Society*, vol. 13, no. 5, pp. 529–539, 1967.

[19] T. R. Prabhakar, "A singular integral equation with a generalized Mittag-Leffler function in the kernel," *Yokohama Mathematical Journal*, vol. 19, pp. 7–15, 1971.

[20] S. G. Samko, A. A. Kilbas, and O. I. Marichev, *Fractional Integrals and Derivatives: Theory and Applications.* Gordon and Breach, 1993.

[21] R. R. Nigmatullin, S. I. Osokin, and V. A. Toboev, "NAFASS: Discrete spectroscopy of random signals," *Chaos, Solitons & Fractals*, vol. 44, no. 4–5, pp. 226–240, 2011.

[22] R. R. Nigmatullin and G. Smith, "Fluctuation-noise spectroscopy and a "universal" fitting function of amplitudes of random sequences," *Physica A: Statistical Mechanics and its Applications*, vol. 320, pp. 291–317, 2003.

[23] R. R. Nigmatullin and I. A. Gubaidullin, "Nafass: Fluctuation spectroscopy and the prony spectrum for description of multi-frequency signals in complex systems," *Communications in Nonlinear Science and Numerical Simulation*, vol. 56, pp. 252–269, 2018.

[24] G. R. de Prony, "Essai expérimental et analytique: sur les lois de la dilatabilité des fluides et sur celles de la force expansive de la vapeur de l'alcool, à différentes températures," *Journal de l'École Polytechnique*, vol. 1, pp. 24–76, 1795. Classical paper introducing the method now known as Prony's method.

[25] A. N. Kolmogorov, *Foundations of the Theory of Probability.* Chelsea, 1950. Original German edition 1933.

[26] E. T. Jaynes, *Probability Theory: The Logic of Science.* Cambridge University Press, 2003.

[27] W. Feller, *An Introduction to Probability Theory and Its Applications, Vol. I.* Wiley, 3rd ed., 1968.

[28] T. M. Cover and J. A. Thomas, *Elements of Information Theory.* Wiley, 2nd ed., 2006.

[29] A. N. Kolmogorov, "Three approaches to the quantitative definition of information," *Problems of Information Transmission*, vol. 1, pp. 1–7, 1965. English translation (Int. J. Computer Mathematics, 1968) DOI: 10.1080/00207166808803030.

[30] A. Rukhin, J. Soto, *et al.*, "A statistical test suite for random and pseudorandom number generators for cryptographic applications," Tech. Rep. SP 800-22 Rev.1a, NIST, 2010.

[31] D. W. Allan, "Statistics of atomic frequency standards," *Proceedings of the IEEE*, vol. 54, no. 2, pp. 221–230, 1966.

[32] D. R. Brillinger, *Time Series: Data Analysis and Theory.* Holt, Rinehart and Winston, 1975.

[33] M. B. Priestley, *Spectral Analysis and Time Series.* Academic Press, 1981.

[34] G. E. P. Box, G. Jenkins, G. C. Reinsel, and G. M. Ljung, *Time Series Analysis: Forecasting and Control.* Wiley, 5th ed., 2015.

[35] E. Ott, *Chaos in Dynamical Systems.* Cambridge University Press, 2nd ed., 2002.

[36] A. I. Khinchin, *Mathematical Foundations of Statistical Mechanics.* Dover, 1949.

[37] U. Frisch, *Turbulence: The Legacy of A. N. Kolmogorov.* Cambridge University Press, 1995.

[38] B. B. Mandelbrot, *The Fractal Geometry of Nature.* W. H. Freeman, 1982.

[39] A. J. McNeil, R. Frey, and P. Embrechts, *Quantitative Risk Management.* Princeton, NJ: Princeton University Press, 2005.

[40] S. Coles, *An Introduction to Statistical Modeling of Extreme Values.* Springer, 2001.

[41] L. de Haan and A. Ferreira, *Extreme Value Theory: An Introduction.* Springer, 2006.

[42] C.-K. Peng, S. Havlin, H. E. Stanley, and A. L. Goldberger, "Quantification of scaling exponents and crossover phenomena in heartbeat time series," *Chaos*, vol. 5, no. 1, pp. 82–87, 1995.

[43] E. Barsoukov and J. R. Macdonald, *Impedance Spectroscopy: Theory, Experiment, and Applications.* Wiley, 2nd ed., 2005.

[44] A. Lasia, *Electrochemical Impedance Spectroscopy and Its Applications.* Springer, 2014.

[45] W. E. Leland, M. S. Taqqu, W. Willinger, and D. V. Wilson, "On the self-similar nature of ethernet traffic," *IEEE/ACM Transactions on Networking*, vol. 2, no. 1, pp. 1–15, 1994.

[46] A. Der Kiureghian and O. Ditlevsen, "Aleatory or epistemic? does it matter?," *Structural Safety*, vol. 31, no. 2, pp. 105–112, 2009.

[47] D. Dua and C. Graff, "UCI machine learning repository." `http://archive.ics.uci.edu/ml`, 2019. University of California, Irvine, School of Information and Computer Sciences.

[48] G. J. Chaitin, "Randomness and Mathematical Proof," *Scientific American*, vol. 232, pp. 47–53, may 1975.

[49] R. Cont, "Empirical properties of asset returns: stylized facts and statistical issues," *Quantitative Finance*, vol. 1, no. 2, pp. 223–236, 2001.

[50] P. Walters, *An Introduction to Ergodic Theory*, vol. 79 of *Graduate Texts in Mathematics*. New York, NY: Springer-Verlag, 2000. Reprint of the 1982 edition.

[51] A. L. Goldberger, L. A. N. Amaral, L. Glass, J. M. Hausdorff, P. C. Ivanov, R. G. Mark, J. E. Mietus, G. B. Moody, C.-K. Peng, and H. E. Stanley, "Physiobank, physiotoolkit, and physionet: Components of a new research resource for complex physiologic signals," *Circulation*, vol. 101, no. 23, pp. e215–e220, 2000.

[52] K. Aki and P. G. Richards, *Quantitative Seismology*. University Science Books, 2nd ed., 2002.

CHAPTER 2

Discrete Geometrical Invariants (DGI)-Based Methods

This chapter presents 3D-DGI (Discrete Geometrical Invariants) plus CAPoNeF (Comparative Analysis of Positive/Negative Fluctuations) methods. The presented new methods are shown to be best in addressing how to compare trendless sequences with each other and how to compress and extract additional information from raw data with randomness.

2.1 PROBLEM FORMULATION – HOW TO READ AND COMPARE TRENDLESS SEQUENCES?

In this chapter, the authors want to prove that a trendless sequence (TLS) can be used as an additional source of information. This additional information can be extracted from random noise with the help of 3D-DGIs (Discrete Geometrical Invariants) and CAPoNeF (Comparative Analysis of Positive and Negative Fluctuations) methods that allow to reduce 3N random data points to 13/10 parameters composed from the combination of integer moments and their intercorrelations up to the fourth order inclusive. Actually, the first method forms a "universal" 13-feature space for comparison of one random sequence with another one. Comparison of these parameters associated with different noise tracks allows to use this set of parameters for comparison with each other on the common platform in order to differentiate the differences and for other purposes associated with "standard"/reference equipment.

Examples taken from different sources should convince the skeptically tuned and attentive reader that this 3D-DGI method is general and, in many cases, it is rather efficient and "universal" in various applications

The idea that the chosen set of random fluctuations (noise) contains a "hidden" source of information now is not new. Part of the researchers tried to extract an information based on some reasonable suppositions. In particular, the basics of the fluctuation noise spectroscopy (including also many useful references) is given in the

DOI: 10.1201/9781003794691-2

book [1] and papers [2, 3]. The approach based on the Mori-Zwanzig formalism are given in the papers [4, 5]. Unfortunately, these approaches include some unjustified suppositions and contain some treatment errors. Therefore, they become useless in analysis of random sequences having different nature. These suppositions and their analysis are listed in paper [6]. The essential results were obtained from analysis of electrochemical noise [7, 8, 9, 10, 11, 12], other type of "noises" as understanding of the earthquakes phenomenon and their quantitative description, analysis of medical data is given in papers [3, 4, 13, 14]. Unfortunately, in spite of these promising attempts the general picture associated with analysis of arbitrary types of random sequences (especially TLS) is far from an 'ideal' one. Researches based on their own (and in many cases unjustified suppositions) experience/intuition try to process many types of random sequences and extract an "information" mixed with treatment/uncontrollable errors. Especially, they have problems with analysis of equipment "noise" (pure TLS) that in many cases is not the Gaussian, uniform and other type of "color" noises that are widely used in the mathematical statistics as an approximation and some imposed model of the real TLS(s).

Therefore, there is a vital task in creation of the reliable processing tool that: (a) should be free from the treatment errors and (b) rather "universal" for application to any TLS that is generated in the given equipment or comes from natural phenomenon. One can compare one type of fluctuations with another one and select the desired and calibrated equipment based on the measurements of "reference" data only in cases when the reliable and universal platform for their comparison can be created.

In this chapter, we propose for a wide group of master/postgraduate students and young researchers the desired and general tool based on the "modeless" reading of random sequences/fluctuations. The creation of this promising method is based on the ideas expressed by Prof. Yu. I. Babenko [15, 16] associated with the generalization of the Pythagorean theorem. One of us (RRN) applied these ideas to random sequences and used the discrete geometrical invariants (DGI) in 2D space for differentiation of different types of olive oils [17] and equipment noise [18] reducing initially 2N random data points to 8 independent parameters representing a combination of integer order moments and intercorrelations up to the fourth order inclusive. One can continue these ideas and consider the complete set of the DGI(s) in 3D space. Application of this method allows to transform the set of arbitrary chosen 3N data points, forming the complete fourth order form to 13 independent parameters forming the feature space of the corresponding dimension. These parameters including the desired combinations of the integer moments and their intercorrelations up to the fourth order inclusive are proved to be very sensitive and general for comparing of the chosen random sequence with another one. Besides, these 13 parameters extracted from the considered random sequence represent themselves a specific "fingerprint" for observing the evolution of the reduced random sequence in time or against another external factor as concentration, electromagnet/acoustic field intensity etc., in 3D space. This reduction procedure reminds a procedure used in the statistical mechanics; when with the help of the Gibbs partition function 3N trajectories of the microscopic particles are reduced to a finite set of thermodynamic parameters. Actually, the 3D-DGI(s) method realizes a similar procedure, i.e., it reduces 3N combination of an arbitrary random data points to 13 quantitative parameters forming the specific feature space. We want to stress here that this reduction procedure does not use any model and can be considered as the "modeless" method. Besides, it does not contain any treatment errors as well and keeps only the measurement/experimental errors that are accompanied by the used equipment

2.2 3D-DGI-METHOD AND DIAGONALIZATION OF THE FOURTH ORDER FORM

In this section, we describe the mathematical details associated with the derivation of the complete DGI in 3D-space. We remind here that preliminary results based on the application of the incomplete DGI form of the fourth order in 3D space is outlined in [19]. Let us consider the power-law form of the fourth order:

$$\begin{aligned}
L_k^{(4)} &= A_{40}^{(1,0)}(y_1 - r_{1k})^4 + A_{40}^{(2,0)}(y_2 - r_{2k})^4 + A_{40}^{(3,0)}(y_3 - r_{3k})^4 - \\
&B_{22}^{(1,2)}(y_1 - r_{1k})^2 \cdot (y_2 - r_{2k})^2 - B_{22}^{(1,3)}(y_1 - r_{1k})^2 \cdot (y_3 - r_{3k})^2 - \\
&B_{22}^{(2,3)}(y_2 - r_{2k})^2 \cdot (y_3 - r_{3k})^2 + \\
&C_{211}^{1(2,3)}(y_1 - r_{1k})^2 \cdot (y_2 - r_{2k}) \cdot (y_3 - r_{3k}) + C_{211}^{2(1,3)}(y_2 - r_{2k})^2 \cdot (y_1 - r_{1k}) \cdot (y_3 - r_{3k}) + \\
&C_{211}^{3(1,2)}(y_3 - r_{3k})^2 \cdot (y_1 - r_{1k}) \cdot (y_2 - r_{2k}) - \\
&\frac{1}{2}D_{31}^{(1,2)}(y_1 - r_{1k}) \cdot (y_2 - r_{2k})\left[(y_1 - r_{1k})^2 + (y_2 - r_{2k})^2\right] - \\
&\frac{1}{2}D_{31}^{(1,3)}(y_1 - r_{1k}) \cdot (y_3 - r_{3k})\left[(y_1 - r_{1k})^2 + (y_3 - r_{3k})^2\right] - \\
&\frac{1}{2}D_{31}^{(2,3)}(y_2 - r_{2k}) \cdot (y_3 - r_{3k})\left[(y_2 - r_{2k})^2 + (y_3 - r_{3k})^2\right].
\end{aligned} \tag{2.1}$$

In expression (2.1) the upper indices define the combination of the variables y_α (α =1,2,3) fixing the location of an arbitrary point $M(y_1,y_2,y_3)$ in 3D-space, the low indices determine the values of the power-law exponents that correspond to the algebraic form of the fourth order. The choice of the sign's combination ($\pm$) before the constants in (2.1) will be explained below. Three random sequences are determined by the values $r_{\alpha k}$ (α =1,2,3; k=1,2,...,N). Expression (2.1) represents itself the *complete* form of the fourth order that contains the combination of three variables associated with an arbitrary point $M(y_1,y_2,y_3)$ and three arbitrary sequences $r_{\alpha k}$ ($\alpha = 1,2,3$). The desired DGI is obtained from the following requirement

$$\frac{1}{N}\sum_{k=1}^{N} L_k^{(4)} = I_4. \tag{2.2}$$

In order to remove in the expression (2.2) the cubic terms, we introduce the variables

$$Y_\alpha = y_\alpha - \langle r_\alpha \rangle, \ \langle r_\alpha \rangle = \frac{1}{N}\sum_{k=1}^{N} r_{\alpha k} \tag{2.3}$$

and nullify the *linear* terms. This requirement helps us to separate the desired variables Y_α from each other and keep only the terms of the second and fourth orders, correspondingly. In order to decrease the number of constants in (2.2) and derive the DGI not depending on some additional constants, one defines three key ratio constants $R^{(\alpha,\beta)}$, with the following combination $(\alpha,\beta) = (1{,}2), (1{,}3), (2{,}3)$

$$R^{(\alpha,\beta)} = \frac{B^{(\alpha,\beta)}}{A} = \frac{C^{\gamma(\alpha,\beta)}}{A} = \frac{D^{(\alpha,\beta)}}{A}, \qquad (2.4)$$
$$A_{40}^{(\alpha)} = A_{40}^{(\beta)} = A_{40}^{(\gamma)} \equiv A,\ \alpha,\beta,\gamma = 1,2,3.$$

It is convenient also to introduce the following notations for the integer moments and their intercorrelations presenting them in the following compact form

$$Q_{\alpha^n\beta^m\gamma^l} = \frac{1}{N}\sum_{k=1}^{N}\left((\Delta r_{3k})^m(\Delta r_{2k})^n(\Delta r_{1k})^l\right) \equiv \left\langle(\Delta r_\alpha)^m(\Delta r_\beta)^n(\Delta r_\gamma)^l\right\rangle, \qquad (2.5)$$
$$\alpha \geq \beta \geq \gamma,\ (\alpha,\beta,\gamma) = 1,2,3.$$

In the result of the introduced notations (2.4) and (2.5), the system of linear equations for the finding of unknown ratios $R^{(\alpha,\beta)}$ from the nullification requirement of the entering linear terms accepts the form

$$\begin{aligned}
&\left[2Q_{221} - Q_{332} + \frac{3}{2}Q_{211} + \frac{1}{2}Q_{222}\right]\cdot R^{(1,2)} + \\
&+\left[2Q_{331} - Q_{322} + \frac{3}{2}Q_{311} + \frac{1}{2}Q_{333}\right]\cdot R^{(1,3)} - 2Q_{321}\cdot R^{(2,3)} = 4Q_{111}, \\
&\left[2Q_{211} - Q_{331} + \frac{3}{2}Q_{221} + \frac{1}{2}Q_{111}\right]\cdot R^{(1,2)} - 2Q_{321}\cdot R^{(1,3)} + \\
&+\left[2Q_{332} - Q_{311} + \frac{3}{2}Q_{322} + \frac{1}{2}Q_{333}\right]\cdot R^{(2,3)} = 4Q_{222}, \\
&-2Q_{321}\cdot R^{(1,2)} + \left[2Q_{311} - Q_{221} + \frac{3}{2}Q_{331} + \frac{1}{2}Q_{111}\right]\cdot R^{(1,3)} + \\
&+\left[2Q_{322} - Q_{211} + \frac{3}{2}Q_{332} + \frac{1}{2}Q_{222}\right]\cdot R^{(2,3)} = 4Q_{333}.
\end{aligned} \qquad (2.6a)$$

The linear system of equations helps to reduce 3 central moments (Q_{333}, Q_{222}, Q_{111}) and 7 intercorrelations of the third order (Q_{332}, Q_{322}, Q_{221}, Q_{211}, Q_{331}, Q_{311}, Q_{321}) to the calculation of three unknown ratios $R^{(\alpha,\beta)}$ only. We should notice also that the combination of the algebraic signs in (2.1) is chosen in that way for the keeping of the partial solution R=1 of system (2.6a) in the case when all three random sequences $r_{\alpha k}$ are identical to each other, i.e., $r_{1k} = r_{2k} = r_{3k}$. It is natural to define it as the case of *spherical* symmetry. If only two sequences coincide with other (for example, $r_{1k} = r_{2k} \neq r_{3k}$), then we deal with the case of the *cylindrical* symmetry. In this case, the linear system (2.6a) is reduced to the couple of linear equations relatively the variables $R^{(1,2)} \neq R^{(1,3)} = R^{(2,3)}$. The number of triple correlations equals four in this case (Q_{111}, Q_{113}, Q_{133}, Q_{333}). The system (2.6a) is simplified and reduced to the couple of linear equations relatively two variables $R^{(1,2)}$ and $R^{(1,3)}$ only

$$\begin{aligned}&(4Q_{111}-Q_{333})\,R^{(1,2)}+\left(2Q_{331}+\frac{1}{2}Q_{333}-\frac{3}{2}Q_{311}\right)R^{(1,3)}=4Q_{111},\\&-2Q_{311}R^{(1,2)}+(4Q_{311}-Q_{111}+3Q_{331})\,R^{(1,3)}=4Q_{333}.\end{aligned} \tag{2.6b}$$

Equations (2.6) facilitate considerably the further calculations. After averaging procedure applied to expression (2.2) the structure of the fourth order form can be rewritten as

$$K_4(Y_1,Y_2,Y_3)+K_2(Y_1,Y_2,Y_3)=I_4. \tag{2.7}$$

As before [17, 18, 19], we chose the value of the invariant I_4 as the double value of the free constant in the left-hand side of (2.7). After some algebraic manipulations (we strongly recommend to a potential reader to reproduce the key expressions given below) the fourth and the second order forms entering to the left-hand side can be presented as

$$\begin{aligned}&K_4(Y_1,Y_2,Y_3)=Y_1^4+Y_2^4+Y_3^4+R^{(1,2)}Y_1Y_2\left[Y_3^2-\frac{1}{2}(Y_1+Y_2)^2\right]+\\&+R^{(1,3)}Y_1Y_3\left[Y_2^2-\frac{1}{2}(Y_1+Y_3)^2\right]+R^{(2,3)}Y_2Y_3\left[Y_1^2-\frac{1}{2}(Y_2+Y_3)^2\right].\end{aligned} \tag{2.8a}$$

$$\begin{aligned}K_2(Y_1,Y_2,Y_3)&=A_{11}Y_1^2+A_{22}Y_2^2+A_{33}Y_3^2+\\&+A_{12}Y_1Y_2+A_{13}Y_1Y_3+A_{23}Y_2Y_3.\end{aligned} \tag{2.8b}$$

The constants A in the expression (2.8b) are defined as

$$\begin{aligned}A_{11}&=6Q_{11}-\left(Q_{22}+\frac{3}{2}Q_{21}\right)R^{(1,2)}-\left(Q_{33}+\frac{3}{2}Q_{31}\right)R^{(1,3)}+Q_{32}R^{(2,3)},\\A_{22}&=6Q_{22}-\left(Q_{11}+\frac{3}{2}Q_{21}\right)R^{(1,2)}+Q_{31}R^{(1,3)}-\left(Q_{33}+\frac{3}{2}Q_{32}\right)R^{(2,3)},\\A_{33}&=6Q_{33}+Q_{21}R^{(1,2)}-\left(Q_{11}+\frac{3}{2}Q_{31}\right)R^{(1,3)}-\left(Q_{22}+\frac{3}{2}Q_{32}\right)R^{(2,3)},\\A_{12}&=-\left(4Q_{21}+\frac{3}{2}Q_{11}+\frac{3}{2}Q_{22}-Q_{33}\right)R^{(1,2)}+2Q_{32}R^{(1,3)}+2Q_{31}R^{(2,3)},\\A_{13}&=2Q_{32}R^{(1,2)}-\left(4Q_{31}+\frac{3}{2}Q_{11}+\frac{3}{2}Q_{33}-Q_{22}\right)R^{(1,3)}+2Q_{12}R^{(2,3)},\\A_{23}&=2Q_{31}R^{(1,2)}+2Q_{21}R^{(1,3)}-\left(4Q_{32}+\frac{3}{2}Q_{22}+\frac{3}{2}Q_{33}-Q_{11}\right)R^{(2,3)}.\end{aligned} \tag{2.9}$$

The constant I_4 (defined by 3 moments and 12 intercorrelations of the fourth order) in the right-hand side of (2.7) is defined as

$$\begin{aligned} I_4 = Q_{1111} + Q_{2222} + Q_{3333} - \left(Q_{2211} - Q_{3321} + \frac{1}{2}Q_{2111} + \frac{1}{2}Q_{2221}\right) R^{(1,2)} - \\ - \left(Q_{3311} - Q_{3221} + \frac{1}{2}Q_{3111} + \frac{1}{2}Q_{3331}\right) R^{(1,3)} - \\ - \left(Q_{3322} - Q_{3211} + \frac{1}{2}Q_{3222} + \frac{1}{2}Q_{3332}\right) R^{(2,3)}. \end{aligned} \tag{2.10}$$

It is interesting to notice that in the case of the spherical symmetry ($r_{1k} = r_{2k} = r_{3k}$) all correlations coincide with each other and the value of I_4 equals zero. The form of the fourth order (2.7) admits the separation of the variables in the spherical system of coordinates. If one accepts the conventional notations:

$$\begin{aligned} y_1 &= \langle r_1 \rangle + R\sin\theta\cos\varphi, \\ y_2 &= \langle r_2 \rangle + R\sin\theta\sin\varphi, \\ y_3 &= \langle r_3 \rangle + R\cos\theta, \\ 0 &\le \theta \le \pi, 0 \le \varphi \le 2\pi, \end{aligned} \tag{2.11}$$

then substitution of these variables into (2.7) (we notice that $Y_\alpha = y_\alpha - \langle r_\alpha \rangle$) leads to the following biquadratic equation relatively the unknown radius $R(\theta,\varphi)$

$$[R(\theta,\varphi)]^4 + \left(\frac{P_2(\theta,\varphi)}{P_4(\theta,\varphi)}\right)[R(\theta,\varphi)]^2 - \frac{I_4}{P_4(\theta,\varphi)} = 0. \tag{2.12a}$$

The desired solution ($R(\theta,\varphi)) > 0$) is written as

$$R(\theta,\varphi) = \left[\frac{\sqrt{P_2^2(\theta,\varphi) + 4I_4 \cdot P_4(\theta,\varphi)} - P_2(\theta,\varphi)}{2P_4(\theta,\varphi)}\right]^{\frac{1}{2}}. \tag{2.12b}$$

The polynomials $P_{2,4}(\theta,\varphi)$ entering in (2.12) are defined by the following expressions

$$\begin{aligned} P_4(\theta,\varphi) &= \sin^4\theta\cdot\cos^4\varphi + \sin^4\theta\cdot\sin^4\varphi + \cos^4\theta + \\ &+ R^{(1,2)}\sin^2\theta\sin\varphi\cos\varphi\left[\cos^2\theta - \frac{\sin^2\theta}{2}(\sin\varphi + \cos\varphi)^2\right] + \\ &+ R^{(1,3)}\sin\theta\cos\theta\cos\varphi\left[\sin^2\theta\sin^2\varphi - \frac{1}{2}(\sin\theta\cos\varphi + \cos\theta)^2\right] + \\ &+ R^{(2,3)}\sin\theta\cos\theta\sin\varphi\left[\sin^2\theta\cos^2\varphi - \frac{1}{2}(\sin\theta\sin\varphi + \cos\theta)^2\right] \end{aligned} \tag{2.13a}$$

$$\begin{aligned} P_2(\theta,\varphi) &= A_{11}\sin^2(\theta)\cos^2(\varphi) + A_{22}\sin^2(\theta)\sin^2(\varphi) + A_{33}\cos^2(\theta) + \\ &+ A_{12}\sin^2(\theta)\sin(\varphi)\cos(\varphi) + A_{13}\sin(\theta)\cos(\theta)\cos(\varphi) + A_{23}\sin(\theta)\cos(\theta)\sin(\varphi). \end{aligned} \tag{2.13b}$$

The last expressions (2.11–2.13) determine the final form of the DGI in 3D-space. It includes three surfaces determined by expressions (2.11). The further analysis shows that the expression (2.12b) equals zero (because $I_4 = 0$) in the case of the coincidence of three compared random sequences ($r_{1k} = r_{2k} = r_{3k}$). The radius $R(\theta, \varphi)$ can contain the complex expression when the integrand in (2.12b) becomes *negative*. It accepts the negative values when the constant I_4 defined by the expression (2.10) becomes *negative*. In this case, it is convenient to rewrite initial expressions (2.11) in the form

$$\begin{aligned}
&y_1 = \langle r_1 \rangle + |R(\theta,\varphi)| \sin\theta \cos\varphi, \\
&y_2 = \langle r_2 \rangle + |R(\theta,\varphi)| \sin\theta \sin\varphi, \\
&y_3 = \langle r_3 \rangle + |R(\theta,\varphi)| \cos\theta, \\
&|R(\theta,\varphi)| = \sqrt{[\mathrm{Re}(R(\theta,\varphi))]^2 + [\mathrm{Im}(R(\theta,\varphi))]^2}, \\
&0 \le \theta < \pi,\ 0 \le \varphi < 2\pi.
\end{aligned} \tag{2.14a}$$

Three other "imaginary" surfaces (when $I_4 < 0$) can be necessary for more detailed analysis

$$\begin{aligned}
&Y_1 = |\mathrm{Im}(R(\theta,\varphi))| \cos(\varphi) \sin(\theta), \\
&Y_2 = |\mathrm{Im}(R(\theta,\varphi))| \sin(\varphi) \sin(\theta), \\
&Y_3 = |\mathrm{Im}(R(\theta,\varphi))| \cos(\theta) \\
&0 \le \varphi < 2\pi,\ 0 \le \theta < \pi.
\end{aligned} \tag{2.14b}$$

One can notice also that expressions representing the desired surfaces can be replaced by three *plane* functions, also. Really, if we subordinate the angular variables to condition

$$\begin{aligned}
&y_1(j) = \langle r_1 \rangle + R(\theta_j,\varphi_j) \cos(\varphi_j) \sin(\theta_j), \\
&y_2(j) = \langle r_2 \rangle + R(\theta_j,\varphi_j) \sin(\varphi_j) \sin(\theta_j), \\
&y_3(j) = \langle r_3 \rangle + R(\theta_j,\varphi_j) \cos(\theta_j), \\
&\varphi_j = 2\pi\left(\frac{j}{N}\right),\ \theta_j = \pi\left(\frac{j}{N}\right),\ R_j \ge 0, j = 0.1..., N,
\end{aligned} \tag{2.15a}$$

then these three functions can serve for identification and quantitative description of the given random/deterministic sequences. The compact numbers of the parameters that determine their behavior remained the *same* and equaled 13. It is interesting to notice also that if three compared random functions completely coincide with each other ($I_4 = 0$), then these functions are reduced to the gravity point

$<r_1>=<r_2>=<r_3>$. If two random functions become identical to each other, i.e., $y_1(j) = y_2(j) \neq y_3(j)$ then in this case the azimuth angle $\varphi = \pi/4$ and these three functions are reduced to a couple of functions

$$
\begin{aligned}
&y_1(j) = \langle r_1 \rangle + \frac{1}{\sqrt{2}} R\left(\theta_j, \frac{\pi}{4}\right) \sin(\theta_j), \\
&y_3(j) = \langle r_3 \rangle + R\left(\theta_j, \frac{\pi}{4}\right) \cos(\theta_j), \\
&\varphi_j \equiv \frac{\pi}{4},\ \theta_j = \pi\left(\frac{j}{N}\right),\ R_j \geq 0, \quad j = 0, 1, \ldots, N.
\end{aligned}
\tag{2.15b}
$$

The curves defined by equations (2.15) facilitate considerably the numerical analysis of initial data, while the surfaces (2.14) give more ocular presentation of the compared sequences in 3D-space. We want to stress here again that the final expressions (2.14) and (2.15) do not use *any* model and are determined completely by the measured data together with their measurement errors. After finishing this section one can say that this method can be applied for reduction of initial data. This reduction procedure can be divided on the following stages:

1. Initially, any available data can be presented in the form of rectangle matrix $[\ N \times M\ ]$, where number N ($j = 0, 1, \ldots, N$ - number of rows) determines the given data points and M (m=1, 2, ..., M - columns) determines the number of the repeated measurements forming in total the statistically significant sampling. As the result of application of 3D-DGI method, we obtain the reduced matrix $[\ M \times S]$, where each column of the reduced matrix ($Pr_{m,s}$: $<y_\alpha>(3)$, $R^{(\alpha,\beta)}(3)$, $A_{\alpha\beta}(6)$, $I_4(1)$; $\alpha,\beta =$ 1,2,3) determines the complete combination of the moments and their intercorrelations (3+3+6+1=13) up to the fourth order inclusive. In the result of application of the 3D-DGI method, we obtain s=1,2,...,S (S=13) distributions that demonstrate the variations of each statistical parameter $Pr_s(m)$ with respect to the number of repeated measurements (m=1,2...,M).

2. The further reduction is possible if one takes into account that each random function $y_s(m) \equiv Pr_s(m)$ is located inside the rectangle $M \times (\text{Range}[y_s(m)])$, where $Range(f) = \max(f) - \min(f)$. For comparison one random function $y_{1,s}(m)$ with another $y_{2,s}(m)$ corresponding to the chosen parameter s (s=1,2,...,S), one can use the following simple formula

$$
\begin{aligned}
&Q_{1,2}(s) = \frac{Range(y_{1,s}) + Range(y_{2,s})}{Range(y_{1,s}, y_{2,s})}, \\
&Range(f_1, f_2) = \max(f_1, f_2) - \min(f_1, f_2).
\end{aligned}
\tag{2.16}
$$

This expression in spite of its simplicity is really effective for comparison of the statistical closeness/"resemblance" of a pair random functions belonging to the given/another sampling participating in comparison operation. Really, if the function $Q_{1,2}(s)$ is located in the interval $[1, 2]$, then the pair random functions are statistically *close* to each other. In this case, as it is easy to notice two rectangles associated with pair of functions cross each other.

In the case, when $Q_{1,2}(s)\in(0,1)$ one can conclude that the pair random functions compared (two rectangles do not cross each other) is statistically *different* from each other. Besides this important parameter (2.16), one can take into account the symmetry of the random function $y_1(m)$. Any random function located in the rectangle $M\times Range[y(m)]$ crosses the line $<Pr(m)>$ coinciding with its mean value. Therefore, for evaluation of the symmetry of a random function, one can introduce the following dimensionless value

$$Sm(y) = \frac{mean(y) - 0.5\cdot(\max(y)+\min(y))}{Range(y)}. \tag{2.17}$$

If the value $Sm(y)$ is located near the zero value ($Sm(y)\approx 0$), then the line $<y>$ divides the rectangle $M\times Range[y(m)]$ on two almost equal parts. In this case $mean(y) = 0.5\cdot(\max(y)+\min(y))$. In other cases, the value $Sm(y)\in[-1/2,\ 1/2]$ determines the measure of asymmetry. We should notice also that in the "ideal" and exceptional case, when the ranges of the compared functions in (2.16) equal zero, we receive uncertainty 0/0 that we put equaled to the unit value. After application of the expression (2.16) for comparing of similar columns (belonging to the same parameter Pr_s) one can obtain finally the vector of the length S=13 that contains information about the statistical closeness of two matrices compared. It is interesting to notice that simple expression (2.16) can be used also for comparison of each successive measurement with another one in the given rectangular matrix $[N\times M]$. If one compares the vectors y_m forming the columns of the initial matrix with each other then, in the result of application (2.16), one can obtain the symmetrical matrix $U(m_1,m_2)$ ($m_{1,2}$ $=1,2,\ldots,M$) with elements located in the interval $0\leq U(m_1,m_2)\leq 2$. Only elements located in the interval $1\leq U(m_1,m_2)\leq 2$ will correspond to a "good" experiment; while the elements from the interval $0\leq U(m_1,m_2)<1$ should be considered as possible "outliers" and correspond to "bad" experiment. The analysis of possible outliers needs a special attention of an experimentalist. They can be associated with errors of the used experimental equipment or they can be associated with unusual phenomenon/discovery that should be take taken into account and analyzed further.

3. How to find the parameters belonging only to one matrix in order to compare them with similar parameters of another tested matrix in cases when the reference/"pattern" matrix is absent? Initially, it is necessary to scale each column to the *same* interval

$$\begin{gathered}\mathrm{Prn}_s = \frac{\mathrm{DPr}_s}{Range(\mathrm{Pr}_s)} \equiv \frac{\mathrm{Pr}_s - \langle \mathrm{Pr}_s\rangle}{Range(\mathrm{Pr}_s)},\\ -\frac{1}{2}\leq \mathrm{Prn}_s\leq \frac{1}{2},\ \langle \mathrm{Pr}_s\rangle = \frac{1}{M}\sum_{m=1}^{M}\mathrm{Pr}_{m,s}.\end{gathered} \tag{2.18}$$

This normalization procedure makes all parameters Pr_s statistically close to each other with mean value equaled zero (mean(Prn_s)=0) and with the $Range(Prn_s)$ =1.

If one integrates expressions (2.18) for each parameter, one can get the statistically different curves JP_s = Integral(Prn_s) for each initial parameter (s=1,2,...,S). The distributions of the ranges of these integral curves P_1=$Range(JP_s)$ together with distribution of asymmetries P_2=$Range(Sm(JP_s))$ calculated with the help of the expression (2.17) give finally the matrix containing $[S\times 2]$ (S=13) rows and two columns only. If we calculate the ranges of these two columns, one obtains finally two values *only* that can characterize the initial matrix $[N\times M]$. If we have a set of matrices $[N\times M]_q$ (q=1,2,...,Q), then this simple and general procedure allows to select the "best" one having minimal values of these two key parameters $P_{1,2}$. It will characterize the stability of the initial sequence and their minimal values will serve as a criterion for selection of the "best" TLS among other TLS(s). We want to notice here that this final stage of treatment of "big" matrices is differed from the procedure used in paper [6]. We will explain this approach that is defined as CAPoNeF (Comparative Analysis of Positive and Negative Fluctuations) below. After finishing this section, one can formulate two conclusions that help to separate the "native" data from the "strange" ones and separate the "best" one from the limited set of selected data.

1. How to separate the "native" data from the "strange" ones if a set of "native" data are known? In this case the following algorithm can be applied.

Step 1. The initial matrix $[N\times M]$ with the help of the 3D-DGI – method is transformed to the reduced matrix $[M\times S]$.

Step 2. Then two or more reduced matrices $[M\times S]_q$ (q=1,2,...) are compared with each other by means of expression (2.16). How to do it? In this case, this expression can be modified as

$$Q_{1,2}(Y,y) = \frac{Range(Y_s^{(q)}) + Range(y_s^{(X)})}{Range(Y_s^{(q)}, y_s^{(X)})}. \tag{2.19}$$

Here, $Y_s{}^{(q)}$ defines the columns of a "native" matrix $[M\times S]_q$ with columns (s; s=1,2,...,13) and $y_s{}^{(X)}$ are referred to columns of the unknown matrix X fixing the same column s. If all compared columns satisfy the condition $1 < Q_{1,2}(Y,y) < 2$, then the columns of the matrix X becomes "native". In the opposite case, when for some columns r we obtain the inequality $0 < Q_{1,2}(Y_r,y_r) < 1$, then one concludes that these set of columns are "strange".

2. How to select the "best" matrix from the limited set of matrices in accordance with criterion described above? In this case, the expression (2.19) is *not* applicable because the initial criterion selecting a "normal" state from "strange" one is absent.

In this case, the first step reducing $[N\times M]_q$ to $[M\times S]_q$ is conserved. Then each selected matrix is normalized in accordance with (2.18) and two key parameters P_1 = $Range(JP_s)$ and P_2= $Range(Sm(JP_s))$ are calculated. The minimal values of these two parameters $P_{1,2}(s)$ taken over all matrix q can characterize the chosen matrix $[NM]_q$. Therefore, the minimal values of these two parameters $[\min(P_{1,2}(s)]_q$ can be used for comparison with other matrices. These parameters will be helpful for selection of the "best" matrix among others that were chosen for this verification by new statistical methods. Let us explain 3D-DGI method on available data.

2.3 THREE APPLICATION EXAMPLES

2.3.1 Example 1. Application of the DGI-method for differentiation of the trans-numbers π and E

The first example is associated with differentiation between two famous trans-numbers as π and **E**.

With the help of this method one can consider the 3D structure of two transcendental famous numbers: (a) $\pi = 3.14159265358979323846264338327 9\ldots$ and (b) Euler constant $\mathbf{E} = 2.71828182845904523536028747135\ldots$. Each of these numbers containing 60000 data points is combined into a set of normalized triads. These 20000 triads are transformed to an "ideal" noise in accordance with the following procedure: (a) the integer part is omitted; (b) each three digits of the infinite sequence form a normalized number in the interval [0,1] after division by 1000. For example, the number of forms the following sequence of numbers falling into the interval [0,1]. π = 3.141 592 653 589 793 238 462 643 383 279... → 0.141, 0.592, 0.653, 0.589, 0.793, 0.238, 0.462, 0.643,... Normalized triads for **E** are obtained similarly. Then, we apply the procedure of reduction to three incident points. This procedure is described in papers [19] and in the book [20]. It allows three types of noise to be obtained from a sequence, i.e., sequences of group-localized maximal ($Yup(x)$), minimal ($Ydn(x)$), and mean values ($Ymn(x)$), correspondingly.

After the application of the reduction procedure to three incident points (the compression parameter b=20, i.e., we compress twenty points in the sequence to their means, maximum and minimum values), each local set $\{b_1, b_2, \ldots, b_{20}\}$ is reduced to three incident points {max(b), min(b) and mean(b)}, giving the desired three sequences. We want to stress also that these three remarkable points are invariant relatively to all possible permutations inside the chosen local set $\{b\}!$. They are shown for the trans-number in Fig. 2.1 and contain 1000 data points only.

Relationship (2.14a) as a "fingerprint" gives three surfaces for the real part of the radial distance when Re$[R(\varphi,\theta))]>0$ and Im$[R(\varphi,\theta)]=0$ and, similarly, it can contain three additional surfaces for the imaginary part of the radial distance, implying that Im $R(\varphi,\theta) > 0$, when the radicand in (2.12a) becomes *negative*. As has been mentioned above, we omit these three additional surfaces for simplicity. Finally, for the identification of the chosen trans-number.

We have three 3D-surfaces and a *minimum* number of 13 parameters (we remind here that 3 parameters identify the gravity center $<r_i>$ (i=1,2,3); 3 parameters $R^{(1,2),(1,3),(2,3)}$ associated with the reduced triple correlations calculated from the linear system (2.6a) ; 6 parameters from (2.6) identifying the correlation of the second order and, finally, the value of the invariant I_4 from relationship (2.10). These 13 indications in the feature space are sufficient in most cases to find the basic differences between compared random sequences. In Figs 2.2–2.4, we show the three desired surfaces corresponding to trans-number. For the real values of the radial distance $R(\varphi,\theta) > 0$. The 13 parameters identifying these surfaces are given in Table 2.1. Similarly, we realize the same procedure for **E**. Compare Figs. 2.6–2.8. for **E** Figs. 2.2–2.4. The comparison of the reduced parameters given in Table 2.1 for the two trans-numbers clearly demonstrates their difference. These differences are seen in the

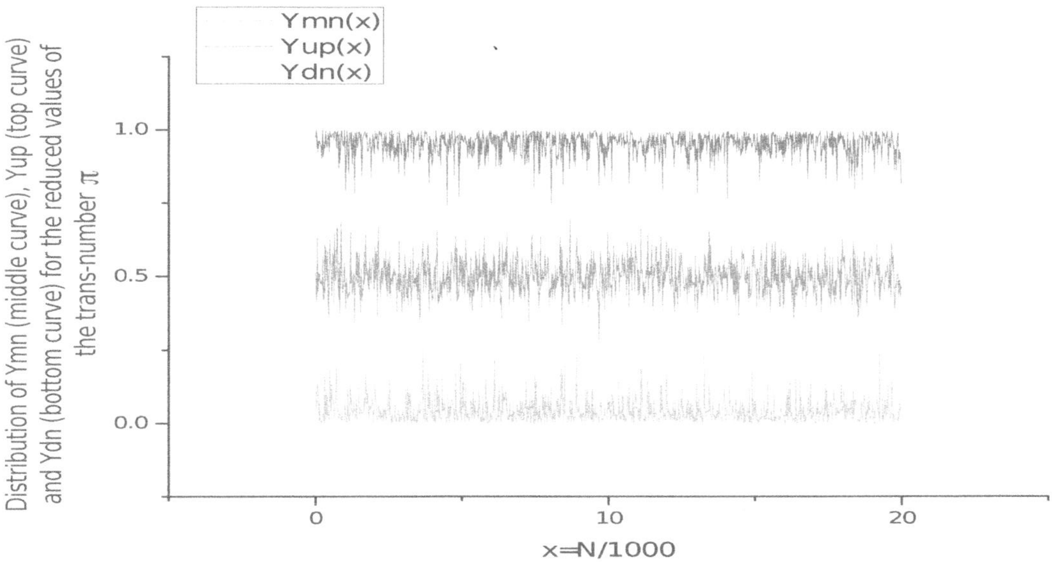

Figure 2.1 Three distributions of the "ideal" noise obtained for the transcendental number π. They are compressed by the factor b=20 and contain only 1000 data points. They form three initial random sequences that can be transformed into three 3D-surfaces (see the expression (2.14a) for identification of their differences and formation of a specific "fingerprint").

values of the parameters and even in their signs. It is not yet clear which parameter is the most sensitive to a possible perturbation (possible attack of the given "ideal" noise). The analysis of these possible perturbations can be a subject for further work.

As one can notice from Table 2.1, the calculated 13 parameters for this trans-numbers are different and, therefore, the applied method can differentiate these numbers from each other very easily.

2.3.2 Application of the DGI-method for description of earthquakes (EQs) signals

Close analysis of the 3D-DGI-method that has been outlined above allows to consider the *simplified* version, when one apply a *similar* digitization for the corresponding angles $\varphi_j = 2\pi \cdot j/N$ and $\theta_j = \pi \cdot j/N (j = 0, 1, ..., N)$. Actually, this simplified consideration allows to obtain specific "projections" of all three surfaces embedded to the plane j/N. In this case, we have three deterministic curves $y_{1,2,3}(j)$ that are defined as:

$$
\begin{aligned}
&y_1(j) = \langle r_1 \rangle + R(\theta_j, \phi_j) cos(\phi_j) sin(\theta_j), \\
&y_2(j) = \langle r_2 \rangle + R(\theta_j, \phi_j) sin(\phi_j) sin(\theta_j), \\
&y_3(j) = \langle r_3 \rangle + R(\theta_j, \phi_j) cos(\phi_j), \\
&\phi_j = 2\pi(\frac{j}{N}), \;\; \theta_j = \pi(\frac{j}{N}), R_j \leq 0, j = 0, 1, ..., N.
\end{aligned}
\tag{2.20}
$$

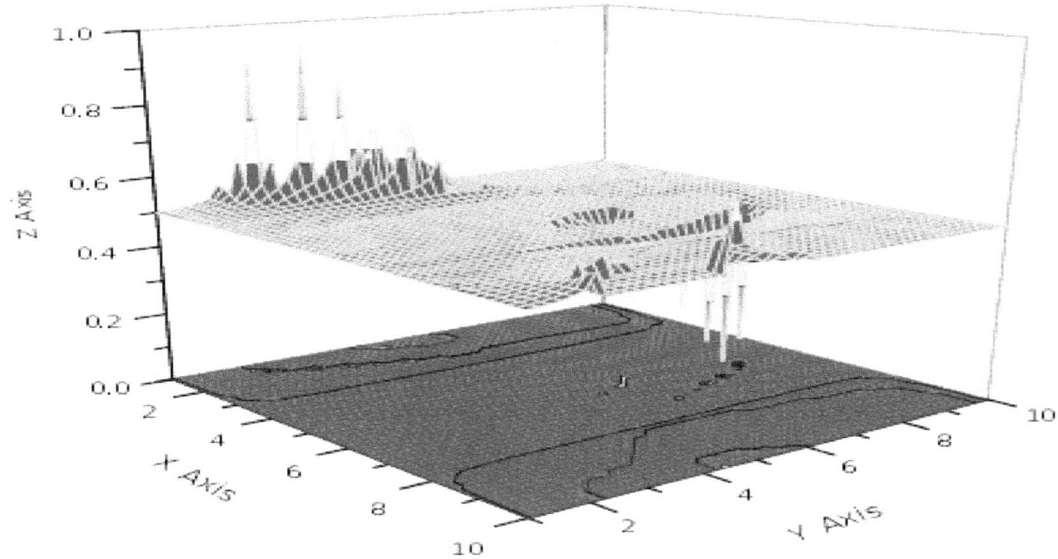

Figure 2.2 This is the first 3D surface formed by the curve $y_1(\varphi, \theta)$ for trans-number π. The scale for the angles ϕ [OX axis] and θ [OY axis] is given in the discrete numbers of the visible data points. Each angle is digitized as $\varphi_j = 2\pi \cdot (j/10)$, $\theta_j = (\pi \cdot j/10)$ and $j=0,1,\ldots,10$.

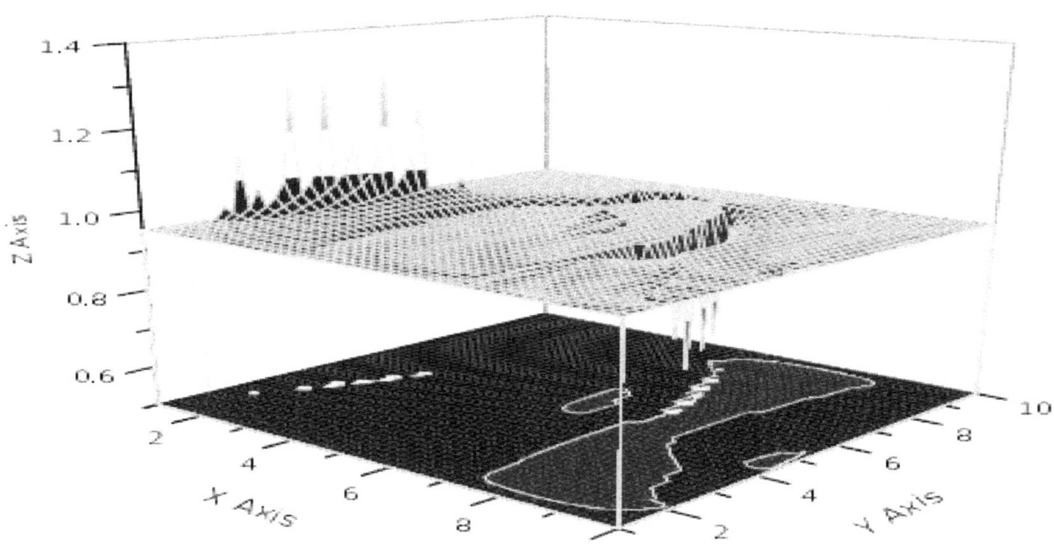

Figure 2.3 Here we display the surface formed by the curve $y_2(\varphi, \theta)$. The scale for the angles φ [OX axis] and θ [OY axis] is given in the numbers corresponding to the discrete data points (j=0,1,..10).

These three functions can serve for identification and quantitative description of the given random/deterministic sequences. The compact number of parameters

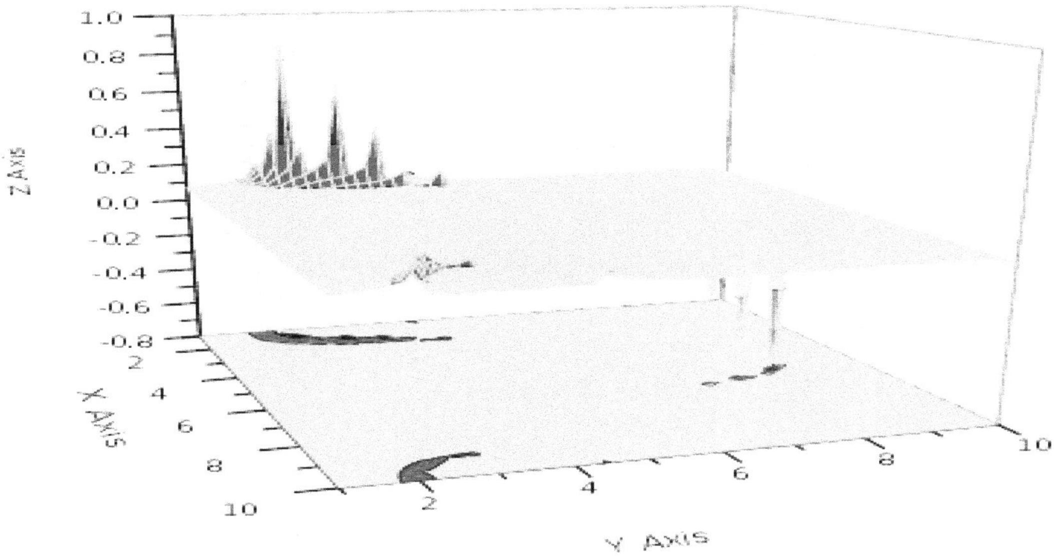

Figure 2.4 Here we display the surface formed by the curve $y_3(\varphi, \theta)$. The scale is similar to that of the previous figures.

TABLE 2.1 The comparison of the 13 basic parameters that correspond to the transnumbers π and E.

Parameter definition	Trans-number π	Trans-number E
$<r_1>$	0.49858	0.49959
$<r_2>$	0.95417	0.9529
$<r_3>$	0.04474	0.04367
$R^{(1,2)}$	8.0119	2.6854
$R^{(1,3)}$	6.2287	2.7388
$R^{(2,3)}$	40.133	−3.3575
$A_{1,1}$	−0.02353	0.00311
$A_{2,2}$	−0.19443	0.00503
$A_{3,3}$	−0.16048	0.0045
$A_{1,2}$	0.04525	−0.0178
$A_{1,3}$	0.03694	−0.02339
$A_{2,3}$	0.13709	−0.00428
I_4	−2.8159E-4	−1.3277E-5

that determine the behavior remain the *same* and the number is 13. The authors decided to apply this simplified analysis to the differentiation and recognition of available EQ (earthquake) data. As it is known, the direct quantitative description of EQ perturbations represents itself an *unsolved* and complex problem. Recently, one of us (RRN) together with other geophysicists was able to apply the NAFASS approach for describing the envelopes of some set of EQs [14]. However, in many cases it is *not* sufficient. It is desirable to have a unified quantitative platform, which enables a comparison of various EQs recorded from different sources. The approach proposed here can serve as a promising tool for the solution of this problem. Referring

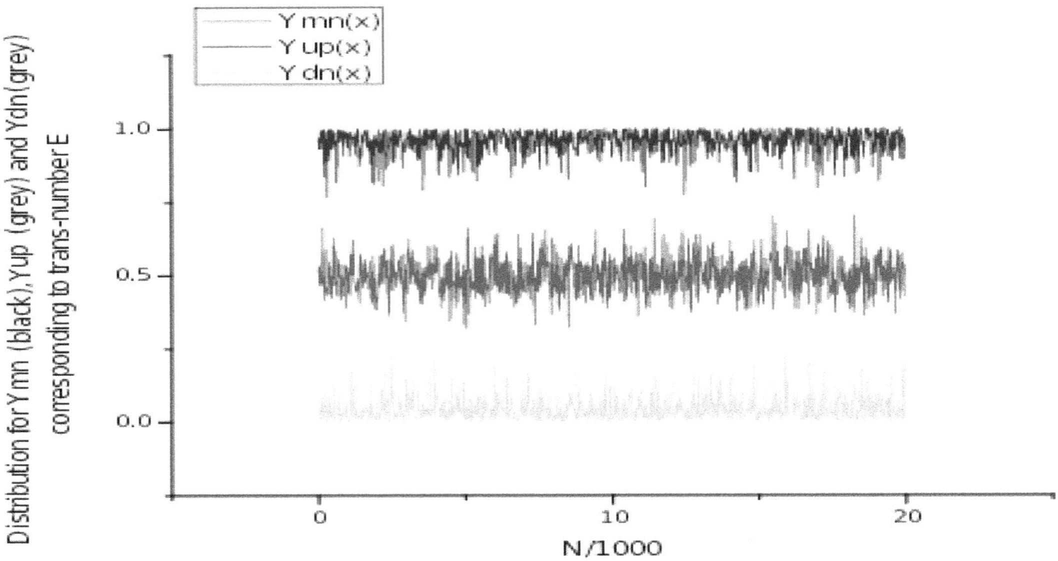

Figure 2.5 Three distributions of the initial "ideal" noise obtained for the transcendental number **E**, with the initial sequence compressed by the factor b=20 and containing again 1000 data points. They form three initial random sequences that can be transformed into three 3D-surfaces for identification of their differences and formation of a specific "fingerprint". Visually, this noise looks like the "ideal" noise depicted in Fig. 2.1.

a potential reader to the reference [14] for details. The authors decided to compare six randomly taken EQ signals recorded from one source (selected EQs signals were obtained from Karagay Bulak station, the Kyrgyz Republic). We are not going to consider this problem in detail, for us it is important only to demonstrate a possibility of applying the DGI-method (expressed in its simplest form [19]) to this complex phenomenon as a quantitative description and possible classification of the complex EQs signals.

The applied algorithm remains partly the same as was used for the trans-numbers described above. We reduced the chosen EQ signal to three signals selecting again their sequences (in terms of the localized maximal ($Y_{up}(x)$), mean ($Y_{mn}(x)$), and minimal ($Y_{dn}(x)$) values), correspondingly. The compression factor $b = 5$. A typical triple EQ signal compressed by a factor $b = 5$ is shown in Fig. 2.9. Application of expressions (2.19) allows the desired $y_{1,2,3}(j)$ projections ($j = 1, 2, N = 50$) to be calculated. These three functions, associated with specific projections of three surfaces on the common plane, can be considered to be the quantitative identifiers of the given EQ-1 signal. These function-projections are shown in Fig. 2.10. Similarly, one can calculate other EQ signals (2–6). Their corresponding projection functions together with their signals are depicted in Figs. 2.11–2.15, and their parameters collected in Table 2.2. Preliminary analysis of these figures allows it to be concluded that the functions $y_{1,2,3}(j)$ for the EQs -1,3,4,6 have the *same* statistical nature. The other two EQs -2,5 are different and should be analyzed separately. Therefore, even the simplified version of the DGI approach, realized in a planar projection, allows a *naive* user (i.e.,

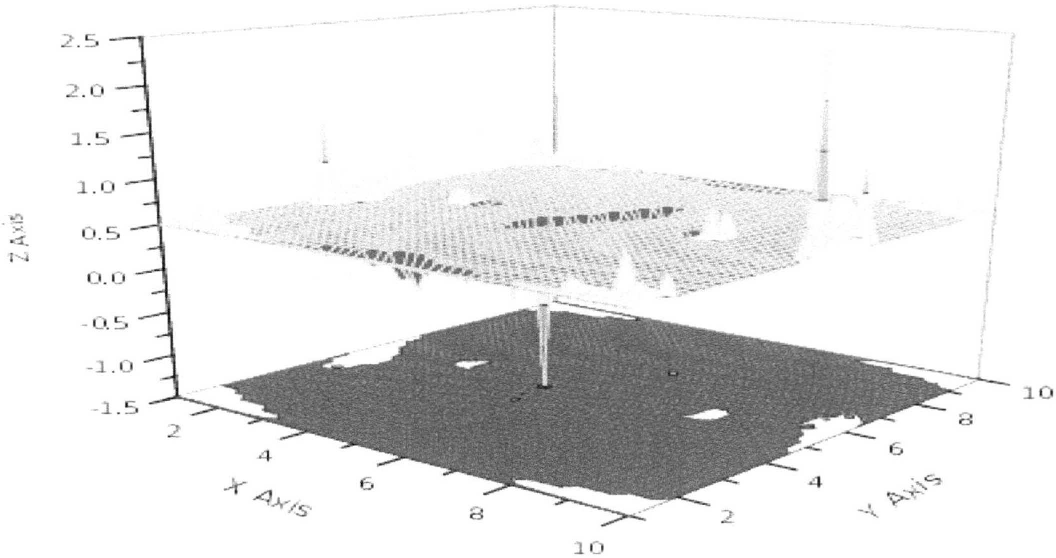

Figure 2.6 This figure corresponds to the surface Z=$y_1(\varphi,\theta)$ for the trans-number **E**. The scale for the angles φ [OX axis] and θ [OY axis] is given in the numbers of the visible data points. Each angle is digitized as X Axis $= \varphi_j = 2\pi\cdot(j/10)$, Y Axis$=\theta_j= \pi\cdot(j/10)$, j=0,1,...,10.

for a person that does not have any specialized geophysical education) to differentiate the EQs and select those that are similar with close statistical characteristics. It implies that this routine work can be realized with the help of a computer program containing the DGI algorithm.

2.3.3 Example 3. Comparison of the TLS(s) measured from electric circuits.

Description of electric circuitry. (Sometimes it is useful to describe the electric gadgets, especially for young researchers for their possible reproduction).

ELVIS II workbench by National Instruments was used as ADC (analog digital converter). ELVIS II has 1.25 MS/s (Mega samples per second) maximum sampling rate and 16-bit resolution [21]. As other ADC(s), ELVIS II contains some circuits for initial signal processing. The simplified block diagram of the channel for digitizing of an analog signal is shown in Fig. 2.16. This workbench represents itself an analog input circuitry. In Fig. 2.16 the abbreviation "MUX" defines a multiplexer. It routs one Analog Input (AI) channel at a time to the ADC, because ELVIS II uses one ADC for several AI(s). PGIA serves as an abbreviation for the Programmable Gain Instrumentation Amplifier (PGIA). The PGIA amplifier attenuates a signal to obtain the maximum resolution of the ADC. As Analog Input Lowpass Filter is a mandatory block of any ADC, it is necessary to suppress aliasing phenomenon. Any signal digitization contains anti-aliasing filtering. The passband of the used filter starts from "zero" Hz to the upper cutoff frequency that equals to the sampling frequency divided by two. The signal that is passed only through the anti-aliasing filter is defined as the "unfiltered" signal. One can change the lower and upper cut-off frequencies

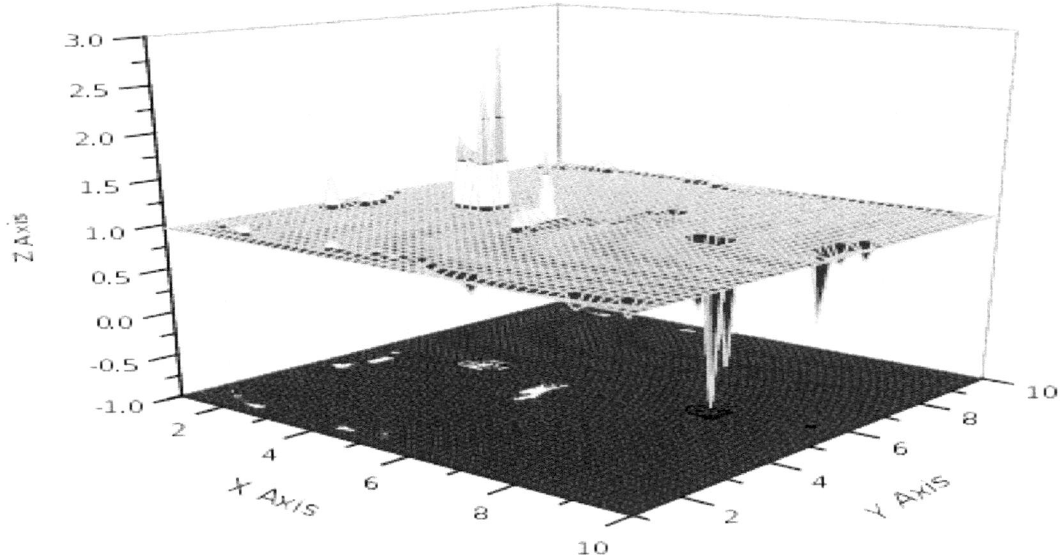

Figure 2.7 This figure corresponds to the surface Z=$y_2(\varphi,\theta)$ for the trans-number **E**. The scale for the angles φ [OX axis] and θ [OY axis] is kept the same. Each angle is digitized as X Axis $= \varphi_j = 2\pi\cdot(j/10)$, Y Axis=$\theta_j = \pi\cdot(j/10)$, j=0,1,. . .,10. This figure should be compared with Fig. 2.3, where the same surface Z=$y_2(\varphi,\theta)$ for the trans-number is shown.

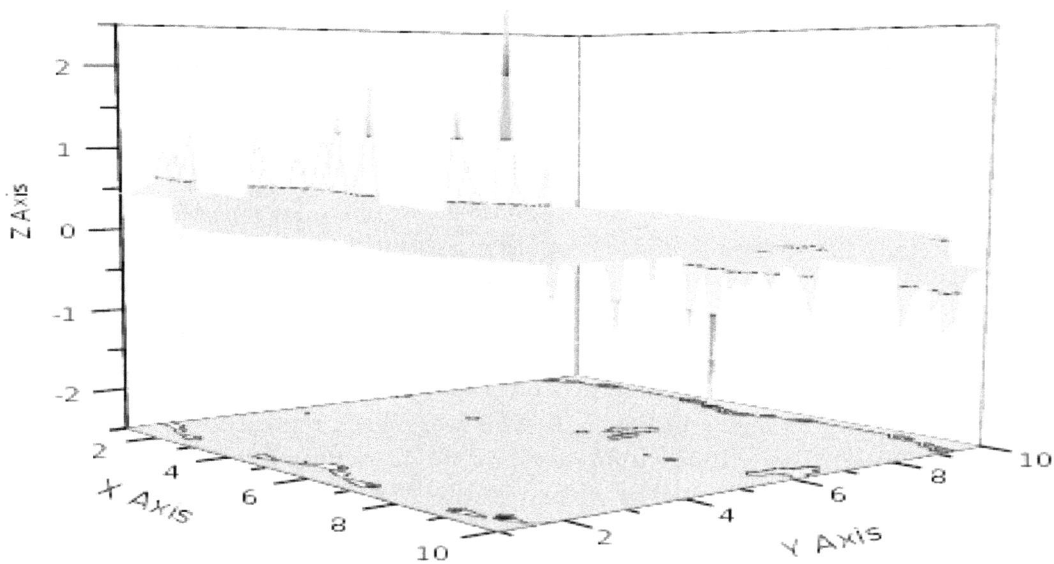

Figure 2.8 Here we display the surface formed by the curve $y_3(\varphi,\theta)$. The scale and axes notations are similar to previous figures.

to the desired values. After realization of this procedure, a digitized signal should be passed through a digital filter. If a signal is passing through an additional digital

TABLE 2.2 The comparison of the 13 basic parameters that correspond to the EQs -1–6.

Parameter definition	EQ-1	EQ-2	EQ-3	EQ-4	EQ-5	EQ-6
$<r_1>$	−0.00685	−0.54278	−0.02754	0.53838	−0.56875	0.51126
$<r_2>$	0.15427	−0.03818	0.13593	1.0115	−0.06519	1.0299
$<r_3>$	−0.16776	−1.07100	−0.19439	0.03459	−1.07987	−7.7699E-4
$R^{(1,2)}$	−11.675	1.41378	0.64384	0.81474	−1.79039	0.54332
$R^{(1,3)}$	−2.6388	0.28944	1.08751	1.3083	1.06854	1.1826
$R^{(2,3)}$	−1.6658	4.21355	0.55153	−0.07425	1.41171	0.13465
$A_{1,1}$	5.2946	8.14784	0.32308	1.9778	12.24669	4.8296
$A_{2,2}$	4.5826	−11.39183	0.59027	8.2579	9.36705	13.466
$A_{3,3}$	0.43886	1.68236	0.65184	4.1815	4.74627	5.2402
$A_{1,2}$	11.17	2.12666	−0.09848	−4.1677	25.24948	−2.6723
$A_{1,3}$	−1.1022	10.16391	−0.80483	−7.2016	−17.47757	−9.7788
$A_{2,3}$	−2.463	−28.90426	0.00492	4.5317	−19.20899	4.0414
I_4	1.1754	−2.82897	0.04084	5.2987	4.35078	15.882

The EQs having similar statistical nature are bolded.

filtering, one can define it as the filtered signal. In this paper, the filtered signals are not considered. We connected 50 Ohm resistor between analog input and the "ground". It simulates the output impedance of the circuit that can be connected to the given ADC. The input signal range has been programmed and finally the PGIA has a gain equals unit value. Such connection provides the conditions that the ADC itself becomes the *main* source of a noise in the presented circuitry. Four different types of ELVIS II were used in the experiment. In this chapter, they are defined as "ELVIS II-L" (L = 1,2,3,4). Fifteen (m=1,2,3...,M; M=15) successive measurements were realized using the selected type of the embedded ADC belonging to the chosen ELVIS II-L. Each of these measurements was carried out with the fixed 50 Ohm resistor at the input (as it was mentioned earlier) and $2 \cdot 10^5$ data points were recorded with a sampling frequency 10 kHz.

Description of data treatment algorithm: From the result of the procedure described in the previous section, one can obtain 4 matrices. Each rectangular matrix includes $N = 2 \cdot 10^5$ data points and M=15 successive measurements corresponding to the chosen "ELVIS II -L" (L =1,2,3,4) workbench. The problem is that one can find the differences between them and choose the "best" one among them. In order to solve this problem, one can propose the following algorithm. It is divided on the following steps.

S-1. In order to decrease the computational time associated with treatment of long sequences ($N > 10^5$), one can suggest the procedure of reduction to three incident points that was described above and successfully used earlier in papers [19, 22]. This procedure helps to divide initial TLS on three independent parts – Yup_j, Ymn_j, Ydn_j, where number of data points is compressed j=1,2,...,$[N_c = N/b]$, and the compression parameter is equaled b=200, the rectangular brackets [...] determine the integer part extraction operation. These three obtained random sequences $[Yup_j,\ Ymn_j,\ Ydn_j\ ]$ describe the distribution of the "up", "mean", and "down" amplitudes, accordingly.

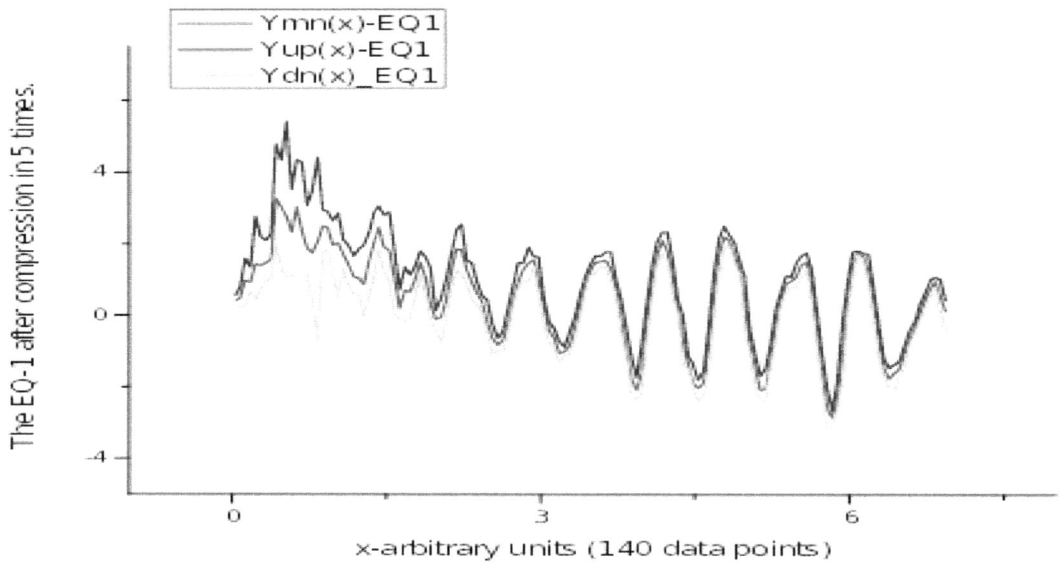

Figure 2.9 The recorded signal corresponding to the EQ-1. After compression by a factor of 5 (b=5) the reduced signal contains 140 data points only. (Ymn - middle curve; Yup - upper curve; Ydn - bottom curve.)

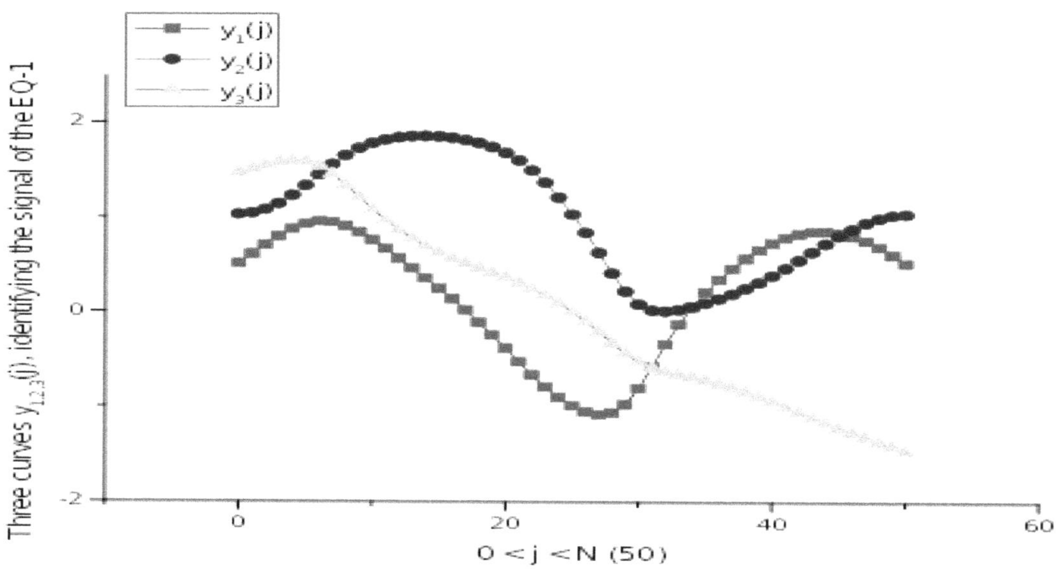

Figure 2.10 Three curves $y_{1,2,3}(j)$ that were calculated with the help of the expression (2.19) and correspond to the EQ-1 signal in 3D – DGI projection space. All 13 quantitative parameters are collected in Table 2.2.

This procedure helps to obtain three similar TLS(s) from the initial sequence and it is illustrated in Fig. 2.17. where three reduced TLS(s) for the first sequence (ELVIS II -1) are shown.

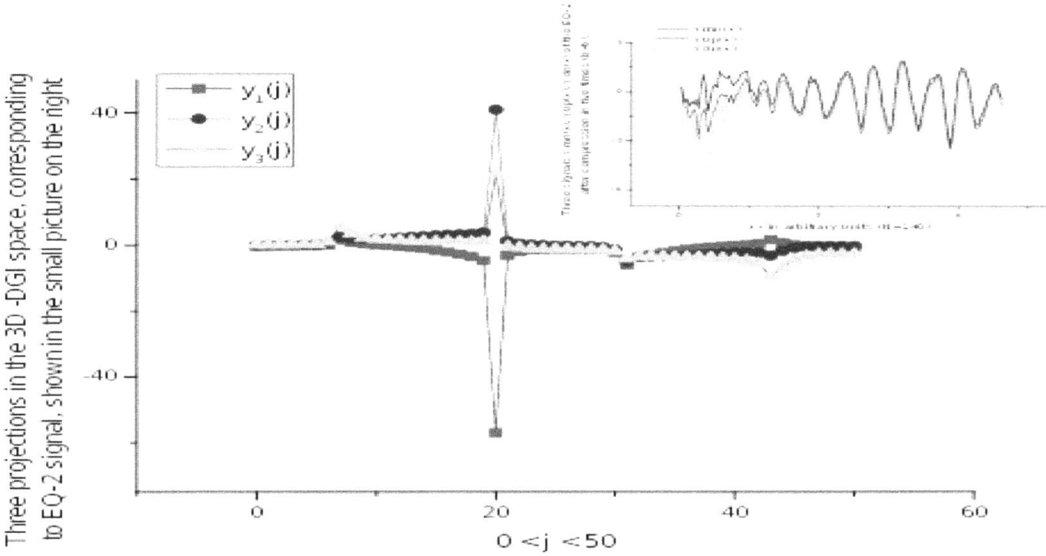

Figure 2.11 Three curves corresponding to the projections of the EQ-2 signal. As one can notice this signal is *different* from the signal shown in Fig. 2.9. Therefore, the usage of the $y_{1,2,3}(j)$ projections helps to differentiate them, at least qualitatively.

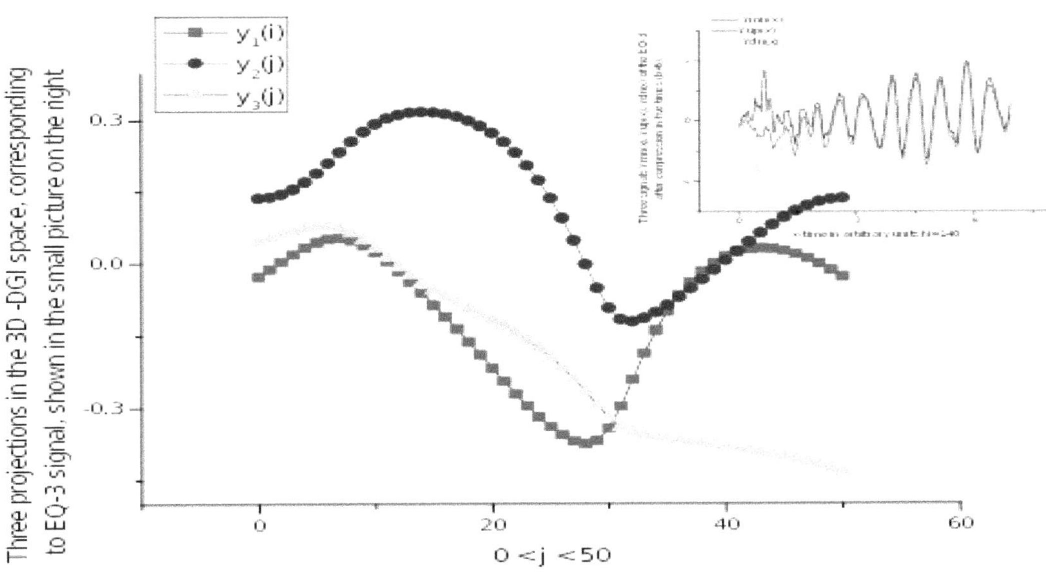

Figure 2.12 Comparison of this figure with Fig. 2.10 allows a preliminary conclusion about the statistical similarity of the EQ-1 with EQ-3 to be made, i.e., the corresponding projections $y_{1,2,3}(j)$ are very similar to each other.

S-2. From the result of application of the first step, one can prepare $3[N\_c=1000 \times M=15]$ matrices corresponding to each "Elvis II -L" (L =1,2,3,4). Then these three matrices are reduced to one matrix having the size $(M=15 \times S=13)$ with the

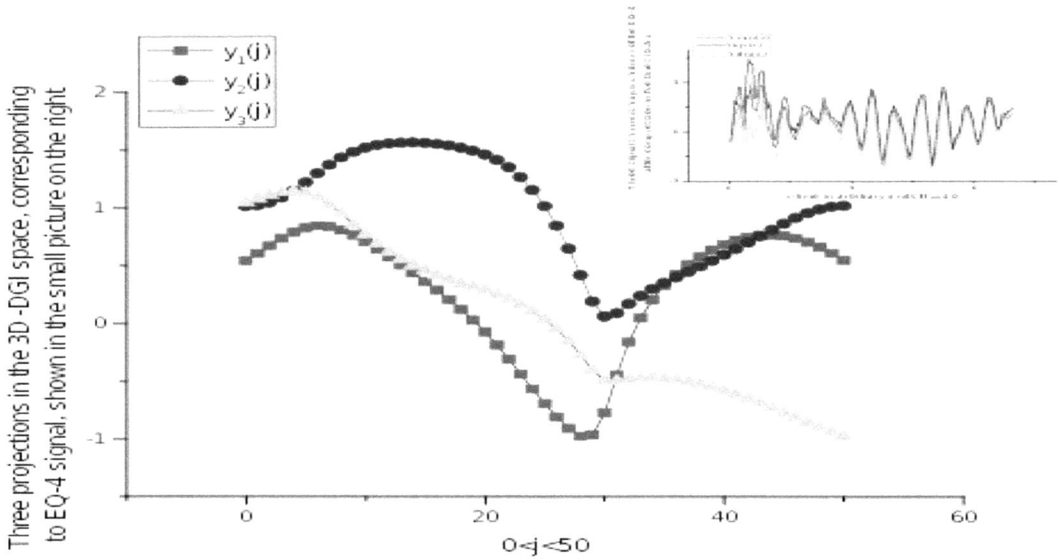

Figure 2.13 Plots for $y_{1,2,3}(j)$ for the EQ-4 signal. In spite of their quantitative difference, the EQs-1,3,4 have the same statistical nature.

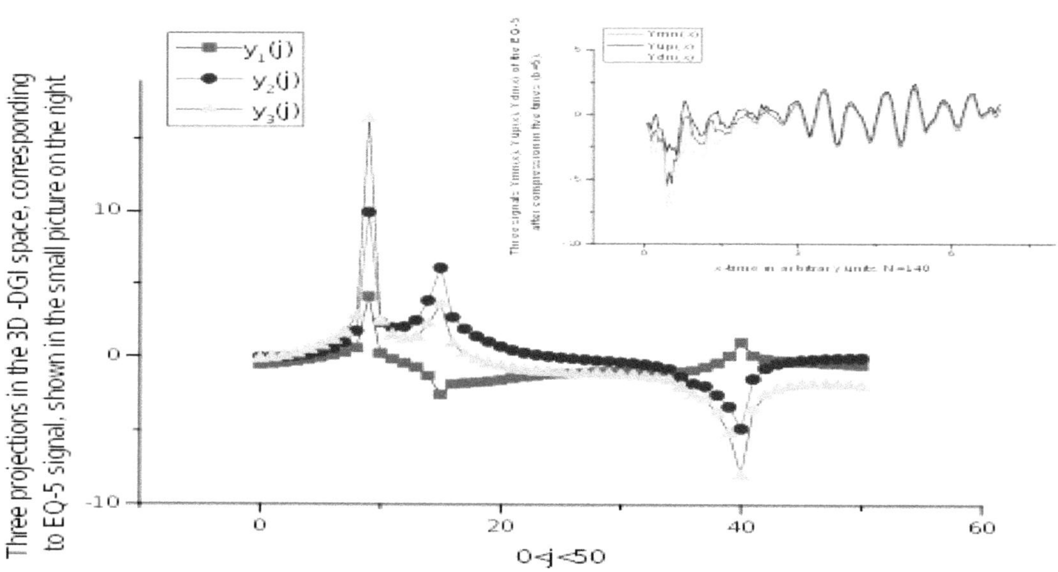

Figure 2.14 The projections $y_{1,2,3}(j)$ of the EQ-5 have quite different behavior to those of EQ-1,3,4,6 and the nature of this signal should be analyzed separately.

help of 3D-DGI method, described above. Each calculated set combining 13 parameters corresponds as minimum to three surfaces that are defined by equation (2.14a). The most significant surface is associated with the behavior of the radius modulus $|R(\theta,\varphi)| = \sqrt{[\mathrm{Re}(R(\theta,\varphi))]^2 + [\mathrm{Im}(R(\theta,\varphi))]^2}$. It is depicted in Fig. 2.18 for three reduced curves shown in the previous Fig. 2.17.

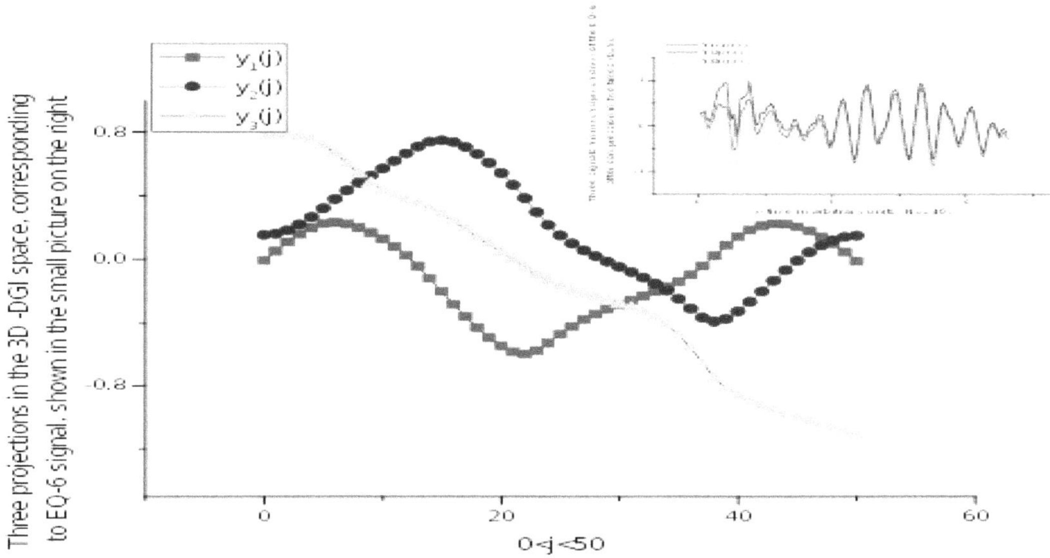

Figure 2.15 The behavior of these projections $y_{1,2,3}(j)$ corresponding to EQ-6 are, in spite of their differences in amplitudes, similar to EQs-1,3,4. Therefore, this analysis of the different EQs enables them to be classified as similar or different.

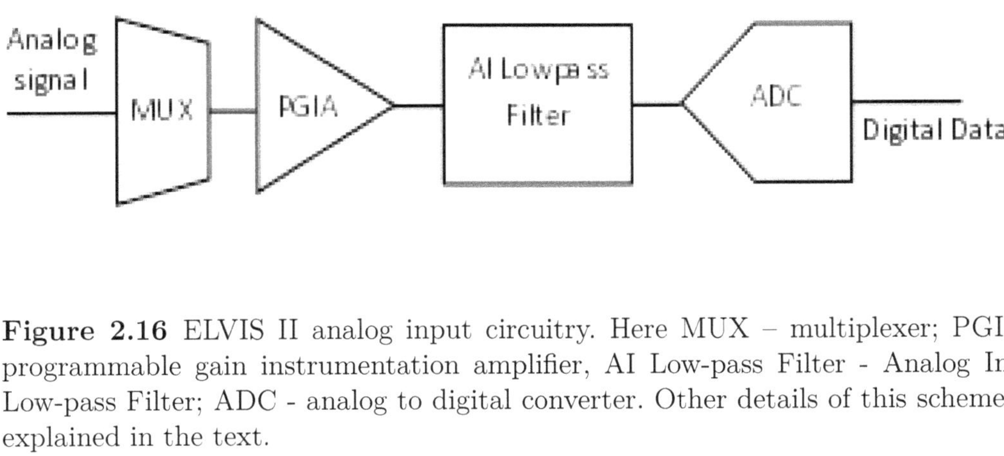

Figure 2.16 ELVIS II analog input circuitry. Here MUX – multiplexer; PGIA – programmable gain instrumentation amplifier, AI Low-pass Filter - Analog Input Low-pass Filter; ADC - analog to digital converter. Other details of this scheme are explained in the text.

In order to save place for the key figures we show only the basic stages and demonstrate the differences between parameters obtained with the help of the expression (2.16).The necessary comments are given in Table 2.3. The final stage allows to select the "best" ELVIS II -L based on the range values of the integrals (depicted in Figure 2.19). Figure 2.20 demonstrates the behavior of the asymmetrical coefficients for all workbenches. The range value among selected samplings allows to select the "best" Elvis II workbench. These two key parameters are listed in Table 2.4. Other

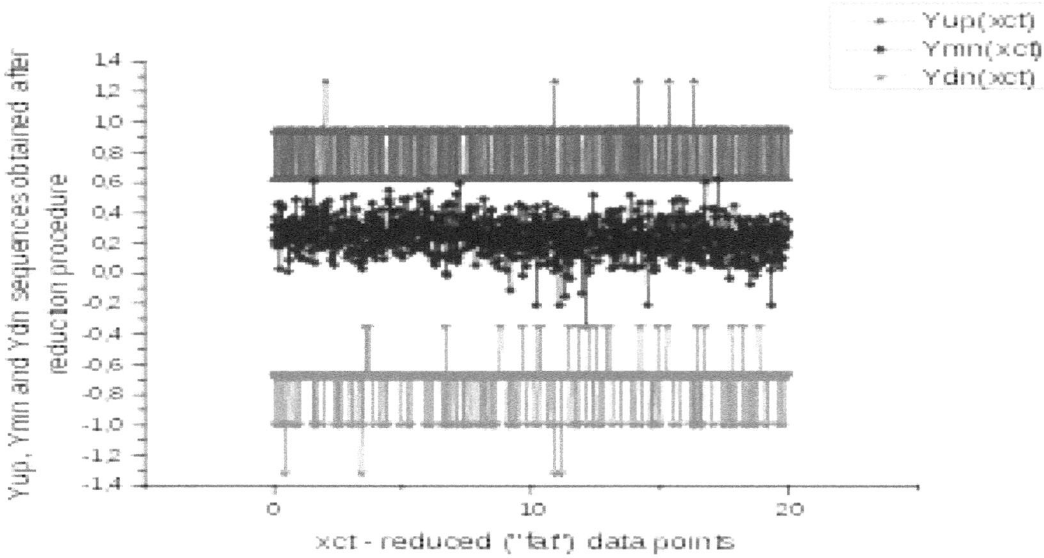

Figure 2.17 This figure demonstrates the result of application of the reduction procedure to three incident points. In the result of this procedure we obtain three TLS(s). The upper and down distributions $Yup(x)$, $Ydn(x)$ describe the distributions of the "up" and "down" amplitudes. The TLS $Ymn(x)$ corresponds to distributions of the mean values.

TABLE 2.3 The comparison of similar columns obtained with the help of the expression (2.16) for all 13 nonnormalized parameters.

(E-1-E-2)	(E-1-E-3)	(E-1-E-4)	(E-2-E-3)	(E-2-E-4)	(E-3-E-4)
0.43676	**0.19879**	1.2805	**0.12805**	**0.72951**	**0.22895**
0.39943	**0.16599**	1.0712	**0.08564**	**0.80189**	**0.18836**
0.2704	**0.12796**	**0,91776**	**0.06911**	**0.54511**	**0.14546**
1.0966	1.3844	1.6007	1.1938	1.2278	1.4019
1.111	1.2847	1.517	1.5039	1.2453	1.5155
1.1336	1.3775	1.3336	1.4025	1.2653	1.5234
1.0987	1.3176	1.4593	1.3384	1.2559	1.7405
1.1001	1.3913	1.5714	1.2535	1.2226	1.5232
1.1246	1.2474	1,5192	1.5837	1.2593	1.4406
1.1012	1.3645	1.6297	1,1954	1.1387	1.3812
1.1121	1.2403	1.379	1.6331	1.3413	1.5438
1.4622	1.5436	1.2641	1.2879	1.1517	1.1096
1.087	1.457	1.7088	1.2074	1.2206	1.6099

four figures (Figs. 2.21–2.24) show four 3D-surfaces that use 13 parameters from the integral curves shown in Fig. 2.20. In order to use these curves for calculation of the desired 3D "effective" surfaces, each integral curve is dissected into four parts: (a) the first three points can be referred to the center of the gravity point <x>, <y>, <z>, other three can be associated with triple correlations ($R_{1,2,3}$), other six (see the expression (2.9)) for characterization of pair correlations and the last parameter can be associated with the value of the invariant I_4 (see the expression (2.10)). This

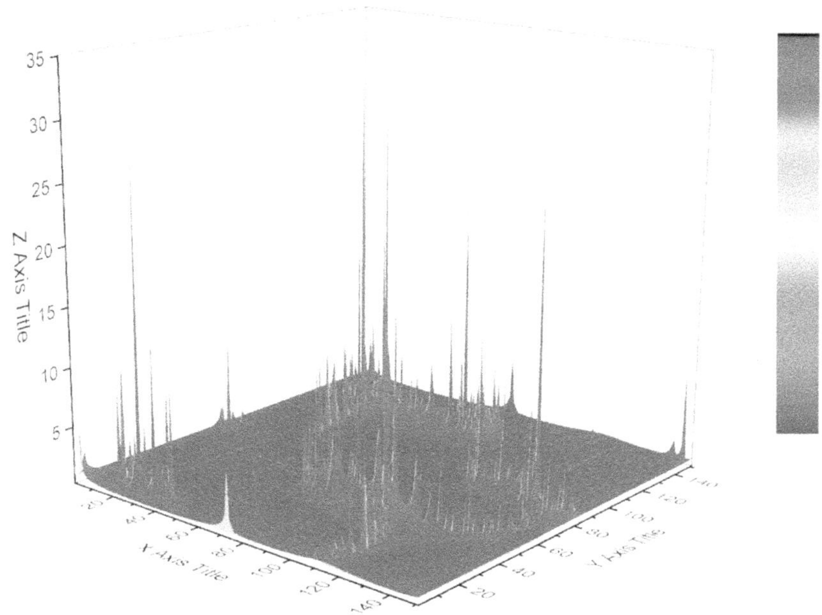

Figure 2.18 With the help of 3D-DGI method one can transform the previous plane TLS(s) depicted in Fig. 2.17 to the 3D-surface. Here, we show the dependence of the radius $|R(\theta,\varphi)|$ modulus with respect to spherical angles. Each angle has 150 data points that are shown clearly in this figure.

TABLE 2.4 Summary of curve integral and the value of asymmetry for 4 workbenches.

The type of the ELVIS II	Range of the Integral curve	Range the values of asymmetry
E - 1	8.49494	1.22538
E - 2	10.56565	0.76423
E - 3	8.87181	1.35162
E - 4	**7.04651**	**1.25806**

procedure helps to replace 15 surfaces for $|R(\theta,\varphi)|_m$ $(m = 1, 2, ..., 15)$ by one "effective" 3D-surface associated with integral curve and characterizing the given TLS data belonging to the chosen ELVIS II-L.

2.3.4 Discussion of the preliminary results

In this chapter, we obtained the complete form for the 3D-DGI surface. This surface contains 13 parameters that determine the desired feature space. These parameters, in turn, include the combination of the moments and intercorrelations up to the fourth

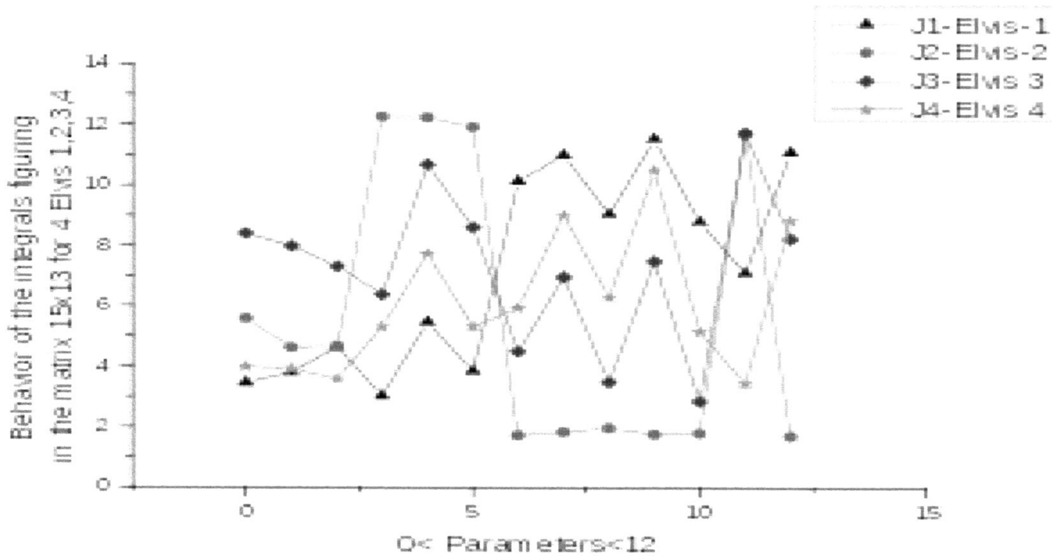

Figure 2.19 The behavior of the integrals that appear in the result of the reduced procedure. One can divide these values into 4 parts (the first three parameters associated with gravity center), other 3 with parameter R, 6 parameters to pair correlations A and the final parameter is identified with the value of the invariant I. It helps to construct the modulus $|R(\theta, \varphi)|$ for an "effective" surface and replace the set of 15 surfaces corresponding to each sampling (m=1,2...,M; M=15). These three surfaces for 4 workbenches are shown below.

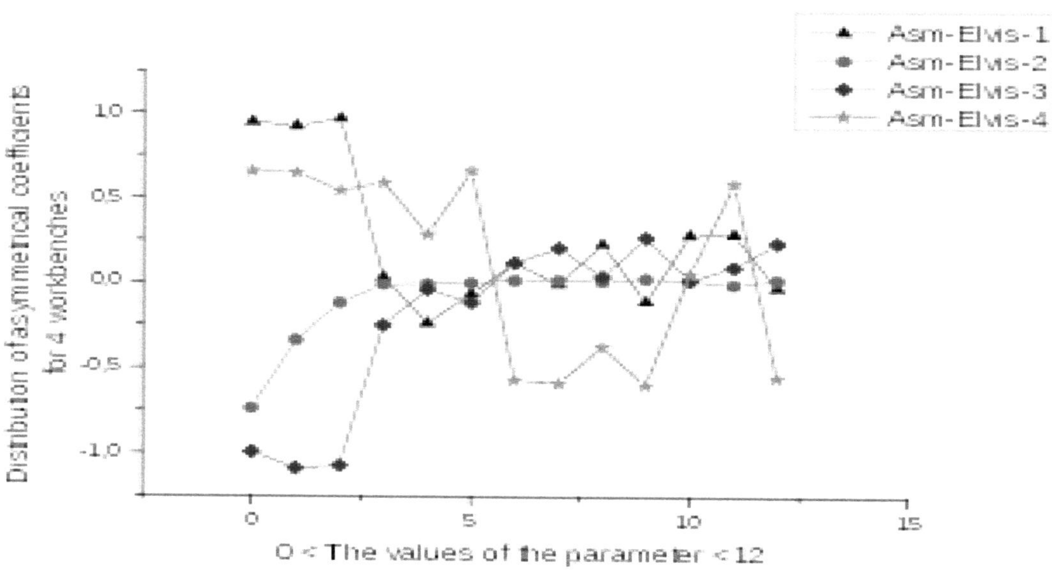

Figure 2.20 The behavior of the distributions $Sm(y)$ corresponding to the expression (2.17) for all 4 workbenches.

Figure 2.21 This reduced surface for ELVIS II-1 that is based on the system parameters corresponding to Fig. 2.18. This surface replaces all set of 15 surfaces corresponding to each selected sampling.

order inclusive. From mathematical point of view, the 3D-DGI method is derived in the result of diagonalization of the fourth-order form. As is it known that the modern statistics is based on diagonalization of the quadratic forms only and the moments of the third and fourth order were *not* taken into account. This method is applicable in the consideration of three arbitrary random sequences having N data points. The selection of an *optimal* triple combination associated with the chosen random sequence represents a problem and merits a further research. Here, it is reasonable to use the procedure of reduction to three incident points. However, other options are also possible. Besides the description of 3D-DGI method, the general reduction procedure is suggested that was proved to be helpful in comparison of "big" rectangular matrices with each other. If all data can be dismembered into two parts – "friend-or-foe", the simplest expression (2.16) becomes efficient, when this division onto reference/tested data) is absent.

Remark 1. *Comment to Table 2.3.*

The bold lines show the differences between each couple of matrices (1513) characterizing each workbench "ELVIS-L" (L=1,2,3,4). As one can notice from this table the minimal differences are observed between the couple (ELVIS-1,4). Other compared pairs have the same 3 different parameters referring to 3 gravity centers (GC) as the most sensitive to possible differences. The data of this table correlate with Figs. 2.6–2.9. The similar pair of workbenches (1,4) has only one difference parameter. Another pair (2,3) has different parameters close to zero, it means that the ratio in the expression (2.16) close to uncertainty 0/0, when minimal of numerator is less than the

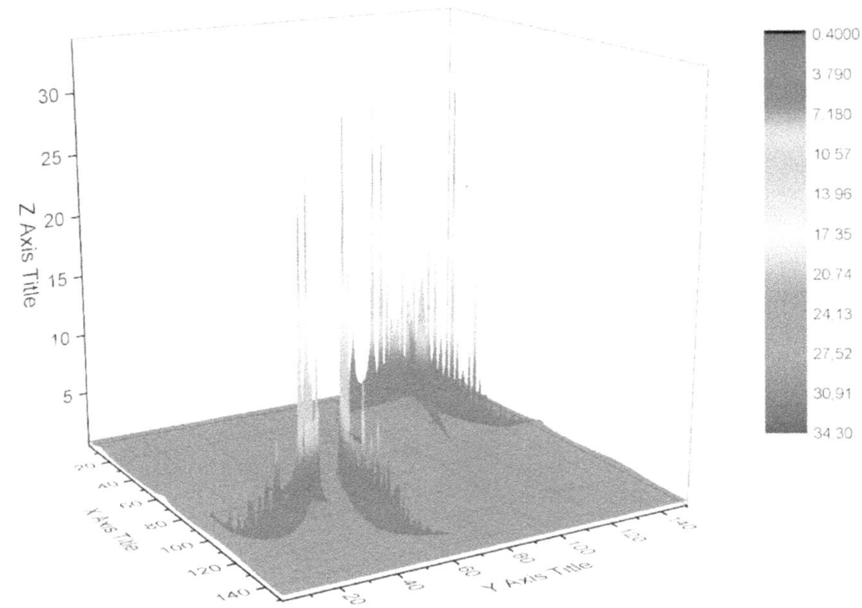

Figure 2.22 This plot constructed from the curve J2-ELVIS II-2 (depicted in Fig. 2.3) dissected preliminary into the combination of 13 data points: gravity center-(1-3); R-triple correlations (4-6); A – pair correlations (7-12) and the value of the invariant I_4 (1) can serve as a specific "fingerprint" characterizing the TLS(s) of the ELVIS II-2. It can be used as a specific visual identifier for all measured matrix $(2 \cdot 10^5)$ data points 15 successive measurements.

minimal value of the corresponding dominator. They are close to each other. This observation is confirmed by the corresponding Figs. 2.21 and 2.23.

As it follows from Table 2.4, the "best" workbench corresponds to E-4 (bold curve) after that the others can be listed as E-1,3,2, correspondingly. We think that this simple selection scheme can be applied to other data expressed in the form of their rectangular matrices.

In these cases, one can use the procedure based on the expressions (2.17) and (2.18). The attentive reader may notice that this approach can be applied successfully to the analysis of different images as well. Really, let us suppose that some rectangular matrix corresponds to an initial "image". With the help of the procedure described above, one can reduce the initial image to an "effective" surface containing for its construction 13 parameters only. This effective surface can serve as a specific "fingerprint" differentiating the chosen "image" from another one. Besides, this effective image helps an operator/robot in making the right decision. This simple idea for its justification merits the further research. The TLS data obtained from four ELVIS

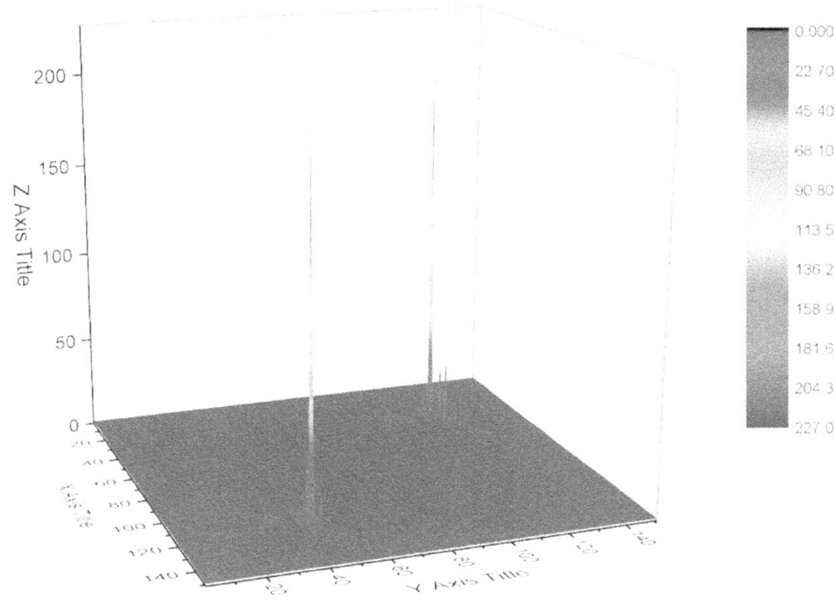

Figure 2.23 The reduced surface constructed from the curve J3-ELVIS II-3. The same procedure is realized for the curve J2-Elvis II-2 depicted in the previous figure.

II workbenches showed the effectiveness of the proposed approach and allowed to select the "best" one (see Table 2.4).

The authors do believe that potential researchers will receive a new original and rather general tool for treatment and comparison of different data, especially associated with rectangular matrices having "large" sizes. This method enables to suggest more accurate scheme for classification of different "color" noises including Gaussian, heat, and flicker-noises based on the unified platform, containing integer moments and intercorrelations, up to the fourth order, inclusive. One can stress here that other platforms of such kind *cannot* be realized, because the analytical separation of the polynomial forms exceeding the fourth order becomes *impossible*. Therefore, the 3D-DGI method representing itself a "specific area" for comparison of random functions is general and unique. This method opens a new horizon in analysis of different random fluctuations/functions and because of its importance merits further research.

To the end of this section, the following are to be noted:

- It can differentiate the real image from the image created by the artificial intellect.
- If these 13 proposed parameters are not sufficient, then one can propose the 3D-DGI of the third order. This variant will be given in *Reading Randomness II*.

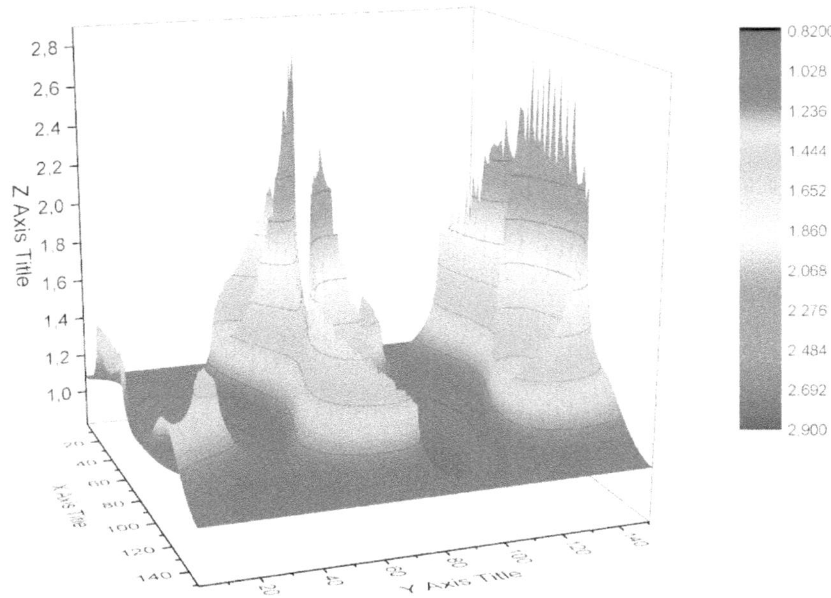

Figure 2.24 The reduced surface constructed from the curve J4-ELVIS II-4. It is interesting to note that the pair of surfaces (2,3) and (1,4) are similar pictorially.

- The series of the trans-numbers can be considered as an "ideal" noise and can be used as note-book for cryptographic purposes.

2.4 THE CAPONEF-METHOD AS AN ALTERNATIVE OF THE 3D-DGI-METHOD

Any reader who read this part of the Chapter 2 has a right to pose the following question: Is the 3D-DGI method really unique and original as the authors claim? The answer for this question will be the following. In the sense of connection of different moments and their intercorrelations up to the fourth order inclusive, the 3D-DGI method is really unique. However, instead of these 13 parameters, one can propose another set of parameters that it follows from conventional analysis of the given TLS. Why not to consider the positive and negative fluctuations as a specific "struggle"/competition between them? The initial attempts of this method were considered in Chapter 9 of the book [20]. Let us define this "struggle" principle as the CAPoNeF (Comparative Analysis of Positive/Negative fluctuations) approach. We introduce a set of 10 parameters (that is free from treatment errors also) and show how to apply them in analysis of some rectangular matrix $[N \times M]$ representing a set of similar measurements.

One can suggest the following ten key and relatively independent from each other parameters that are derived from any TLS y_j (j=1,2,...N) and used for a quantitative description of positive/negative fluctuations:

TABLE 2.5 This table shows the compression of the initial matrix (TX_2_"0"_real part) to the size [(M=*50* × P=*10*)].

p_1	p_2	p_3	p_4	p_5	p_6	p_7	p_8	p_9	p_{10}
0,01898	16,06	0,02204	1364,95	−6,86E-4	4	6,8511	5,80726	−0,45154	130,345
−0,07052	16,6	1,12104	1354,39	−0,03377	18	6,80257	6,49561	0,87444	126,888
0,072	16,46	−0,344	1336,46	0,01045	−10	6,71075	6,67485	1,15573	138,005
0,06838	15,99	−0,1468	1382,39	0,00459	4	6,93833	6,90705	−0,78356	134,156
−0,14242	15,78	−0,3952	1301,86	0,01252	6	6,53665	6,78913	−1,71884	153,63
−0,05504	15,03	−0,2799	1405,15	0,00931	−12	7,04987	7,15859	0,63415	130,688
0,18724	17,44	−0,8945	1290,85	0,02564	−2	6,48709	5,37345	1,68132	138,627
0,0187	15,91	0,8526	1362,07	−0,02679	20	6,83843	7,18201	−0,20517	135,744
−0,2009	17,67	1,0918	1319,85	−0,03089	16	6,63284	4,32157	−1,49699	135,438
0,00778	15,16	−0,0156	1306,42	5,132E-4	−12	6,55685	7,55793	−0,2155	135,993
0,16276	16,14	0,5145	1349,39	−0,01594	−18	6,77492	5,10632	1,17498	131,182
−0,03866	16,16	−0,2427	1270,38	0,00751	−8	6,38012	6,16455	0,51495	124,682
−0,10692	16,22	0,79384	1356,45	−0,02447	10	6,8111	5,13789	−0,41785	137,007
0,0581	16,8	−0,3362	1287,64	0,01001	−4	6,46808	5,25176	−0,77172	130,966
0,02692	16,15	−0,2439	1368,66	0,00755	−2	6,87047	5,52875	−0,40619	130,954
−0,07344	16,63	1,3569	1348,66	−0,0408	8	6,7746	6,39889	0,8938	128,767
0,05446	16,4	−0,5089	1345,85	0,01552	−10	6,75793	6,66246	1,07702	139,481
0,0622	16,05	−0,2944	1380,6	0,00917	14	6,92991	6,83158	−0,81669	135,686
−0,13342	15,82	0,06684	1314,12	−0,00211	4	6,59705	6,81697	−1,66924	143,713
−0,05116	14,99	−0,1877	1402,63	0,00626	−20	7,03697	7,16314	0,60288	128,192
0,19062	17,35	−0,7312	1290,65	0,02107	−12	6,48544	5,36504	1,70383	142,97
0,02926	16	0,66148	1361,41	−0,02067	22	6,83494	7,10687	−0,25984	136,573
−0,19474	17,72	1,14948	1316,07	−0,03243	22	6,61422	4,37963	−1,55098	133,512
0,01054	15,12	−0,2211	1308,12	0,00731	−12	6,56563	7,72239	−0,1687	133,705
0,17726	16,41	0,41548	1344,4	−0,01266	−28	6,75054	4,96043	1,1639	129,135
−0,03368	16,12	−0,1926	1264,13	0,00598	−4	6,34862	6,21487	0,54076	121,332
−0,11598	16,3	0,791	1350,26	−0,02429	12	6,78033	5,13462	−0,5134	132,783
0,05828	16,79	−0,0466	1280,09	0,00139	−2	6,42963	5,2257	−0,71548	132,053
0,03854	16,03	−0,5671	1361,16	0,01769	0	6,83359	5,41717	−0,27569	129,575
−0,06694	16,49	1,3639	1335,16	−0,04135	0	6,70698	6,3829	0,88095	129,363
0,05246	16,37	−0,6349	1344,33	0,01939	−14	6,75064	6,56117	1,0148	137,457
0,06756	16,2	−0,2551	1377,65	0,00787	4	6,91549	6,723	−0,83132	132,046
−0,12228	15,59	−0,0054	1310,86	1,745E-4	4	6,58009	7,0483	−1,62489	145,943
−0,0516	15,04	−0,357	1401,18	0,01186	−12	7,0303	7,18585	0,64829	127,007
0,18582	17,4	−0,4516	1291,85	0,01298	−18	6,49087	5,46797	1,64677	144,676
0,03348	16,16	0,75304	1364,48	−0,0233	22	6,85095	6,91289	−0,2656	138,232
−0,19276	17,66	1,42552	1313,41	−0,04036	18	6,60147	4,42075	−1,53917	137,373
0,0075	15,18	−0,535	1314,3	0,01762	−12	6,59755	7,63669	−0,11374	132,102
0,18464	16,94	0,33072	1348,18	−0,00976	−26	6,77044	4,74618	1,23049	129,059
−0,02348	16,13	−0,1630	1271,49	0,00505	−6	6,38536	6,45906	0,45667	128,16
−0,1219	16,32	0,5038	1353,38	−0,01544	14	6,79522	5,21169	−0,56689	134,552
0,05674	16,69	0,07652	1278,4	−0,00229	0	6,42103	5,37347	−0,75475	123,575
0,05198	15,85	−0,834	1366,28	0,02631	−2	6,85937	5,3907	−0,21467	129,551
−0,07394	16,82	1,4279	1332,53	−0,04245	−4	6,69485	6,19078	0,86359	129,393
0,03718	16,56	−0,5944	1351,54	0,01795	−4	6,78696	6,40901	0,90403	136,974
0,0693	16,23	−0,2886	1377,55	0,00889	12	6,91518	6,82866	−0,83996	130,204
−0,11712	16,15	0,1442	1323,9	−0,00447	−4	6,64684	6,79367	−1,53285	142,444
−0,05756	15,18	−0,2649	1408,58	0,00872	−10	7,06728	7,15325	0,69212	125,858
0,17768	17,22	−0,2954	1300,03	0,00858	−12	6,53085	5,63733	1,70435	145,286
0,03432	16,23	0,48136	1376,21	−0,01483	28	6,90891	6,81509	−0,33642	134,644

Comment to the Table 2.5. The first line of the reduced Table corresponds to the parameters p_1-p_{10} shown on the previous Figs. 2.25–2.29.

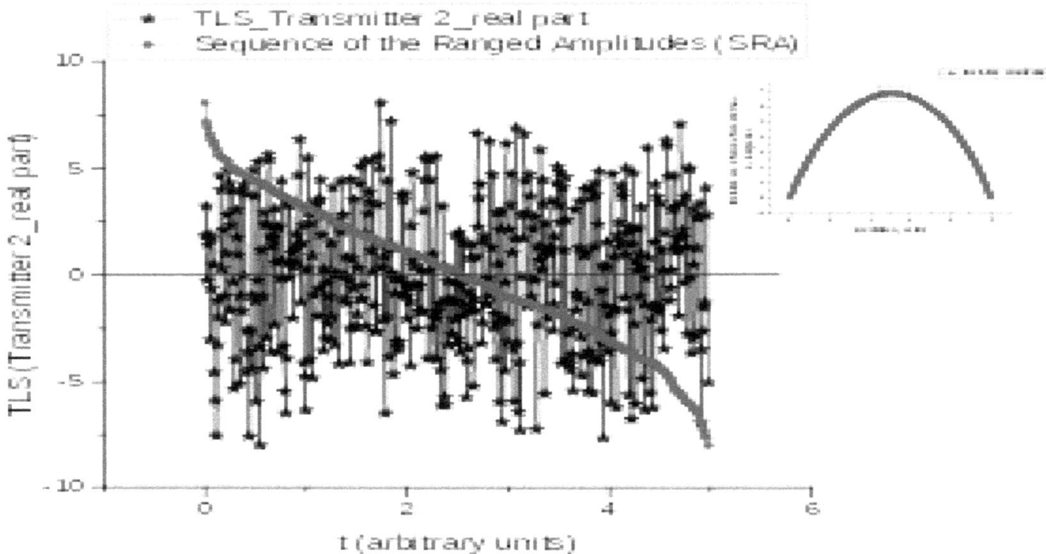

Figure 2.25 This figure demonstrates a typical behavior of TLS selected for the transmitter TX_2 (angle 0)_real part. We show also SRA (marked by point line) and in the small figure it is shown the bell-like curve and the maximal parameter p_7 of its distribution is enclosed in a rectangle. One calculates also the number of the roots (extract the mean frequency p_{10}) and mean phase.

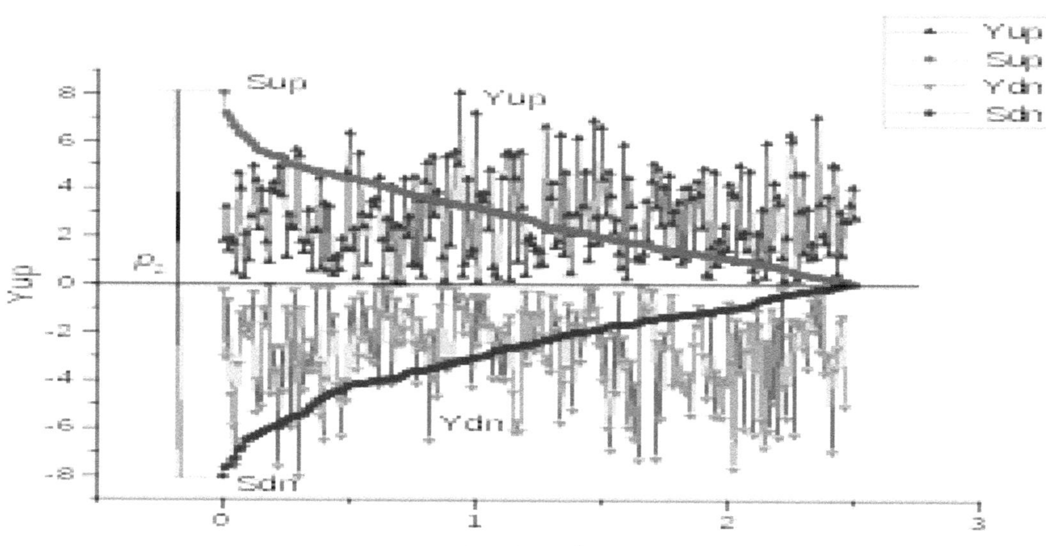

Figure 2.26 In order to see other quantitative parameters, we should divide fluctuations on positive/negative parts and separate the range between them expressed with the help of parameter p_2.

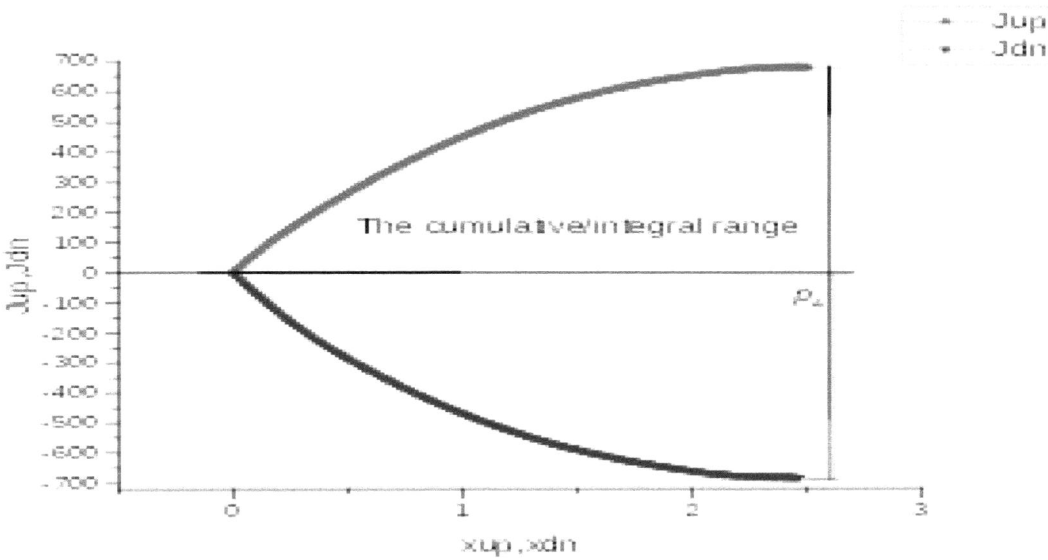

Figure 2.27 The integration of the Sup, Sdn curves shown on the previous figure leads to the formation of a specific "fork" and the range of the fork (parameter p_4) serves as sensitive parameter to the presence of the "strange" influence to a set of "native" arrays.

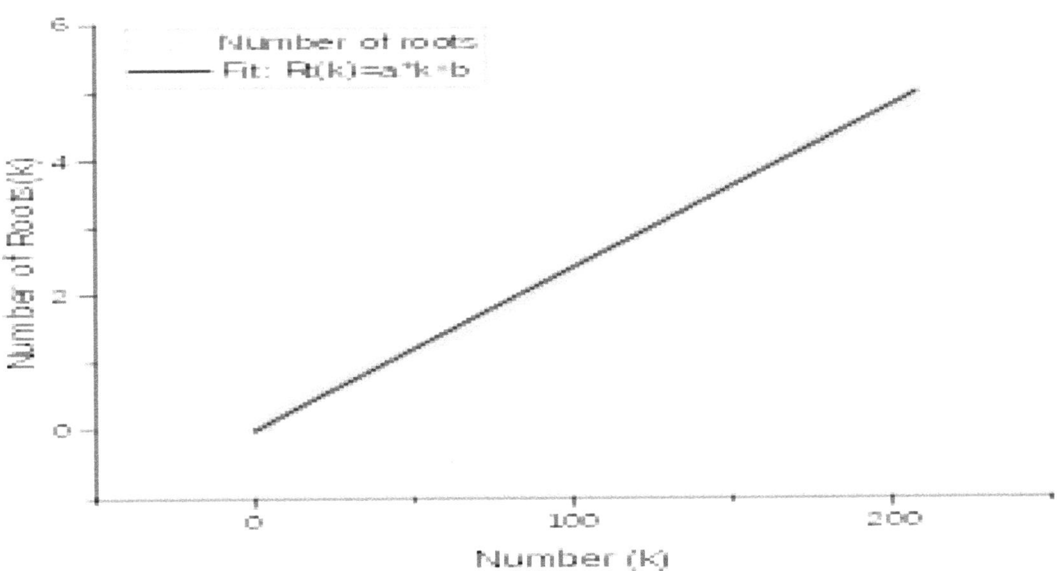

Figure 2.28 This important plot shows the basic relationship that is formed by number of the roots. In many cases this relationship shows to the straight line (especially for rather long TLS(s) with $N > 500$. This relationship allows to calculate the mean frequency and mean phase, correspondingly.

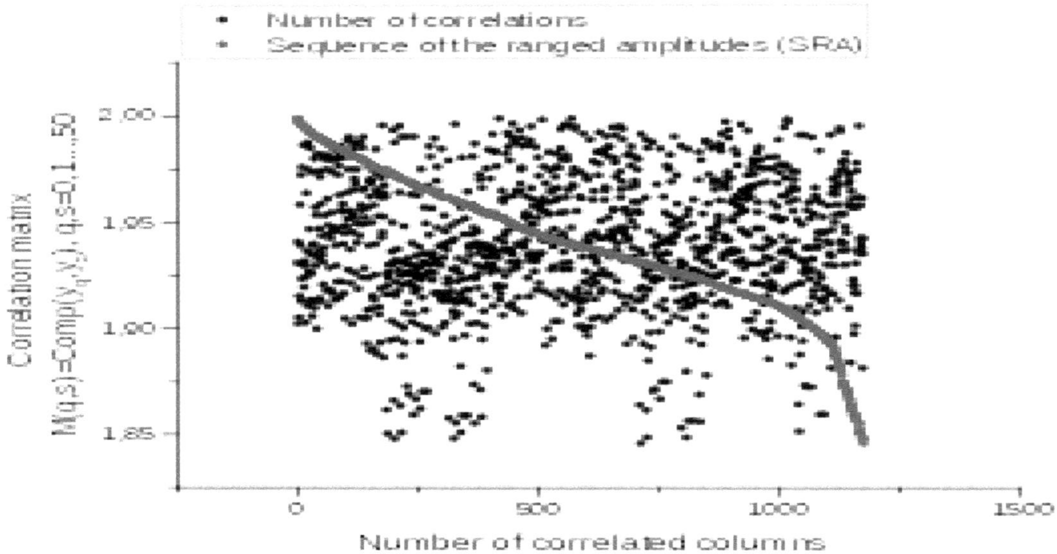

Figure 2.29 Application of the expression (2.19) to analysis of external correlations between columns obtained for the matrix TX_2_"0"_real part. As one can see for the number of pairs equaled $N(N+1)/2 = 1175$ the external correlations are rather high and located in the interval (1.8–2.0). The same tendency are observed for other matrices.

1. $p_1 = \langle y \rangle = \frac{1}{N}\sum_{j=1}^{N} y_j$, the mean value of the TLS that determines a specific balance between positive and negative fluctuations. For "ideal" TLS, this balance should be located near zero value.

2. $p_2 = Range(Dy) = \max(Dy) - \min(Dy)$, which defines the range of the given sequence, where $Dy_j = y_j.-.<y>$, j=1,2,...N. This value is always positive and corresponds to the maximal intensity of the given TLS. Again, for "ideal" TLS this parameter determines a certain stability of the chosen sequence and, therefore, in comparison of two randomly taken, TLS the more stable sequence should accept a *minimal* value of p_2.

3. $p_3 = Range(|Dy|) = \max(Dy_+) - |\min(Dy_-)|$, which defines the relative contribution of amplitudes that are located in the opposite sides of the TLS, where Dy_+ corresponds to the positive part of the TLS and Dy_- includes the negative values of the TLS. The equality $Range(|Dy|) \cong 0$ corresponds to an "ideal" balance between positive and negative amplitudes. In the opposite case, when $Range(|Dy|) > 0$ *or* $Range(|Dy|) < 0$ (one can detect a set of specific "spikes/outliers" of positive (negative) amplitudes in the given TLS relatively to each other. Hence, for an "ideal" TLS, this parameter should be close to zero values.

4. $p_4 = Range(Sm) = \max(Sm_+) - \min(Sm_-)$ with $Sm_\pm = \sum_{j=1}^{N_+,N_-} Dy_{j,\pm}$, which defines the range of sums that evaluates the cumulative effect of the given

TABLE 2.6 The comparison of 6 tables with the help of the expression (2.19).

TX(2-2) "0"-"/3" Real part.	TX(2-2) "0"-"/3" Imag. part.	TX(3-3) "0"-"/3" Real part.	TX(3-3) "0"-"/3" Imag. part	TX(2-3) "0"-"0" Real part.	TX(2-3) "0"-"0" Imag part.	TX(2-3) "/3"- "/3" Real part.	TX(2-3) "/3"- "/3" Imag. part.
1,61208	1,83064	1,67831	1,6856	1,84	1,82073	1,74059	1,97775
0,53201	**0,657**	**0,62762**	**0,61713**	1,04806	1,06494	**0,89362**	1,78607
1,60692	1,84096	1,43456	1,44352	1,71282	1,64586	1,90203	1,73987
0,44	**0,58408**	**0,59312**	**0,59124**	**0,72416**	**0,75175**	**0,63231**	1,02174
1,81182	1,88756	1,72414	1,70435	1,84035	1,76095	1,89868	1,74392
1,75	1,72222	1,75758	1,64706	1,79487	1,84848	1,84	1,77778
0,43742	**0,58173**	**0,59057**	**0,58803**	**0,72031**	**0,74831**	**0,63086**	1,02098
1,78295	1,86591	1,70941	1,71639	1,84506	1,8718	1,73828	1,8899
1,78781	1,96394	1,88646	1,90104	1,62202	1,60694	1,5394	1,71616
1,53419	1,67629	1,71464	1,83077	1,76153	1,85247	1,6093	1,65924

Comments to Table 2.6. In this table we bold the key parameters $p_{2,4,7}$ that have values less than the unit value. These values differentiate the small differences between the same transmitters but oriented on different values (the first 4 columns). This method differentiates also the differences between different transmitters (the last 4 columns) but oriented on the same angles. The values close to the unit values are underlined.

fluctuations. It is proved to be very effective together with the independent parameter p_2. For "ideal" TLS, this range should accept a minimal value also.

5. $p_5 = Sm(y) = \langle y \rangle - \frac{1}{2}(\max(y) + \min(y))$, which reflects a possible asymmetry between positive and negative fluctuations with respect to their mean value. If p_5 is close to zero, then one concludes that the TLS is completely symmetric, in other cases when $p_5 > 0$, (< 0), it is possible to evaluate the value of asymmetry. For "ideal" sequence, this parameter should be close to zero value, as well.

6. $p_6 = DN = N(x_+) - N(x_-)$, which determines the number of amplitudes located on the abscissa axis. If $p_6 > 0$ or $p_6 < 0$, then the number of positive amplitudes $N(x_+)$ exceeds the number of negative amplitudes $N(x_-)$ or vice versa. The range p_6 for an "ideally equalized" TLS tends to zero value.

7. $p_7 = \max(Bd)$. If all amplitudes locate in descending order $y_1 > y_2 > ... > y_N$ and then this sequence of these ranged amplitudes is integrated with respect to its mean value $<y>$, then one receives the bell-like curve. This curve can be fitted approximately by the expression $Bd_j \cong A(x_j - x_0)^{\alpha}(x_N - x_j)^{\beta} + B$, containing 4 fitting parameters (A, B, α, β). The maximum of this integrated curve clearly indicates the boundary between positive and negative fluctuations. For "ideal" TLS parameter, p_7 should accept a minimal value also. From statistically point of view, this bell-like curve corresponds to maximal value of fluctuation that is calculated for the given interval $\Delta = x_N - x_0$.

8. $p_8 = Range(JDy) = \max(JDy) - \min(JDy)$, this parameter evaluates the contribution of the cumulative fluctuations that are obtained after summation of positive and negative fluctuations, accordingly. The summation formula is

written as $JDy_j = JDy_{j-1} + y_j, (j = 1, 2, ..., N), JDy_1 = 0$ and N is the number of data points in the considered sequence. The minimal value of this parameter signifies about the minimal range/stability of the given fluctuations.

9. $p_9 = \langle JDy \rangle$ is also an important parameter. It serves for evaluation of contribution of the cumulative fluctuations. For "ideal" TLS, the contributions of the positive/negative cumulative fluctuations should be equal to each other and, therefore, the parameter p_9 tends to zero.

10. $p_{10} = \langle \omega \rangle$ is an important parameter for evaluation of the mean frequency of the initial fluctuations that cross the horizontal abscissa axis. One of us (RRN) noticed that in the most cases for rather relatively long ($N > 1000$ data points) TLS, the distribution of the roots can be approximated by a segment of the straight line $r_k = a + b \cdot k$, where the integer value k determines the number of the calculated roots. In this simple case, the mean frequency $<\omega>$ and the corresponding phase $<\Phi>$ can be found from the condition: $\cos(\langle \omega \rangle r_k - \langle \Phi \rangle) = 0$ or, equivalently, $\langle \omega \rangle r_k - \langle \Phi \rangle = \frac{\pi}{2} + k \cdot \pi$. From the last equality, it follows that $\langle \omega \rangle \cdot b = \pi$ and these parameters are in inverse proportional dependence to each other and, therefore, for "ideal" TLS the contributions of positive/negative fluctuations should be approximately equal to each other that are achieved for relatively large values of b. Therefore, for TLS close to an "ideal" one parameter p_{10} can accept relatively small values.

We should notice here that the proposed the CAPoNeF-method is the generalization of the Struggle Principle that was described earlier in Chapter 9 of [20]. After publication of this principle and its verification on many available data, it becomes possible to select the most sensitive parameters as $p_{2,4,7,10}$ that enable to solve the problem "friend-or-foe". In many cases these 10 parameters are sufficient for detection of possible differences based on expressions (2.16–2.19). For additional verification of these parameters that can serve as an alternative method to the DGI-method described above, one of us (RRN) received some radiometric data for its verification and demonstration.

These data are associated with TLS(s) obtained from similar transmitters (fabricated by the same producer) and these TLS(s) were recorded by one fixed receiver with known characteristics. From the result of these experiments, one can obtain the TLS(s) recorded on the fixed antenna receiver. The problem can be formulated as follows: is it possible to differentiate one transmitter from another one (based on their TLS(s)) if these transmitters are involved in the common electric coupling conditions? Their differentiation should be based on their specific TLS only. Will CAPoNeF-method be applicable in this case? The figures given below demonstrate the basic steps of the applied algorithm. We deliberately show these figures for young researchers for their better understanding of the basic content of this chapter. We received the data recorded from two transmitters: TX=2,3 oriented with respect to each other on angles "0" and "$\pi/3$" with their real and imaginary parts, accordingly. Therefore, each record represents itself a matrix $[(N{=}500 \times M{=}50)]$ and we have 4 matrices (TX-2,3; angles $(0, \pi/3)$, real/imaginary parts) for each selected transmitter.

Remarks to Chapter 2

For the advanced readers that got over the content of the Chapter 2, the authors want to highlight the following:

1. The researchers receive new tool for reading the trendless sequences of *any* nature. Actually, the common platform for comparison of TLS(s) are created and these 13 parameters that are given by DGI method and 10 parameters that are added by thew CAPoNeF are sufficient for reducing the initial data and present them in the compact form for their comparison.
2. The authors strongly recommend to advanced readers to reproduce the basic expressions (2.8), (2.9), and (2.13) for better understanding the proposed method. The derivation of these expressions can be recommended for readers as exercises.
3. We think that potential readers have access to different trendless data as acoustic noise, EM noise from different devices, optical, medical (ECG - Electrocardiograph data etc.) for verification of the proposed methods and for writing their original publications, based on this original tool.

BIBLIOGRAPHY

[1] S. F. Timashev, *Flicker-Noise Spectroscopy: Information in Chaotic Signals.* Moscow: Fizmatlit, 2007.

[2] S. F. Timashev and Y. S. Polyakov, "Review of flicker noise spectroscopy in electrochemistry," *Fluctuation and Noise Letters*, vol. 7, no. 2, pp. R15–R47, 2007.

[3] S. F. Timashev, S. A. Demin, O. Y. Panischev, Y. S. Polyakov, A. Y. Kaplan, and Y. A. Nefedov, "Flicker-noise spectroscopy as a tool for the personalized medicine of the future," *Uchenye Zapiski Kazanskogo Universiteta. Seriya Fiziko-Matematicheskie Nauki*, vol. 154, no. 4, pp. 161–177, 2012. [In Russian].

[4] R. Yulmetyev, P. Hänggi, and F. Gafarov, "Stochastic dynamics of time correlation in complex systems with discrete time," *Physical Review E*, vol. 62, no. 5, pp. 6178–6194, 2000.

[5] R. Yulmetyev, P. Hänggi, and F. Gafarov, "Quantification of heart rate variability by discrete nonstationary non-Markov stochastic processes," *Physical Review E*, vol. 65, no. 4, p. 046107, 2002.

[6] R. R. Nigmatullin and A. S. Vorobev, "The "universal" set of quantitative parameters for reading of the trendless sequences," *Fluctuation and Noise Letters*, vol. 18, no. 4, p. 1950023, 2019.

[7] V. A. Tyagai, "Faradaic noise of complex electrochemical reactions," *Electrochimica Acta*, vol. 16, pp. 1647–1654, 1971.

[8] G. C. Barker, “Faradaic reaction noise,” *Journal of Electroanalytical Chemistry and Interfacial Electrochemistry*, vol. 82, pp. 145–155, 1977.

[9] S. Bozdech, K. Krischer, D. A. Crespo-Yapur, E. Savinova, and A. Bonnefont, “$1/f^2$ noise in bistable electrocatalytic reactions on mesoscale electrodes,” *Faraday Discussions*, vol. 193, pp. 187–205, 2016.

[10] R. R. Nigmatullin, S. Martemianov, Y. K. Evdokimov, E. Denisov, A. Thomas, and N. Adiutantov, “New approach for PEMFC diagnostics based on quantitative description of quasi-periodic oscillations,” *International Journal of Hydrogen Energy*, vol. 41, no. 29, pp. 12582–12590, 2016.

[11] E. Astafev, “Electrochemical noise measurement methodologies of chemical power sources,” *Instrumentation Science & Technology*, vol. 47, no. 3, pp. 233–247, 2019.

[12] E. A. Astafiev and A. Ukshe, “Peculiarities of hardware for electrochemical noise measurement in chemical power sources,” *IEEE Transactions on Instrumentation and Measurement*, vol. 68, pp. 4412–4418, 2019.

[13] B. Gutenberg and C. F. Richter, *Seismicity of the Earth and Associated Phenomena.* Princeton, NJ: Princeton University Press, 1949.

[14] R. R. Nigmatullin, A. S. Vorobev, K. S. Nepeina, and P. N. Alexandrov, “Fractal description of the complex beatings: How to describe quantitatively seismic waves?,” *Chaos, Solitons & Fractals*, vol. 120, pp. 171–182, 2019.

[15] Y. I. Babenko, *The Power Law Invariants of the Point Sets.* St. Petersburg: Professional Publ. House, 2014.

[16] Y. I. Babenko, *Power Relations in a Circumference and a Sphere.* Landisville, NJ, USA: Norell Press, Division of Norell Press Inc., 1997.

[17] R. R. Nigmatullin, A. S. Vorobev, H. C. Budnikov, A. V. Sidelnikov, A. D. Badikova, and E. I. Maksyutova, “The usage of unremovable artefacts for the quantitative “reading” of nanonoises in voltammetry,” *New Journal of Chemistry*, vol. 43, pp. 6168–6178, 2019.

[18] R. R. Nigmatullin, A. S. Vorobev, A. R. Nasybullin, A. D’Orazio, G. Maione, P. Lino, and M. Grande, “Advanced and sensitive method by discrete geometrical invariants for detection of differences between complex fluids,” *Communications in Nonlinear Science and Numerical Simulation*, vol. 73, pp. 265–274, 2019.

[19] R. R. Nigmatullin, “Discrete geometrical invariants in 3D space: How three random sequences can be compared in terms of “universal” statistical parameters?,” *Frontiers in Physics*, vol. 8, p. 76, 2020.

[20] R. R. Nigmatullin, P. Lino, and G. Maione, *New Digital Signal Processing Methods: Applications to Measurement and Diagnostics.* Cham: Springer, 2020.

[21] National Instruments, *NI Educational Laboratory Virtual Instrumentation Suite II (NI ELVIS II) Series User Manual.* National Instruments, Austin, TX, USA, 2016. Part Number: 374629C-01.

[22] R. R. Nigmatullin, R. A. Giniatullin, and A. I. Skorinkin, "Membrane current series monitoring: Essential reduction of data points to finite number of stable parameters," *Frontiers in Computational Neuroscience*, vol. 8, p. 120, 2014.

CHAPTER 3

Non-Orthogonal Combined Fourier Analysis of Smoothed Signals (NOCFASS) Methods

In this chapter, we demonstrate *how* to extend the possibilities of the conventional F-analysis (Fourier analysis) and adapt it for quantitative description of multi-periodic signals recorded from different complex systems. The basic idea lies in filtration property of the Dirichlet function that allows finding the leading frequencies (having the predominant amplitudes) and the calculated shortcut frequency band allows to fit the initial random signal with high accuracy (with the value of the relative error less than 5%). This modification defined as NOCFASS-approach (Non-Orthogonal Combined Fourier Analysis of the Smoothed Signals) can be applied to a wide class of different signals having multi-periodic structure. Verification on mimic data shows the merits and demerits of the proposed approach in comparison with the previously shown NAFASS approach developed earlier by one of the co-authors RRN in his papers [1, 2, 3]. As real random "signals"/sequences, we chose the *economic* data related to distribution of the prices for strategic materials as Au, Ag, and U. Besides, we consider the series of the transcendental numbers of and E (Euler constant) and irrational number $2^{1/2}$ for detection of their differences (that can present an interest for the modern cryptography). In addition, we want to consider also some real geophysical data, which can present a practical interest for a wide circle of potential researchers. We *do* suppose that the proposed NOCFASS-approach having new attractive properties as the simplicity of realization, agility to the problem formulated will find a wide propagation in the modern signal processing area. It can serve as the modified and "universal" tool for the fitting of the LF (low-frequency) trends obtained from the initial trendless sequences by numerical integration procedure, as well.

DOI: 10.1201/9781003794691-3

3.1 INTRODUCTION AND FORMULATION OF THE PROBLEM

As it is well known, the Fourier transformation for the calculation of the desired frequencies occupies the *central* place in the modern data treatment procedure of an arbitrary periodic signal containing the oscillating/harmonic components [4, 5, 6]. However, the numerous modifications of the conventional F-transformation as complete ensemble empirical mode decomposition (CEEMD) [7, 8, 9, 10, 11, 12], multiple signal classification (MUSIC) [13, 14, 15, 16] did not help essentially in fitting of the signals associated with their multi-periodic structure and characterizing different complex phenomena, where the simple fitting formula for quantitative description of the recorded signals is *absent*, especially in cases when *a priori* information is *not* available. In order to fit the desired signal in the frame of the conventional DFT, we need number of frequencies that are close to number of data points $N/2$ (because of the symmetry of the F-transform relatively the ends of the given interval). It creates a certain problem for identification of the calculated F-spectrum, containing excess number of frequencies. In other words, this problem lies in the well-known transformation of the desired and discrete signal $y(t)$ to the frequency domain, associated with aliasing phenomenon [17]. However, the Fourier series can be also used for the fitting purposes after application of the DFT, if, after that, the inverse discrete Fourier transformation (IDFT) is applied. The main drawback of this approach is that the DFT returns the doubled number of amplitudes, compared with the data points of the input signal. It means that the calculated number of the fitting parameters exceeds/(or should be equal) to number of input data points. Therefore, this excess number of frequencies is *not* suitable for the fitting of the discrete signal $y(t)$. From our point of view, it is the main reason why DFT is *not* used usually for the fitting purposes.

Other approaches for transformation of the desired signal targeted on improving the frequency resolution of the DFT, such as MUSIC, CEEMD [7, 8, 9, 10, 11, 12, 13, 14, 15, 16] require knowing *a priori* approximate frequencies location in the transformed signal and have the high computational cost.

As an alternative approach, the NAFASS [1, 2, 3] was proposed presumably for the fitting purposes of an arbitrary signal $y(t)$. It uses for the fitting the segment of the random series (3.1)(1) (shown below) having an arbitrary distribution of the finite number of frequencies with fixed number of amplitudes. Advantage of NAFASS lies in the high accuracy of the proposed iteration procedure and a small number of frequencies forming the desired amplitude-frequency response (AFR). The details of application of the NAFASS approach are discussed in the recent book [18]. Nevertheless, it has few drawbacks. First, it has high computational cost especially for the segments containing large number of data points ($N > 5000$). Second, the NAFASS has high sensitivity to small signal deviations; if the given signal is changed, the calculated amplitudes and frequencies obtained in the frame of the NAFASS-approach can be changed significantly. However, NAFASS can be used for the fitting and calculations of the desired AFR with some precautions. In order to overcome these drawbacks in paper [1], the independent iteration schemes for calculation of the inoculating set of frequencies were used. It helps to create a confidence "fork" of the location of the calculated distribution of frequencies obtained with the help of iteration scheme and then find the optimal frequency branch from the minimization of the fitting error procedure.

In this chapter, we propose a simple algorithm that based on modification of the main formula for the digital Fourier transformation [17]. We want to show *how* to obtain the reduced/shortcut F-spectrum from the complete frequency set that helps to fit the desired temporal signal with high accuracy (the relative error does not exceed 5%). The given signal can contain many periods or beatings and even can have random structure. However, thanks to filtration properties of the Dirichlet function one can extract the desired/leading harmonic frequencies and then find the rest of the amplitudes by means of the linear least squares method (LLSM). We deliberately do *not* use the orthogonal trigonometric functions and symmetrical properties of the F-transformation because the main advantage of the proposed approach, defined as the Non-Orthogonal Combined Fourier Analysis of the Smoothed Signals (NOCFASS), is to find the shortcut/reduced spectrum that provides the fit of the given signal with rather high accuracy. The found spectrum provides the accurate fit of the desired temporal signal. Besides, it is fast, simple in realization and compact because the proposed algorithm containing only four steps (shown below) that can be implemented easily in microcontrollers or other portable devices. As one can be explained below the NOCFASS-approach is rather universal tool. With the help of the segment of the F-spectrum found, one can fit a wide set of signals obtained from a set of trendless sequences. This peculiarity will facilitate considerably the reduction procedure of a big set of data that can be reproduced with the help of the calculated F-spectrum and the finite set of the amplitudes, forming the desired amplitude-frequency response (AFR). Some details of this interesting procedure will be given below.

3.2 THE BASIC EXPRESSIONS AND PROPOSED ALGORITHM

The basic problem lies in the presentation of an arbitrary discrete signal $y(t)$ in the form of a linear combination of the harmonic modes. These modes do *not* follow to the dispersion law $\Omega_k = 2k/T$ as it is conventionally accepted for the F-transformation and the number of these modes should be minimal. This minimal number is selected from the criterion of the minimal value of the fitting error. It should not exceed the given value 5%. As one can see below, the shortcut dispersion law is expressed in the form $\Omega_k = ck+d$. The NOCFASS approach differs from the previously proposed NAFASS approach [1, 2, 3], because it cannot find and then fit the *true* dispersion law and therefore, it compensates this defect by an additional excess of frequencies associated with small values of amplitudes that participate in the fitting procedure. However, this defect is balanced by its fastness and simplicity in applications of the NOCFASS approach in different micro-devices that are used in treatment of different complex signals. We should stress also that the linear dispersion law $\Omega_k = ck+d$ used in the NOCFASS approach allows considering the behavior of the signal *out* of the given temporal interval limits. Mathematically, any multi-periodic signal can be expressed by the formula

$$
\begin{aligned}
&y(t_n) \equiv y_n \cong F(t_n) = A_0 + \sum_{k=0}^{K-1} \left[Ac_k \cos(\Omega_k t_n) + As_k \sin(\Omega_k t_n)\right] \equiv \\
&\equiv A_0 + \sum_{k=0}^{K-1} \left[Amd_k \cos(\Omega_k t_n - \varphi_k)\right], \\
&Amd_k = \sqrt{Ac_k^2 + As_k^2},\ \varphi_k = \tan^{-1}\left(\frac{As_k}{Ac_k}\right), \\
&t_n = t_0 + \left(\frac{n}{N-1}\right)\Delta t,\ \Delta t = t_{N-1} - t_0,\ n = 0, 1, ..., N-1.
\end{aligned} \tag{3.1}
$$

Here, Ω_k contains an unknown dispersion law *not* coinciding with $\Omega_k = ck$ ($c = 2\pi/T$; the letter T defines a period of the signal that in the most cases is not known). This dispersion law is conventionally accepted in the F-transformation and follows from the orthogonality of the modes used in this transformation. In order to "percept" the advantages of the proposed approach and peculiarities that can be used in the further calculations, we present the digital/direct Fourier transformation (DFT) in the simplest form that was used in the book [17]

$$
\begin{aligned}
&X_m = \sum_{n=0}^{N-1} y_n \exp(-i(b_m + a)n), \quad b_m = \frac{2\pi}{N}m, \\
&m = 0, 1, ..., N-1.
\end{aligned} \tag{3.2}
$$

In comparison with the conventional formula, we introduce the *shifting* parameter a (the reason of this modification will be explained later) and at $a = 0$ it coincides with the conventional expression accepted in [17]. We deliberately omitted the normalization coefficient in (3.2) because the main aim is to find accurately the finite set of the shortcut frequencies Ω_k that enables to fit the given signal $y(t_n)$ in (3.1) with high accuracy. The unknown coefficients Ac_k, As_k can be found easily with the help of the linear least squares method (LLSM). In order to understand deeper expression (3.2) and its filtration properties let us *suppose* that the unknown signal $y(t_n)$ represents itself a linear combination of the modes

$$
y(t_n) = \sum_{k=1}^{K} Md_k(t_n),\ Md_k(t_n) = A_k \exp(i(\omega_k t_n - \varphi_k)). \tag{3.3}
$$

Substitution of expression (3.3) for the fixed mode $Md_k(t)$ into (3.2) can be evaluated easily. After some algebraic calculations, we obtain the following expression

$$
\begin{aligned}
&X_m = \exp(i\Phi(k, m, N))\, D(k, m, N) \\
&\Phi(k, m, N) = \frac{N-1}{2}\Delta\omega_{k,m} + \omega_k t_0 - \varphi_k, \\
&D(k, m, N) = A_k \frac{\sin\left(\frac{N}{2}\Delta\omega_{k,m}\right)}{\sin\left(\frac{1}{2}\Delta\omega_{k,m}\right)}, \quad \Delta\omega_{k,m} = \frac{\Delta t}{N-1}\omega_k - (a + b_m).
\end{aligned} \tag{3.4}
$$

Here, the function $\Phi(k,m,N)$ determines the generalized phase and the function $D(k,m,N)$ is defined as the Dirichlet function [17] that gives the maximal contribution at $\Delta\omega_{k,m} \cong 0$. From the condition, viz., $\max(|X_{\mathrm{m}}|) = NA_k$ at $\Delta\omega_{k,m} \cong 0$ one can find easily the desired frequency Ω_k. However, we want to stress here that an attempt to find the accurate values of the frequencies Ω_k (k=1,2,...,K) from the linear combination of the modes leads to some *inevitable* distortions. Really, let us consider the modulus of the function $|X_{\mathrm{m}}|$ for linear combinations of two unknown modes $k_{1,2}$

$$|X_m| = \begin{bmatrix} A_1^2 D^2(N,k_1,m) + A_2^2 D^2(N,k_2,m) + \\ +2A_1A_2 D(N,k_1,m)\cdot D(N,k_2,m)\cos\left(\Phi(N,k_1,m) - \Phi(N,k_2,m)\right) \end{bmatrix}^{1/2}. \tag{3.5}$$

Analysis of this expression shows that the accurate calculation of two frequencies $\omega_{k_{1,2}}$ from the resonance conditions written for each mode $\Delta\omega_{k_{1,2},m} \cong 0$ becomes *impossible*. This is because, besides of this main condition, the amplitudes of the local Dirichlet functions $A_{1,2}$ and partly the phases of the modes shown in the third term of (3.5) can distort the values of the unknown frequencies that are needed to be evaluated. The mimic example of this phenomenon showing these distortions is considered in the third part. What we should do in the situation when the frequencies are *not* known, however, it is necessary to select the optimal range of frequencies Ω_k that can fit the given signal $y(t_n)$ by the fitting function $F(t_n)$ with high accuracy? We remind here that the value of the relative error evaluating the fitting procedure quantitatively is evaluated with the help of the formula

$$\mathrm{Rel}\,Err(\%) = \left(\frac{stdev\,(y(t) - F(t))}{mean\,|y(t)|}\right)\cdot 100\%, \tag{3.6}$$

and this expression should not exceed the upper limiting value fixed initially as 5%. Let us come back to expression (3.2) again. If we put the shifting parameter $a = 0$, then we obtain the symmetrical F-spectrum for $|X_m|$ located on the ends of the given interval because of the well-known property $X_m = (X_{N\text{-}m})^*$ [17]. However, if we put $a = 0$, then we obtain two branches for $|X_m|$ located *symmetrically* relatively the middle of the total interval (m=0,1,...,N-1) with maximal value of m_{Mx} near the value $N/2$. This simple property allows selecting accurately (symmetric to the left-hand branch) the right-hand branch of the F-spectrum that is needed for selection and further calculation of the reduced spectrum Ω_k, which is suitable for the fitting purposes. Therefore, we deal with the reduced or truncated values of m from the interval

$$mc_m = mc_0 + \frac{m}{Nc}\left(m_{Nc} - mc_0\right),\; mc_0 \cong \frac{N}{2},\; m = 0, 1..., Nc. \tag{3.7}$$

The final parameter $Nc < N$ is chosen from the condition that the selected distribution of the amplitudes is located presumably in the interval:

$$mean\left(|X_m|\right)\cdot\kappa \le |X_m| \le \max\left(|X_m|\right). \tag{3.8}$$

Here, the correction multiplier κ in the left-hand side is found empirically and depends on the total number of data points. If N is rather small, then $\kappa < 1$ and its value close to its minimal value $\geq \min(|X_m|)$. It implies that the contribution of the amplitudes located in the interval $([\min(|X_m|), mean(|X_m|)]$) becomes important for the fitting purposes. However, for the large number of data points ($N \geq 1000$), the value of the correction parameter $\kappa > 1$. When the corresponding interval for the amplitudes has been found, then the upper value of the reduced spectrum Ω_k is found at k=Nc. We should remember also that the total frequency set of the digital F-spectrum lies in the interval

$$\begin{aligned} &\omega_0 \cong \frac{2\pi}{Range(t)} < \omega_m < \omega_{Mx} \cong \frac{2\pi}{dt}, \\ &Range(t) = \max(t) - \min(t),\ dt = \min(t_{i+1} - t_i), \end{aligned} \tag{3.9}$$

where ω_{Mx} is the maximal value of the discretization frequency. The procedure described above, allows to find the low limit coinciding approximately with the value ω_0 and truncate the upper limit of the total spectrum by the Ω_{Nc} value. Analysis of available data taken from different sources confirms the behavior of the shortcut spectrum Ω_k presented in the form of the straight line $\Omega_k = ck + d$ that facilitates the prediction purposes of the given signal located out of the initial interval. The limits of this spectrum (ω_0, Ω_{Nc}) can be corrected with the help of NAFASS-approach [1] that uses the original iteration procedure. Finally, we receive very simple algorithm that can be used for the fitting of multi-periodic signals containing different beatings. Concluding this section, we express the proposed 4-step simple algorithm in the following form:

Step 1. Calculation of the total spectrum from (3.2) for all available values of m from the interval (0, N-1) together with the total frequency range with the help of (3.9).

Step 2. Calculation of admissible range of "m" with the help of expression (3.7). The limiting value of Nc is found from the condition (3.8).

Step 3. Calculation of the shortcut spectrum belonging to the range

$$\omega_0 \cong \frac{2\pi}{Range(t)} < \Omega_k \leq \Omega_{Nc}, \tag{3.10}$$

and set of the basic modes $f_k(t_n) = \{\cos(\Omega_k t_n), \sin(\Omega_k t_n)\}$ in (3.1).

Step 4. The fit of the given signal $y(t_n)$ with the help of the LLSM and calculation the AFR – $Amd_k(\Omega_k)$ and $\varphi_k(\Omega_k)$ with respect to the found shortcut spectrum Ω_k.

3.3 VERIFICATION ON MODEL/MIMIC DATA

In order to stress a "universality" of the NOCFASS approach, we deliberately select the combination of the frequencies with the help of the formula

$$\begin{aligned} &\Omega_k = C_1\lambda_1^k + C_2\lambda_2^k,\ k = 0, 1, ..., K = 4, \\ &\lambda_1 = -2,\ \lambda_2 = 2,\ C_1 = -0.125,\ C_2 = 2.125. \end{aligned} \tag{3.11}$$

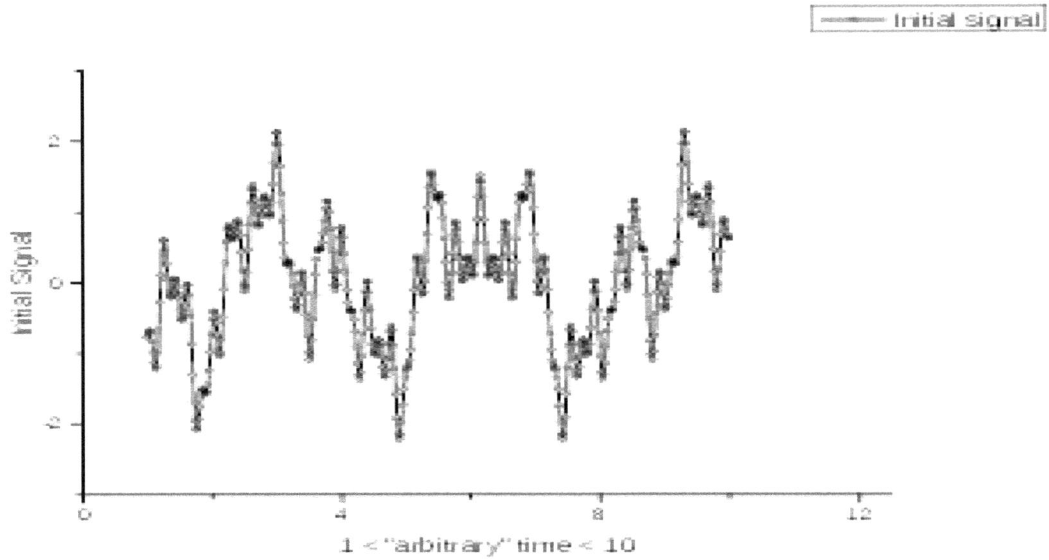

Figure 3.1 Initial signal obtained with the help of frequency distribution (3.11) and distributions of amplitudes and phases from (3.13).

This dispersion law differs from the conventional F-expression $\Omega_k = ck$ ($c = 2\pi/T$) and can be considered as the multi-frequency spectrum. The first five frequencies that are needed for creation of the desired signal form the following set

$$\Omega_0 = 2,\ \Omega_1 = 4.5,\ \Omega_2 = 8,\ \Omega_3 = 18,\ \Omega_4 = 32. \tag{3.12}$$

These values are necessary for calculation of the initial signal $y(t)$ with the following modulus of amplitudes and phases

$$Amd_k = (\lambda_1 + \lambda_2\Omega_k)^{-1/4},\ \varphi_k = C_1\Omega_k,\ k = 0, 1, ..., 4. \tag{3.13}$$

Substitution of expressions (3.12) and (3.13) into (3.1) allows to receive the test signal. It is shown in Fig. 3.1. Then we realize successfully the 4-step procedure for the calculation of the shortcut spectrum and receiving the fit of the testing signal and the desired dependencies as $Amd_k(\Omega_k)$ and phase distribution $\varphi_k(\Omega_k)$. They are explained by the set of the Figs. 3.2–3.5 and captions to them. The most remarkable picture is the AFR dependence $Amd_k(\Omega_k)$ depicted in Fig. 3.5b. In Fig. 3.5c, we show the distribution of the phases $\varphi_k(\Omega_k)$. The sequence of the range amplitudes (SRA) for this distribution (shown by black circles) coincides approximately with straight line. It signifies that distribution is close to the *uniform* distribution of the corresponding phases. In spite of the essential differences associated with the behavior Ω_k from (3.11) and its replacement by the linear dependence $\Omega_k = ck+d$ (where d=ω_0=0.698, c= Range(Ω)/K=0.687, K=73), including 73 modes the fit of the initial signal, shown in Fig. 3.4, it is realized with the value of the relative error 4.75%.

Therefore, a sufficient number of amplitudes selected from the interval (3.8) with the value of $\kappa = 0.25$ provides the desired fit. In comparison with the total spectrum defined by expression (3.8) and containing N=500 (m=0,1,...,N-1) frequencies we

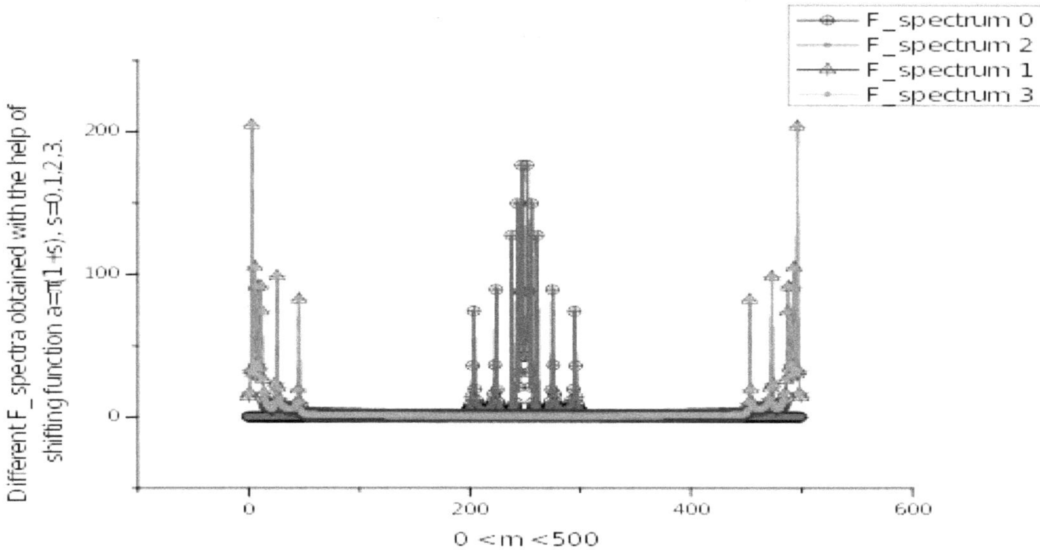

Figure 3.2 Different F-spectra obtained with the help of the shifting function a_s=(1+s), s=0, 1, 2 and 3. This picture helps to identify the location of the leading amplitudes.

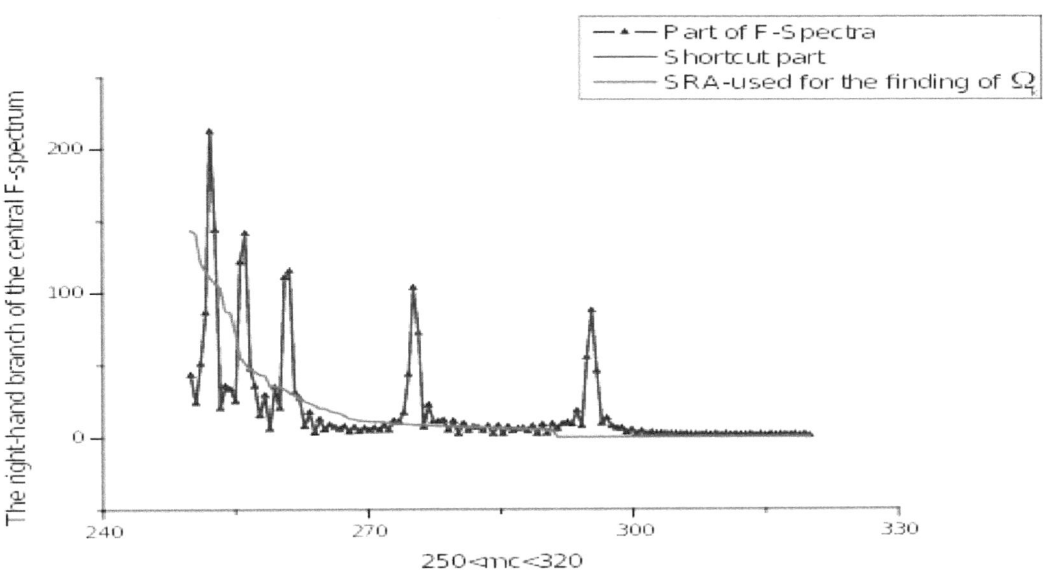

Figure 3.3 The right-hand part of the central F-spectrum 2 (marked in the previous figure by black line). The truncated part of the spectra obtained with the help of expression (3.8) and its sequence of the ranged amplitudes (shown by black line) are used for calculating the shortcut spectrum Ω_k.

decrease sufficiently of this value in 6.8 times! Other high frequencies from (3.9) are *not* essential for the fitting purposes. Analysis of these mimic data allows making some preliminary conclusions:

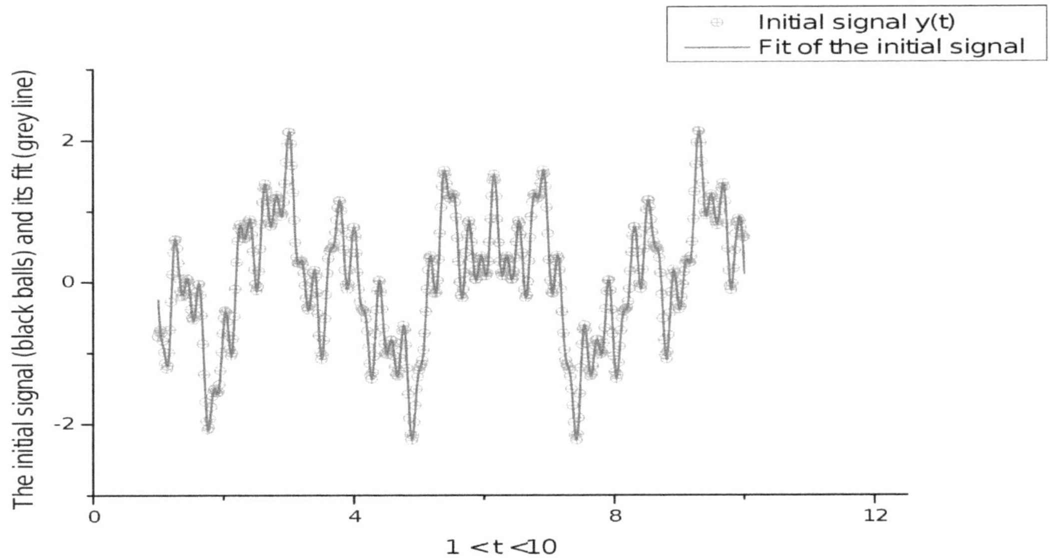

Figure 3.4 The fit of the initial signal $y(t)$ realized with the help of the shortcut spectrum Ω_k. The limiting frequency value is obtained with the help of the previous figure. The value of the relative error equals 4.5%.

1. The NOCFASS approach *cannot* find the true dispersion law and replaces it approximately by the linear spectrum $\Omega_k = ck+d$ with number of frequencies exceeding the initial ones from one side; however, from another side the number of the sufficient/fitting frequencies is much less in comparison with all set of admissible frequencies (equals $N/2$) and given by expression (3.9). It enables to detect the desired frequencies with high accuracy identifying them with the peaks of the Dirichlet function.

2. The most informative picture is connected with dependence $Amd_k(\Omega_k)$ and the distribution of the calculated peaks approximately related with frequencies associated with the initially given harmonic modes. Because of their importance, we compare again the calculated values with exact values given in the parenthesis:

$$\begin{aligned} &\Omega c_0 = 2.0916(2),\ \ \Omega c_1 = 4.1818(4.5),\ \ \Omega c_2 = 8.3622(8), \\ &\Omega c_3 = 18.116(18),\ \ \Omega c_4 = 32.051(32). \end{aligned} \tag{3.14}$$

We want to notice here that the selection of the different distributions for amplitudes and phases (shown in (3.13)) cannot change significantly this result. Their variations lie in the interval (1–5%) and change only the third digit after dividing point in the calculated frequencies.

3.4 ANALYSIS OF ECONOMIC DATA

In order to stress usefulness of the proposed NOCFASS algorithm, we deliberately selected three interesting examples. The first one is related to quantitative description

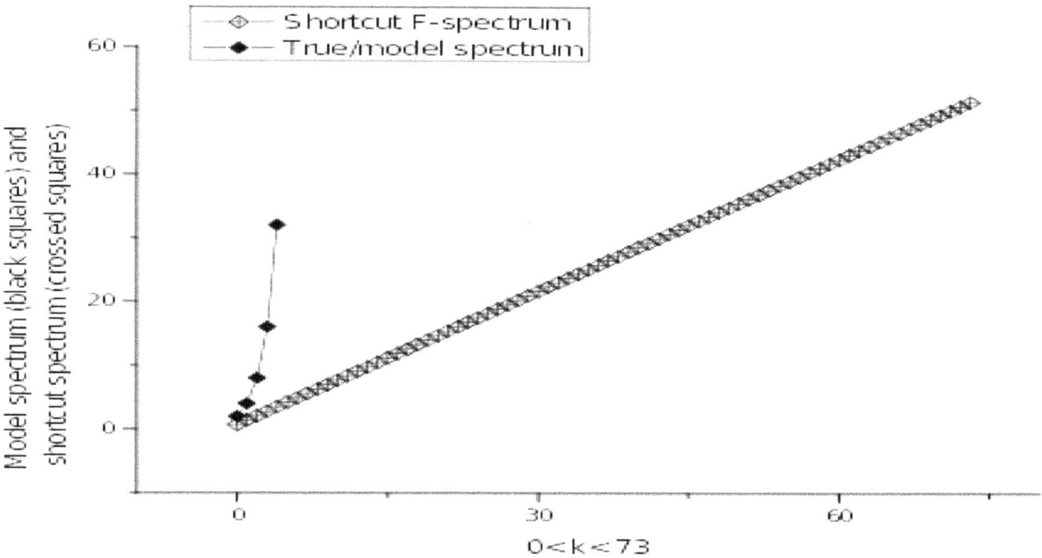

Figure 3.5 This plot demonstrates the differences between true spectra (black squares) and the shortcut spectrum (crossed squares) that is obtained in the result of the 4-step algorithm. The last spectrum contains the excess of frequencies that helps to provide the accurate fit of the signal shown on the previous figure. However, this number is much less (in 6.8 times) in comparison with total spectrum (containing 500 frequencies and coinciding with the total number of data points) defined by expression (3.9).

of the prices associated with strategic materials as gold (Au), silver (Ag), and uranium (U). The distribution of the prices for these materials can be taken from the admissible site "Index Mundi", where one can download the distribution of the gold and silver monthly prices (US cents per Troy Ounce (31.103g.)) covering the 30-year period (September 1989 up to September 2019) and therefore, giving in total 360 data points (unfortunately, the site cannot give more detailed data). For the same period, we downloaded the distribution of the monthly prices for uranium (US cents per American pound (0.454g.)) In order to save place for other pictures we show only two figures for each selected material: the final fit for the given price distribution and the dependence $Amd_k(\Omega_k)$ as the most important characteristics of this research. Figures 3.8, 3.10, and 3.12 demonstrate the perfect fit of the prices distribution (Au, Ag, U) realized with the value of the relative error (0.2–0.3%). The number of the modes that are sufficient for this purpose lies in the interval (80–90). The corresponding AFRs expressed in the form of dependence $Amd_k(\Omega_k)$ are shown in Figs. 3.9, 3.11, and 3.13, correspondingly. Analysis of these plots is similar. The leading modes are concentrated in low-frequency region. The high-frequency modes have approximately the amplitudes with equal values and form a specific plateau.

In this plateau, no any leading mode (or group of the neighboring modes) associated with selected frequency cannot be noticed. The behavior of these AFRs follows approximately to monotone behavior. They are differed only by the values of

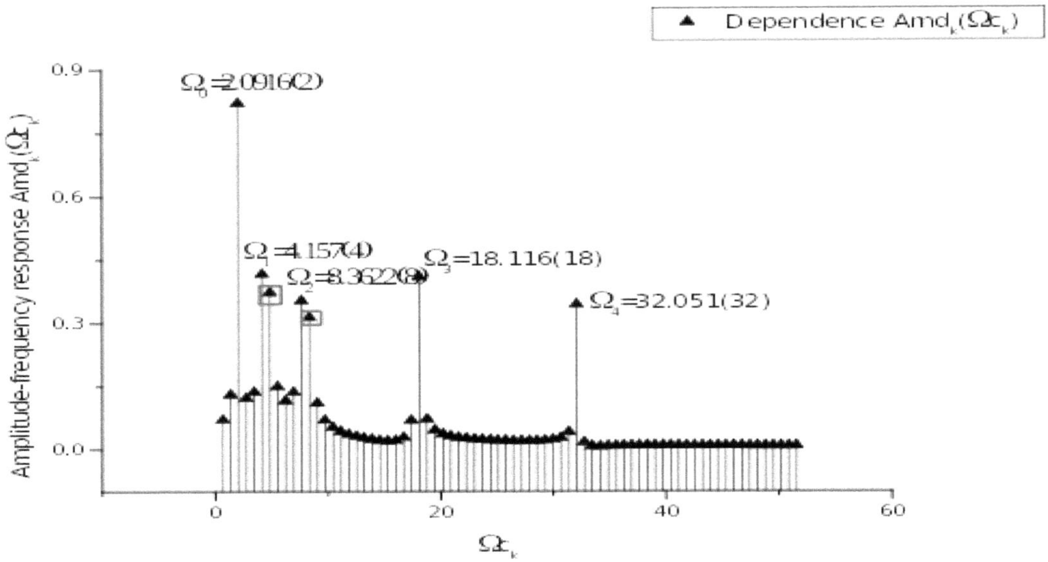

Figure 3.6 This is the central figure of this section. The NOCFASS approach identifies the desired frequencies that are contained in the initial dispersion law (3.11). They are identified with the high accuracy in spite of possible distortions evoked by the nearest lobes of the Dirichlet function and influence of the neighboring modes shown in expression (3.5). The parasite frequencies marked by Grey squares demonstrate the strongest influence of the neighboring modes.

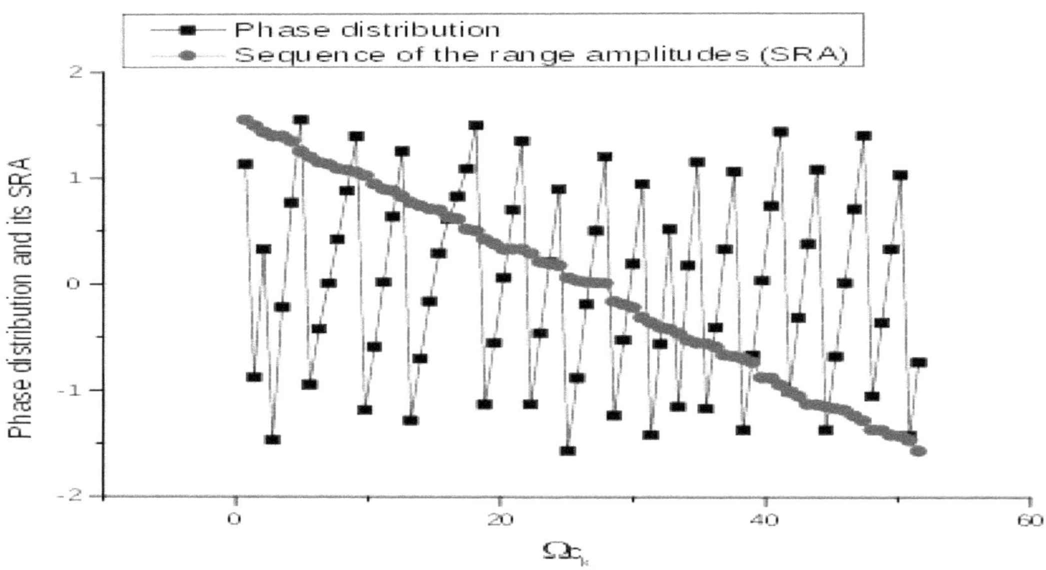

Figure 3.7 Phase distribution characteristic $\varphi(\Omega c_k)$ related to $y(t)$. The sequence of the ranged amplitudes (SRA) (marked by black balls) shows that the distribution $\varphi(\Omega c_k)$ is close to the uniform distribution.

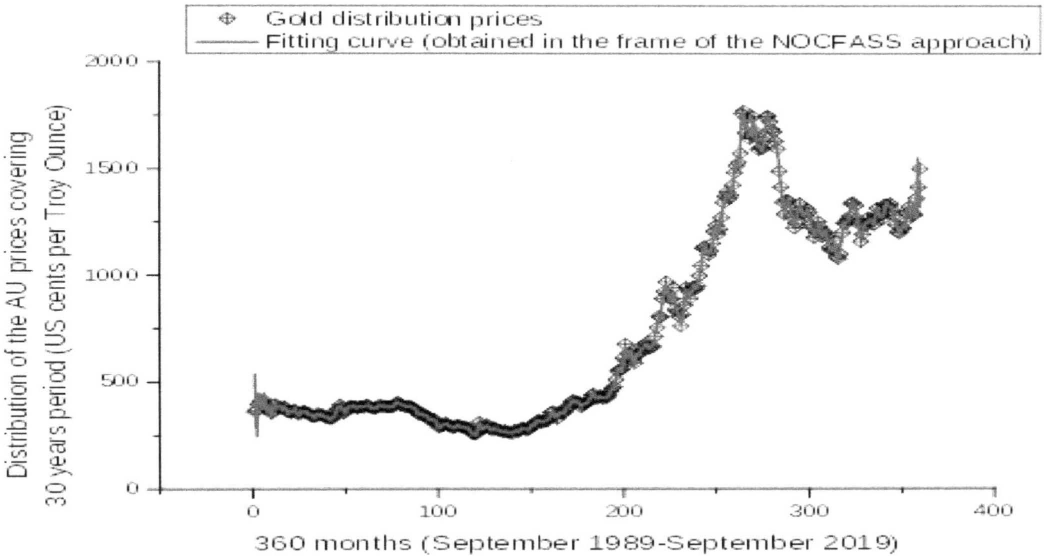

Figure 3.8 The monthly gold (Au) distribution prices (US cents per Troy Ounce) covering 30 years period and its fitting curve (bold black line) obtained in the frame of the NOCFASS approach. In order to obtain this fit (the value of the relative error equals 0.2%) 84 modes are necessary.

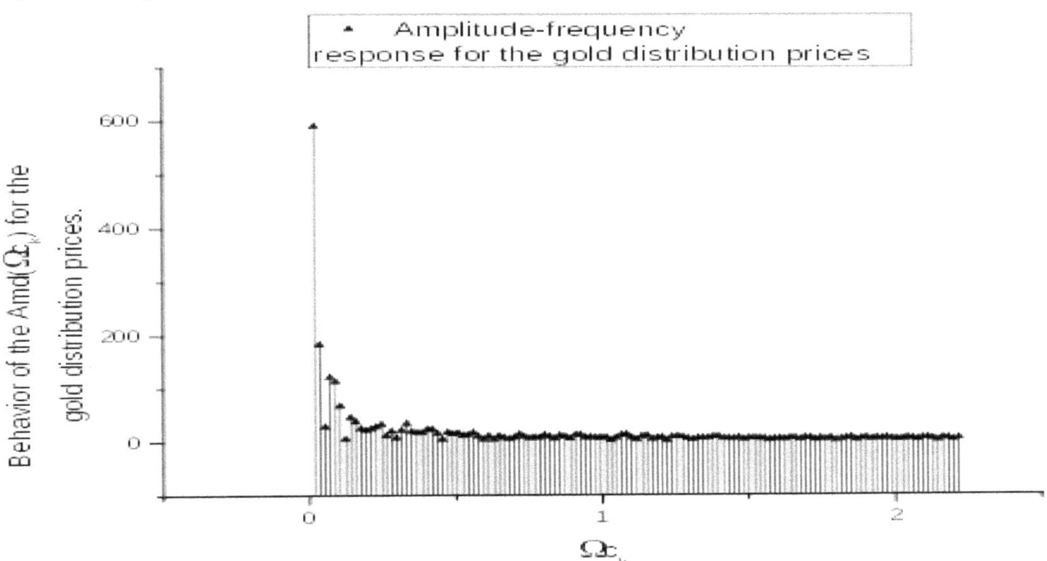

Figure 3.9 Amplitude-frequency response (AFR) for the monthly gold distribution prices. As one can see from analysis of this plot, the AFR of this distribution is "poor" and contains the rectangle-like frequency band. All "resonance" frequencies are concentrated near low-frequency region. No any specific mode can be seen in the "high-frequency" band for the characterization of the fitting curve shown on the previous figure.

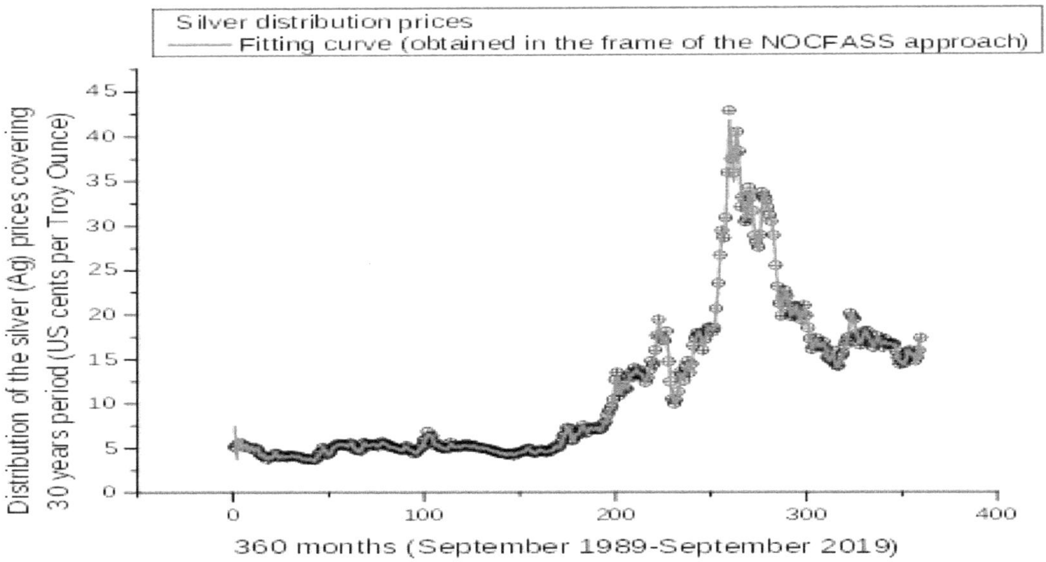

Figure 3.10 The monthly silver (Ag) distribution prices (US cents per Troy Ounce) covering 30 years period and its fitting curve (bold black line) obtained in the frame of the NOCFASS approach. In order to obtain this fit (the value of the relative error equals 0.17%) 90 modes are necessary.

amplitudes located in the low-frequency region and the limiting value of the "high-frequency" band form a specific plateau.

3.5 ANALYSIS OF THE TRANSCENDENTAL/IRRATIONAL NUMBERS SEQUENCES

From our point of view, the series of transcendental/irrational numbers can be used as an "ideal" noise for cryptographic purposes. Really, these digits combined in three, four and etc., digits and chosen from [K(left), K(right)] interval (that is known only for two sides located in the ends of the transmission line) form a segment of a "uniform" noise that can be used for cryptographic purposes. The problem that can arise from analysis of these uniform noise sequences can be formulated as follows: *How to differentiate one segment of the given uniform noise from another one*?

A possible solution of this problem lies in a specific formation of a "crypto-key" and its subsequent identification. In order to find the desired and specific for each sequence the crypto-key one recommends the integration operation. The simplest integral uses the trapezoid formula and it is expressed in the form of iterative expression as

$$
\begin{aligned}
&J_j = J_{j-1} + \frac{1}{2}(x_j - x_{j-1}) \cdot (\Delta y_j + \Delta y_{j-1}), \\
&\Delta y_j = y_j - \langle y \rangle, \ \langle y \rangle = \frac{1}{N}\sum_{j=1}^{N} y_j.
\end{aligned} \tag{3.15}
$$

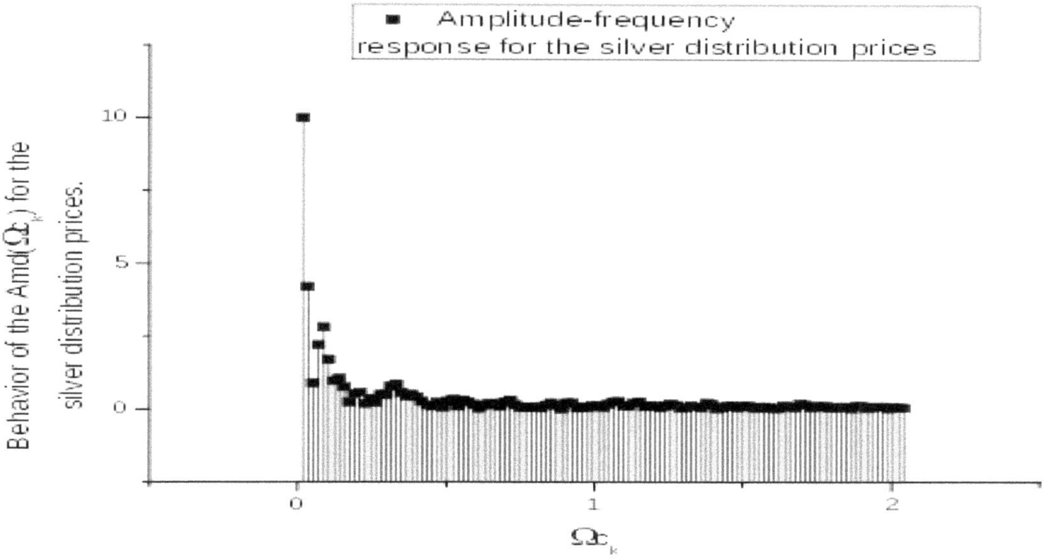

Figure 3.11 Amplitude-frequency response (AFR) for the monthly silver distribution prices. As one can see from analysis of this plot, the AFR of this distribution is similar that is depicted in Fig. 3.9 and "poor" also. The rectangle-like frequency band is seen clearly. All leading frequencies are concentrated again near the low-frequency region. No any specific mode can be seen in the "high-frequency" band for the characterization of the fitting curve shown in the previous figure. However, the order of the magnitude in 60 times less in comparison with distribution shown in Fig. 3.9.

After integration realized with the help of expression (3.15), one can receive a "unique" trend that differentiates one "uniform" segment from another one. Besides, the procedure of reduction to three incident points $(b_1,\ldots,b_p)\rightarrow[\min(b), \text{mean}(b)$, and $\max(b)]$ described in papers [19, 20] helps to increase the speed of the calculation procedure and proves in many cases the self-similarity of the reduced curve to the initial one. In order to prove the statements that was mentioned above, we chose two transcendental famous numbers: as (a) $\pi = 3.14159265358979323846264338 3279\ldots$ and (b) Euler constant $\mathbf{E}= 2.7182818284590452353602 87\ldots$, and one irrational number as (c) $2^{1/2} = 1.4142135623780950488\ldots$

As it was shown in Chapter 2, each of these numbers containing 60000 data points is combined into the set of the normalized triads. These 20000 triads are transformed to an "ideal" noise in accordance with the following procedure: (a) the integer part is omitted; (b) other three digits of the infinite sequence form the normalized number from the interval [0,1] after division on 1000. For example, the number of forms the following sequence of numbers falling into the interval [0,1]. $\pi = 3.141\ 492\ 653\ 689\ 793\ 238\ 462\ 643$ 0.141, 0.492, 0.653, 0.689, 0.793, 0.238, 0.462, 0.643,... Other normalized triads from **E** and $2^{1/2}$ are obtained similarly. In order to leave place for the significant figures only we show a "uniform" noise related to the transcendental number and its "unique" trend. They are shown in Figs. 3.14 and 3.15,

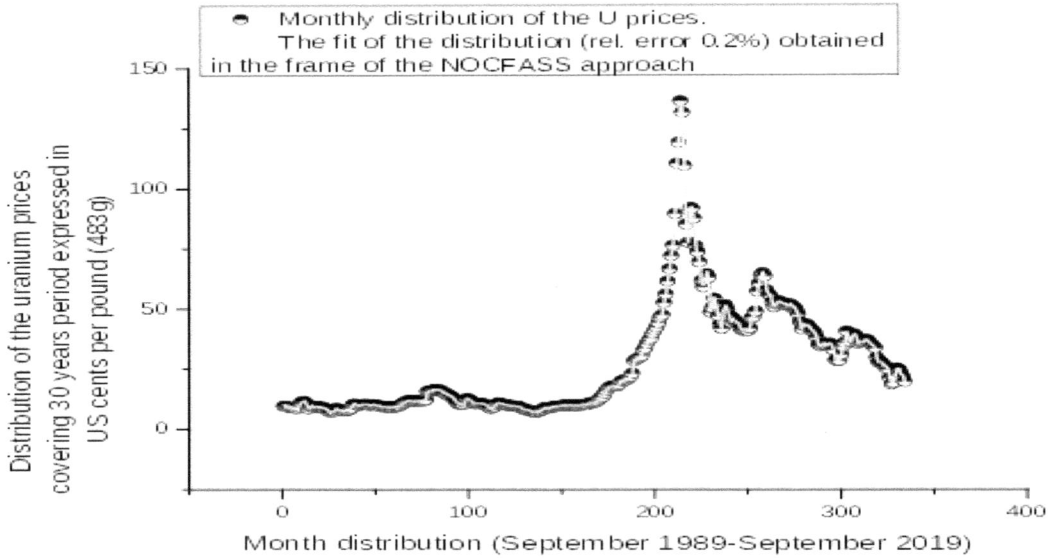

Figure 3.12 The fit of the monthly prices distribution for uranium. The fitting curve obtained in the frame of the NOCFASS approach is shown by the bold black line. The value of the relative error equals 0.2%. The number of modes that are necessary for the fitting is 88. The total number of the data points equals 360.

correspondingly. After application of the reduction procedure to three incident points (the compression parameters b=10) one can obtain three curves (the first curve $Yup(x)$ contains only the maximal values, the down curve $Ydn(x)$ does only the minimal values and the $Ymean(x)$ collects only mean values. These three curves contain 1000 data points only that facilitate considerably the whole treatment procedure. As one can see from comparison of Fig. 3.16 with Fig. 3.15, these three curves (obtained for trans-number π) are similar to the initial trend. For further calculation purposes, we choose only the curve containing the mean values. Omitting the figure related to the demonstration of the whole F-spectrum, we show only two important figures: (a) Fig. 3.17 associated with the fit of the reduced trend and (b) the Fig. 3.18 associated with its AFR. These important figures can be used as the specific "detectors" crypto-keys for identification of the initial sequence shown in Fig. 3.14. In the same manner, we treated the Euler constant **E** and $2^{1/2}$. Their fit and the AFRs are shown in Figs. 3.19 and 3.20 as well as Figs. 3.21 and 3.22, correspondingly.

Comparison of their AFRs depicted in Figs. 3.18, 3.20, and 3.22 shows that in spite of their quasi-monotone and decreasing behavior, the values of their leading amplitudes are *different*. Therefore, they can serve as additional identifiers/crypto-keys alongside with the fitting curves of the reduced trends (shown in Figs. 3.17, 3.19, and 3.21 for the complete description of a "uniform" noise obtained from the digits of the corresponding transcendental/irrational number. This "noise" looks as a random for the third side and contains informative components for two sides located on the ends of the transmission line. This small research allows receiving the answers for the problem posed above. The unique trends together with their AFRs can *differentiate*

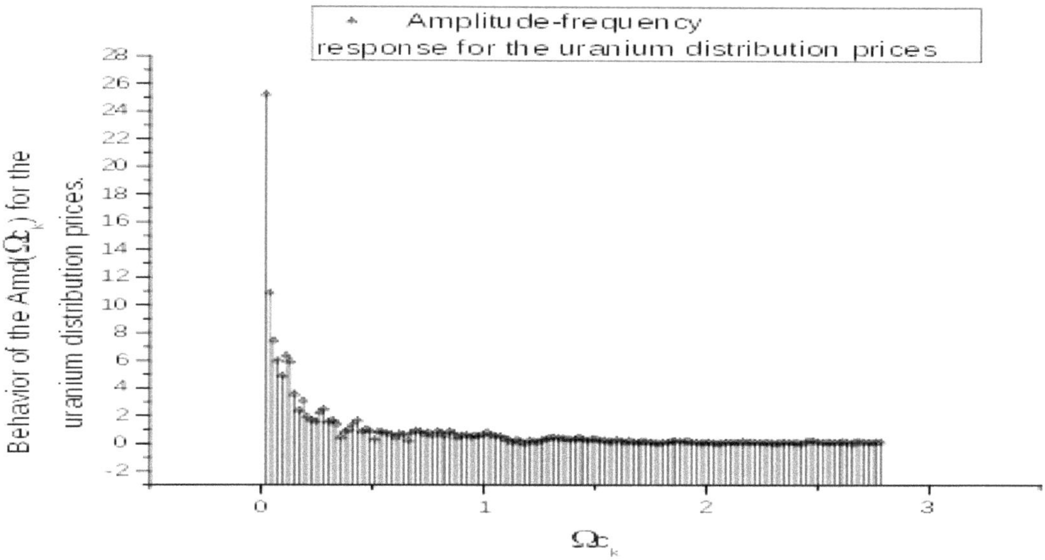

Figure 3.13 Amplitude-frequency response (AFR) for the monthly uranium distribution prices. As one can see from analysis of this plot, the AFR of this distribution is similar that are depicted, correspondingly, in Figs. 3.9 and 3.11. All leading frequencies are concentrated again near the low-frequency region. No any specific oscillating mode can be seen for the characterization of the fitting curve shown on the previous figure. However, they are differed by the values of the amplitudes concentrated in low-frequency region.

one uniform noise from another one. This peculiarity can be used as crypto-keys for the protected communication systems.

3.6 APPLICATION OF THE NOCFASS METHOD TO THE GEOPHYSICAL DATA

We chose deliberately this third example associated with geophysical data as the most difficult for treatment and proper understanding because the content of this manual is aimed presumably for the *advanced* researchers. The data registered by different geophysical equipment/devices are full of "random" time series because our Mother-Earth is a "living" organism and "vibrates" due to many factors that should be taken into account in order to analyze the registered data, representing in many cases a "noise" recorded in the form of trendless sequences. That is why some explanation or preambular is necessary in order to understand deeper a problem that can be solved with the help of the NOCFASS-method.

An observation of ambient seismic noise time series has stimulated by the statistical mechanics approach to earthquake occurrence, applying ideas and methods as scaling laws, universality, fractal dimension, renormalization group, to characterize the physics of earthquakes [21]. However, this approach is useful for areas with repeated clusters of semi-similar *foci* mechanisms. When one interprets data from

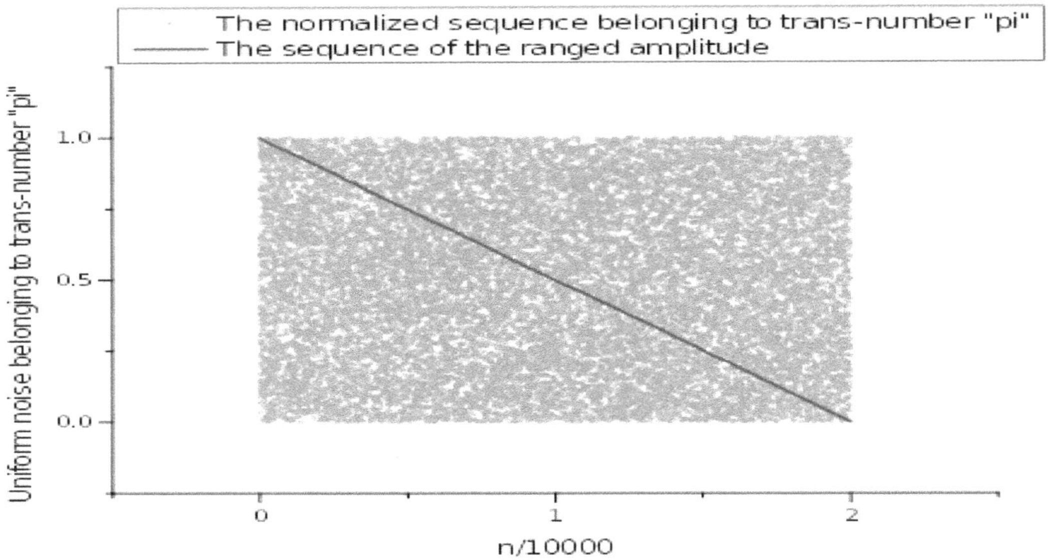

Figure 3.14 The normalized sequence of triads (N=22000) corresponding to the transcendental number "Pi". The light grey straight line shows the sequence of the ranged amplitudes and proves that this triad's distribution is uniform. The same behavior demonstrates the Euler constant "E" and the irrational number $2^{1/2}$ and, therefore, they are not shown.

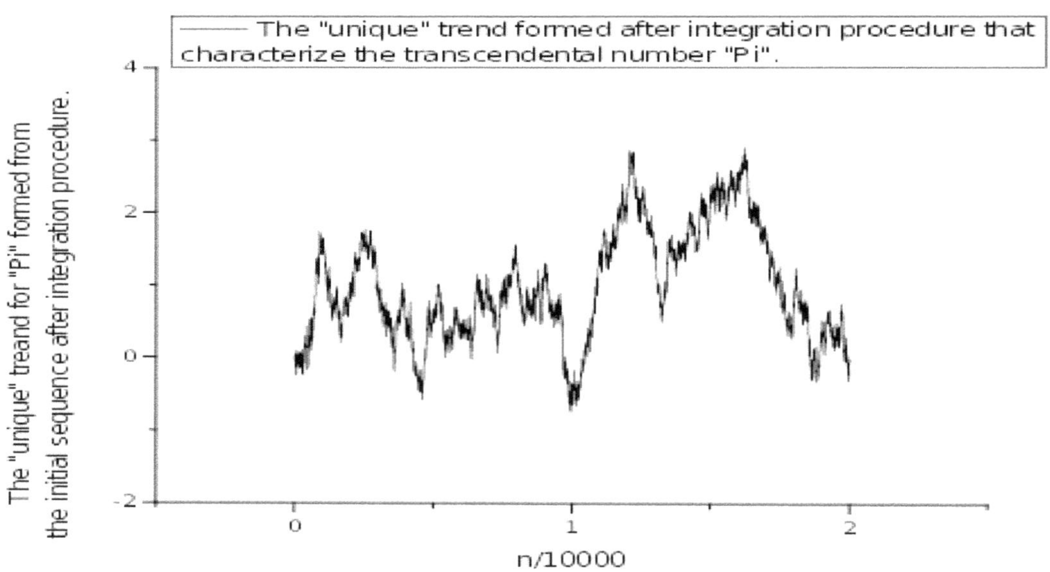

Figure 3.15 The complete trend obtained with the help of trapezoidal formula. It can serve as a specific crypto-key for identification of the sequence shown in Fig. 3.14.

the ambient areas, we need to find specific statistical invariants to describe seismic sequences without quakes appearance. Raw seismic noise-like time series are full of

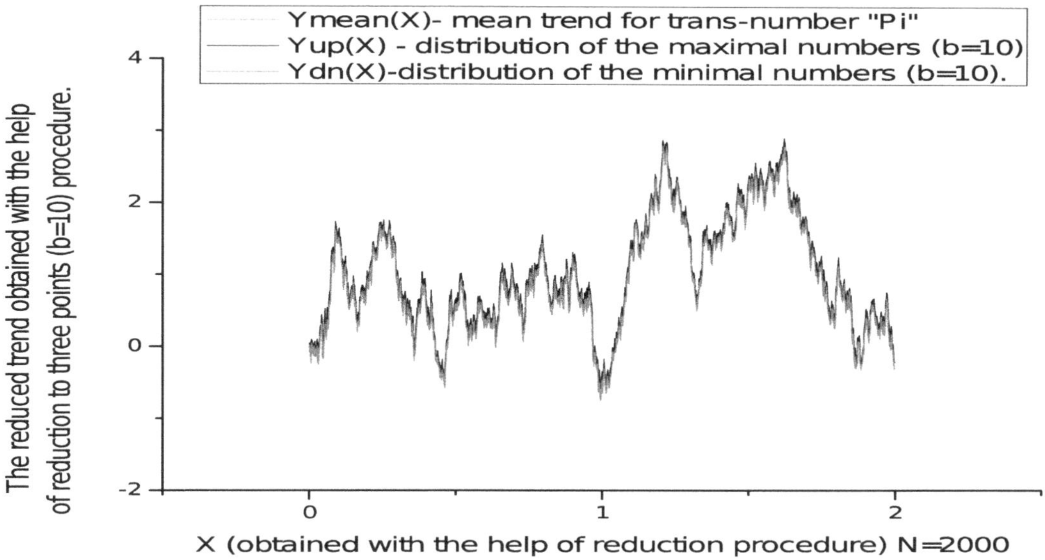

Figure 3.16 If we apply the procedure that defined earlier as the reduction to three incident points and compress the initial trend (see Fig. 3.15 in 10 times then we obtained a similar curve with number of data points N=2000. This procedure facilitates considerably the calculation of the total F-spectrum and the reduced spectrum used for the fitting of the reduced trend. For the fitting purposes, the reduced trend $Ymean(\mathrm{X})$ is used.

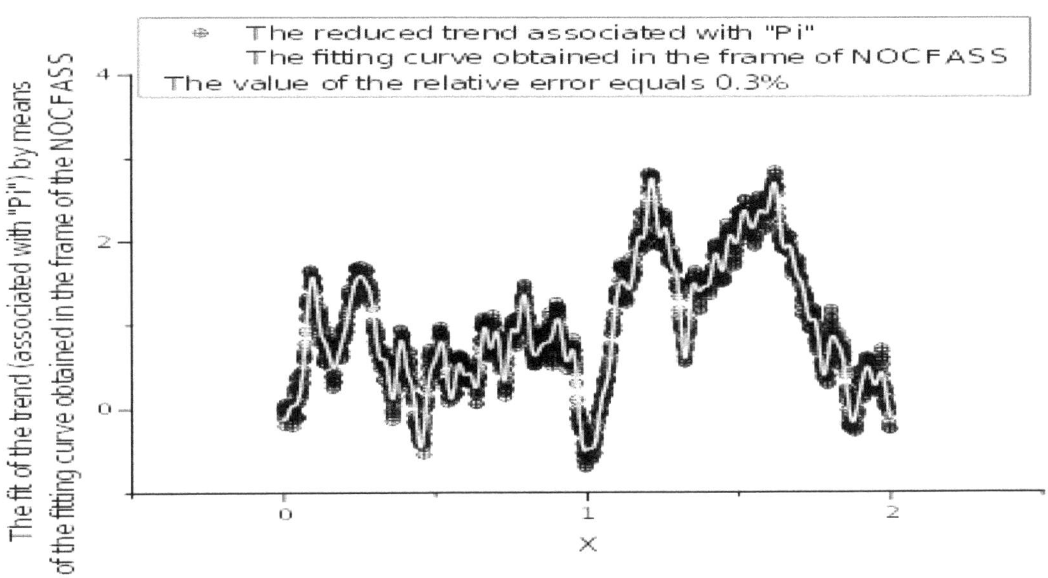

Figure 3.17 The fit of the reduced trend (containing N=2000 data points) for the trans-number "Pi". The value of the relative error equals 0.3%. For this high-quality fit only 50 leading frequencies are necessary.

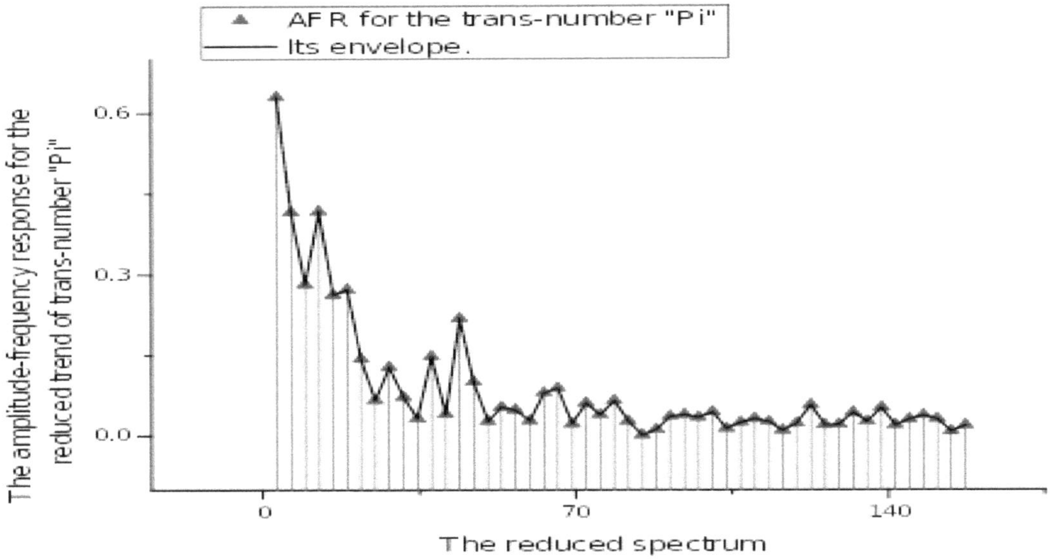

Figure 3.18 The amplitude-frequency response (AFR) defined as the dependence $\mathrm{Amd}(\Omega_k)$ and obtained in the frame of NOCFASS. Only 50 frequencies are needed for description of the reduced trend corresponding to the "Pi" number. This AFR can serve for this number as the second identifier.

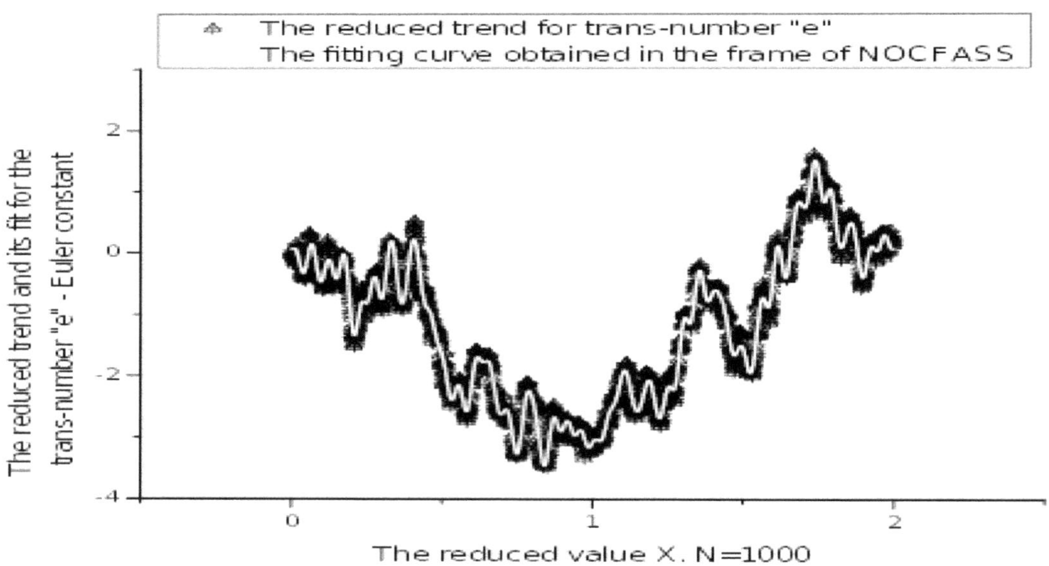

Figure 3.19 The fit of the reduced trend corresponding to the Euler constant "E". The error of the fitting equals 0.25%. For this purpose, only 46 frequencies are necessary.

hidden information. Some of the authors do see some of the rhythms in experimental signals [22, 23]. Most of researchers use the Fourier spectrum to examine the noise-like

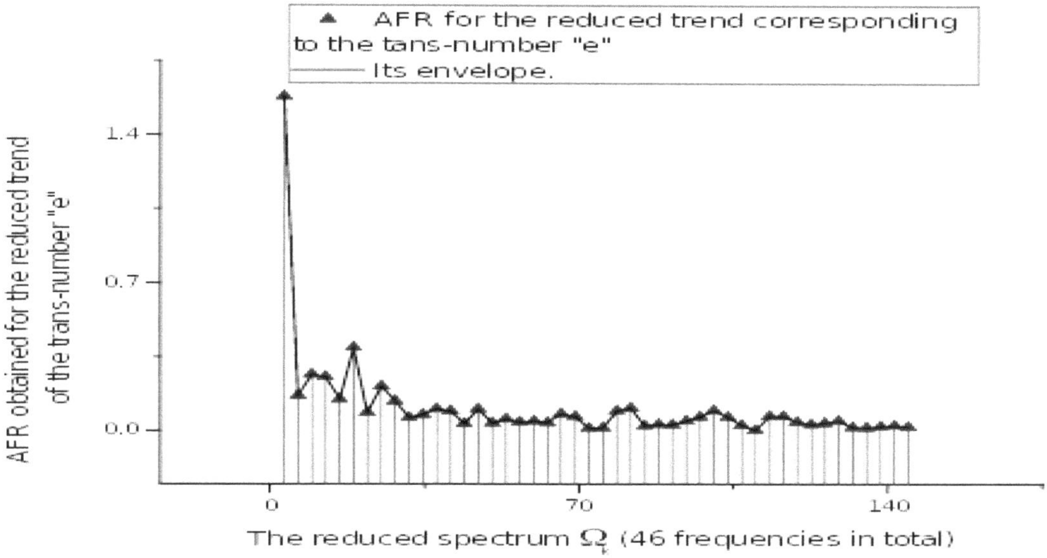

Figure 3.20 The amplitude -frequency response (AFR) defined as the dependence $\mathrm{Amd}(\Omega_k)$ and obtained in the frame of NOCFASS. Only 46 frequencies are needed for description of the reduced trend corresponding to the Euler constant. This AFR can serve for this transcendental number "E" as the second identifier. Compare the ranges of the amplitudes for these two numbers.

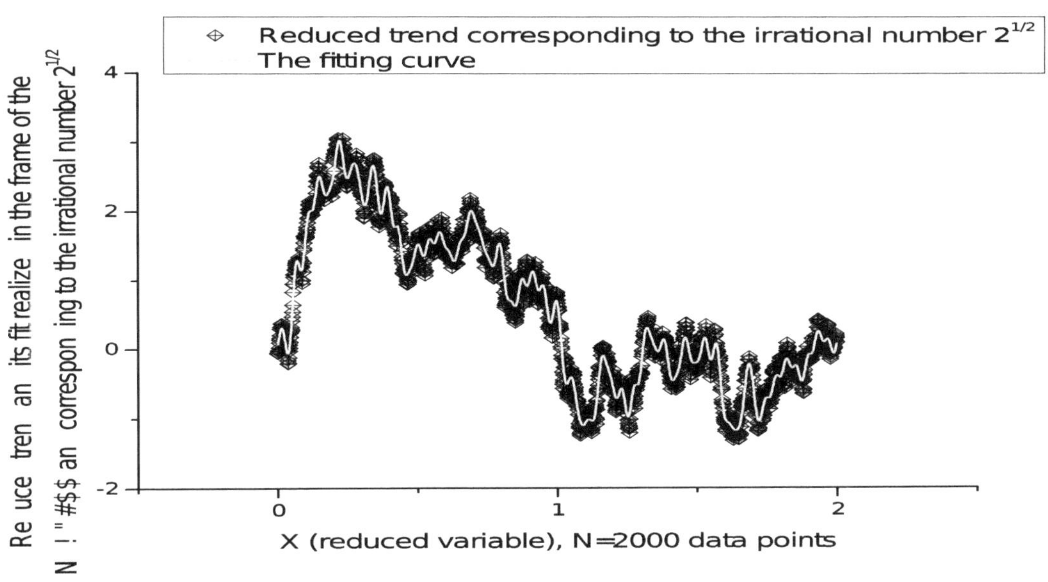

Figure 3.21 The fit of the reduced trend corresponding to the irrational number $2^{1/2}$. The error of the fitting equals 0.2%. For this fit, only 56 frequencies are necessary.

data [24, 25]. However, this way is too strict, because we lose a micro-informational trend, when we move from time space area to frequency domain. The idea to look at the noise-like seismic series as a set of chaotically mixed sequences in time is a new

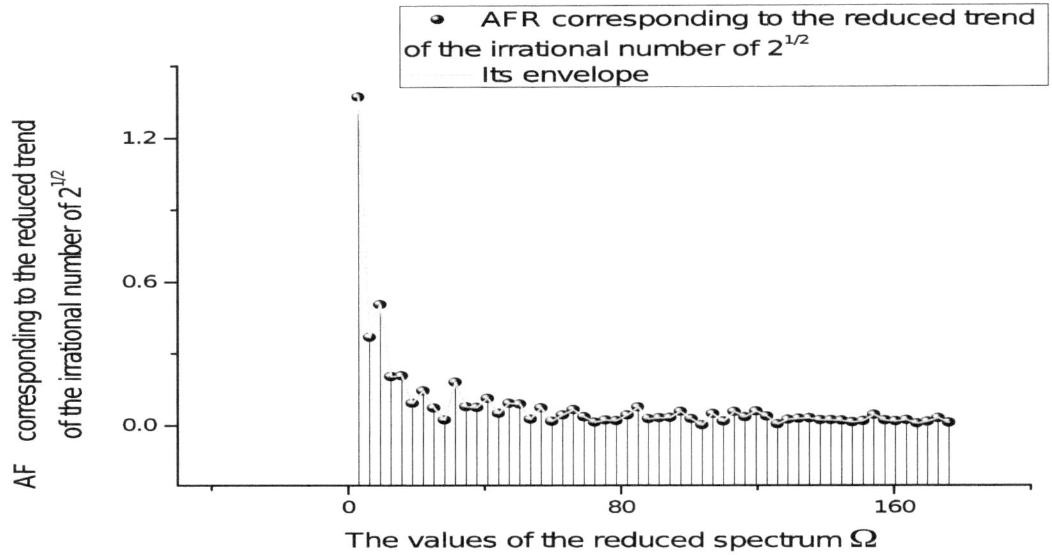

Figure 3.22 The amplitude -frequency response (AFR) obtained in the frame of NOCFASS approach. Only 56 frequencies are needed for description of the reduced trend corresponding to the irrational number $2^{1/2}$. This AFR can serve for this as the second identifier. Comparison of this AFR with similar curves depicted in Figs. 3.18 and 3.20 shows that in spite of their quasi-monotone and decreasing character the values of the leading amplitudes are different.

one. Also, the "normal" noise is suspected to be deleted from the initial recording. But if we have a random noise without a general model (i.e., Gauss) in opposite to [26], this analysis becomes predominant, see, for example, in [27].

We should outline here the *basic* methods that are accepted in geophysics for treatment of a set of seismic signals, which are encrypted as noise-like trendless sequences. The most common existing methods considering a noise, one way or other, as trendless fluctuations, linked to the seismic data, we classify in the Table.

The attentive analysis of these partial methods shows that the universal method of analysis of the TLS is *absent.* That is why the material outlined in Chapter 2 and in this chapter will serve as an example for application of some general and "universal" methods as 3D-DGI, CAPoNeF, and NOCFASS. Before application of the NOCFASS approach, we should give in brief the information about the equipment used.

Prior to this specific data treatment procedure, we design and use a seismic gradient system (SGS) [43], equipped with 18 seismic sensors GS-20DX manufactures by Geospace. Each sensor was connected through a signal conditioning circuit to one of a 32 input channels of analog signal commutator AIMUX32C-2 (N-to-one multiplexer) with an integrated programmable gain amplifier. The output signal was sent to the audio input of a laptop (computer), i.e., ADC (analog-to-digital converter) input of built-in sound card similar to Realtek AC97, etc. The audio ADC of the computer sound card can be used to digitize a signal with a small required dynamic range (digitized signal width less than 14 bits) and a relatively low-frequency signal (no higher than up to 1–10 kHz) [44]. A digitized signal was recorded and saved by software tools of laptop. Added by analog part noise is less than 0.02 % from full signal range. In most cases, it is less than a registered seismic signal. Therefore, one can consider that the output signal of the seismic sensor coincides with the desired/seismic signal. Table 3.1 demonstrates the list of conventional methods used in geophysics.

Previously, we showed that the recordings contain a component of the seismic field, associated with the activation of endogenous sources of elastic waves that form seismic acoustic emission [45]. The nature of such sources is due to the destruction of rocks in the lithosphere of the Earth. The analysis of the deep underground component of the seismic field is connected to the spatial filtering due to the SGS configuration. These components coinciding with sub-vertical oscillations are used for the treatment purposes. If we look at all recordings, they consist of noise-like chaotic trendless sequences. For additional research by the offered method, we use all recordings without any filtering procedures. These recordings associated with the initial data represent themselves the distribution of the seismic fluctuations (measured in Volts) from an arbitrarily chosen channel (SGS geophone) of the gradient system. Therefore, for complete one-day recording, we get 24 files of one-hour length.

The nature of seismic noise at the SGS installation site may differ greatly from the nature of the high-frequency seismic noise taken as the basis. The anthropogenic effect is small and is assumed to be zero at the selected monitoring point Ukok-2. In particular, we note that the recorded seismic signal has a noise-like character. The test analysis shows that, in general, the observed signal has the character of stationary noise, but sporadic disturbances in the stationarity of the noise signal are observed in different time digitizing points.

For further treatment, one divides the analyzed signal fragments into two types: with small-amplitude and -amplitude signal fluctuations (relative number of the saturated value is less and more than 5%, respectively), and analyze them separately. We form two matrices for further processing. The first matrix M_1 corresponds to the small-amplitude signal fluctuations, and M_2 – to the -amplitude signal fluctuations. Mean and standard deviation shows variations which is significantly less than the signal range (-1;+1) for both matrices.

This outcome indicates the absence of any constant source of oscillations such as: interfering noise, induced fluctuations, impulse noise, additive disturbance. Such analysis shows unacceptability standard methods of signal filtering and signal detecting inside trendless fluctuations. The offered approach tends to be compressing and should reflect the fluctuations behavior in time. These careful precautions explaining the influence of many external factors are necessary to understand the nature of the measured fluctuations. Namely, these selected TLS will be treated by the NOCFASS-method. They correspond to Ukok-2 experiment (14th–16th July 2018).

From all available data, we selected as an example initial data that are combined in the form of two rectangle matrices. The first one is $[M_1(N{=}2000)\times(M{=}148)]$ and another one is $[M_2(N{=}2000)\times(M{=}211)]$. In our case, $N{=}2000$ because we take 10-minute discretized sequences for the control. We mentioned above that the values represent the distribution of the seismic fluctuations (measured in Volts) from randomly chosen geophone/channel of the SGS. The first matrix M_1 corresponds to the small-amplitude signal fluctuations, and M_2 – to the -amplitude signal fluctuations

TABLE 3.1 The most common existing methods processing of the seismic noisy fluctuations.

The method	**The limits of its applicability (basics)**	**Reliability (disadvantages)**	**Subject**	**Comments e.g., and References**
Beam-forming and array methods	Adoption of time-windows length and signal-to-noise ratio threshold, that is much averaging in time or the use of array techniques to enhance the hum signal	Stable location parameters for nearby sources for the determined station–location pair. Suits for low frequencies with filtering	Signal energy from the target azimuth direction	[28] Duputel et al., 2009; [29] Kitov et al., 2015; [30] Löer, Riahi and Saenger, 2018
Cross-correlation	Database of known ideal patterns and signal-to-noise ratio threshold	High Cross-correlation coefficient over 0.8 with repeated events	Signals from seismic events encrypted with noise	[31] Stehly, Campillo, and Shapiro, 2006; [32] Adushkin et al., 2015
Green's function	Standard Gaussian distributed stochastic process and statistical probe threshold are used	Estimation with time windows and filtering	Classification of Seismic Noise.	[33] Groos, 2010
Machine learning /Neural networks / least-squares inversion	Multiple synthetic and real data, noise statistics distribution parameters. Test standard estimators: e.g., regression	Indications predictive laws, fitting of multiple damping exponential components, a defined decomposition number	The micro-seismic event monitoring during fracturing or classification of the waves phases.	[34] Mousavi et al., 2016; [35] Chen, 2018; Li et al., 2018; [36] Zhang et al., 2020; [37] Popov and Zamaraev, R.Y., 2020
Wavelet methods	Wavelet database. Comparably small set of existing wavelet models for seismic noise. The initial evaluation of suitable wavelet should be done and justified on model data.	Fix fitting measure impact. The criterion for selection of an optimal/ideal wavelet is not known.	Through a continuous Wavelet transform the features are extracted, yielding a time-frequency-amplitude decomposition.	[38] Learned and Willsky 1995; [39] Riggelsen, Ohrnberger and Scherbaum, 2007
Spectral methods with Fourier analysis	Should define weighting functions as well as reference stations	The selection of the pick probability densities of reference station due to the interference	The monitoring of micro-seismicity with magnitude (M<2) in a densely populated region with poor geology	[40] Plenkers, Husen and Kraft, 2015
Dominant periodicity in time	Long-time series matched with respect to the observed micro-seismicity variations	Alignment with the known processes such as lunar-solar tides of planetary motions	Dominant slow low-frequency processes	[41] Ekstrom, 2001; [42] Lyubushin, A.A., 2009

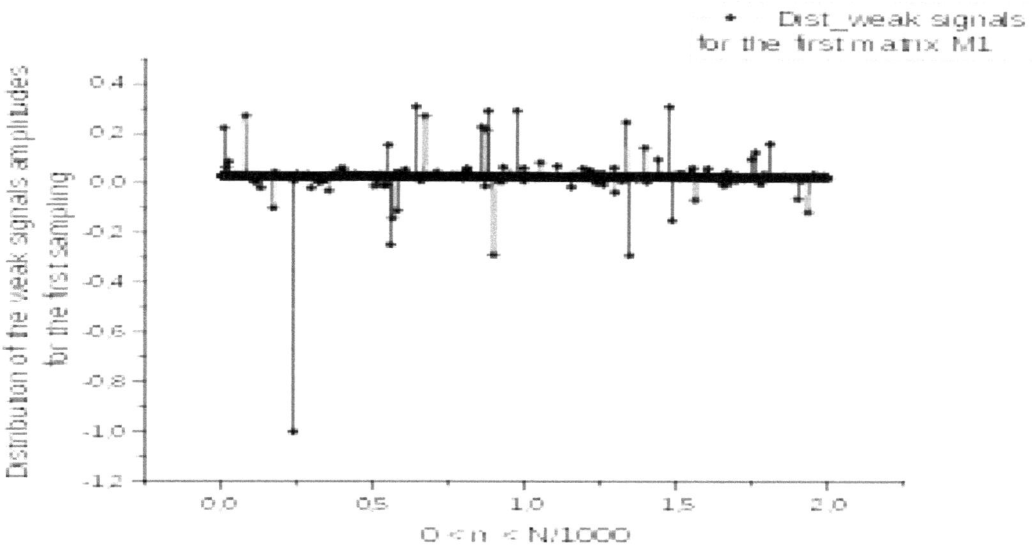

Figure 3.23 Distribution of the weak signals amplitudes for the first column of the initial rectangle matrix M_1.

with the significant values, exceeding the admissible interval [-1, 1]. Initially we tested the stationarity of the mean values and their standard deviations.

For acceleration of the following procedure, we subject these initial data to a reduction procedure containing three incident points. As it has been shown earlier in papers [19, 20] and in the book (Nigmatullin, Lino and Maione, 2020) because of this procedure, one can obtain three similar sequences. For this aim, it is necessary to select from the given small sequence containing b data points only three statistically significant points b_{max}, b_{min}, b_{mean}. These three points conserve their values relatively $(b)!$ permutations. This procedure decreases the subsequent calculations and keeps the similarity to the initial sequence. In Figs. 3.23 and 3.24, we show the initial signals used for this type of treatment.

In Figs. 3.25 and 3.26, we demonstrate the results of the reduction procedure that use this similarity of initial TLS to sequences compressed 10 times. The second important point is the testing of the initial rectangle matrices $M_{1,2}$ and the integrals obtained from them with the help of the recurrence trapezoid formula (3.15).

For evaluation of the quality of experimental measurements and their similarity to each other we use the formula (16) that is explained in details in Chapter 1. This important formula allows us to test *any* initial rectangle matrix that includes in itself the set of successive measurements $y_m(t_n)$ (m=1,2,...,M) performed during the fixed time. These correlations for the initial matrices $M_{1,2}$ and the correlations obtained for their integrated curves (obtained with the help of expression (3.15)) are given in Fig. 3.3.

As one can see from these figures, the integration operation makes the correlation weaker in comparison with the initial TLS. The total number of independent correlations for matrix M_1 equals $M(M$-1)/2 = (147148)/2=10878, while for M_2 = (210211)/2=22155, correspondingly.

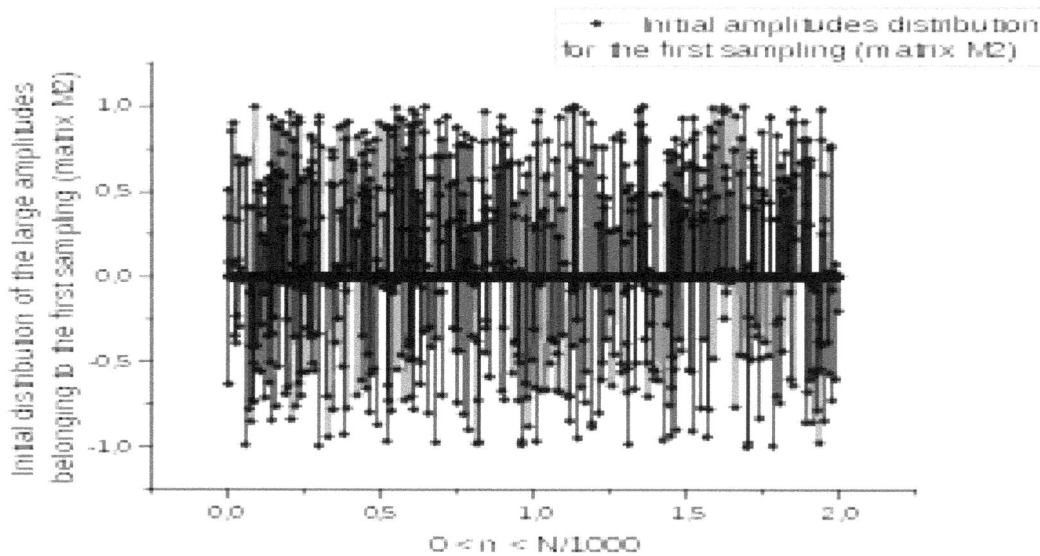

Figure 3.24 Distribution of the significant signals expressed in distribution of amplitudes for the first column of the initial rectangle matrix M_2. As one can see from comparison of two distributions, they are strongly differed from each other.

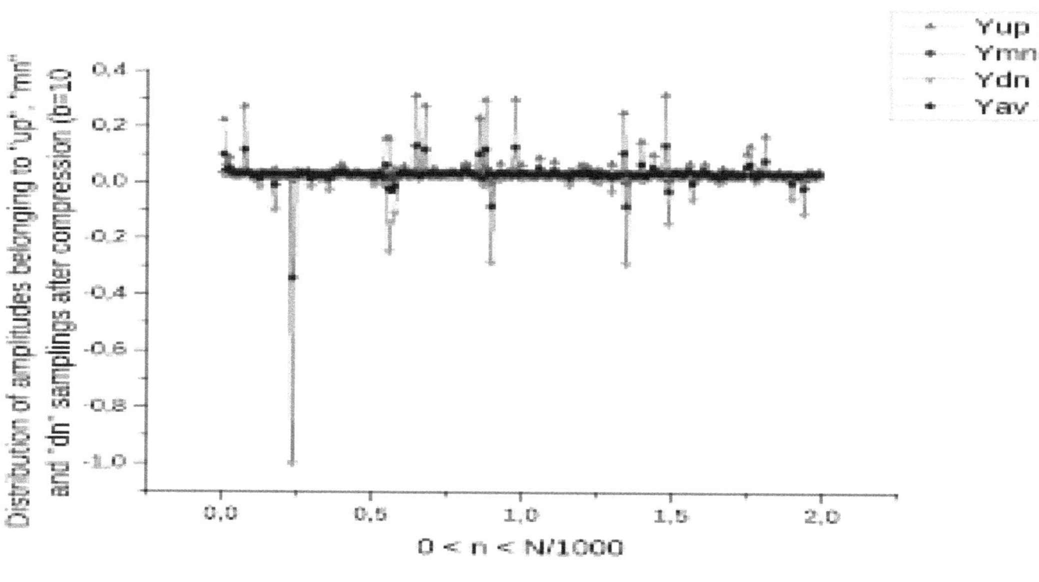

Figure 3.25 Here we demonstrate the results of reduction procedure to three incident points obtained for the first sampling. The value of the compression parameter b=10. The curves $Yup(t_n)$, $Ymn(t_n)$, $Ydn(t_n)$ – demonstrate the distribution of the "maximal", "mean" and "minimal" points, correspondingly. The initial curve and the curve $Yav(t_n) = (Yup(t_n) + Ymn(t_n) + Ydn(t_n))/3$ is strongly correlated with the initial one, depicted in Fig. 3.23. The Pearson correlation coefficient equals 0.95–0.99.

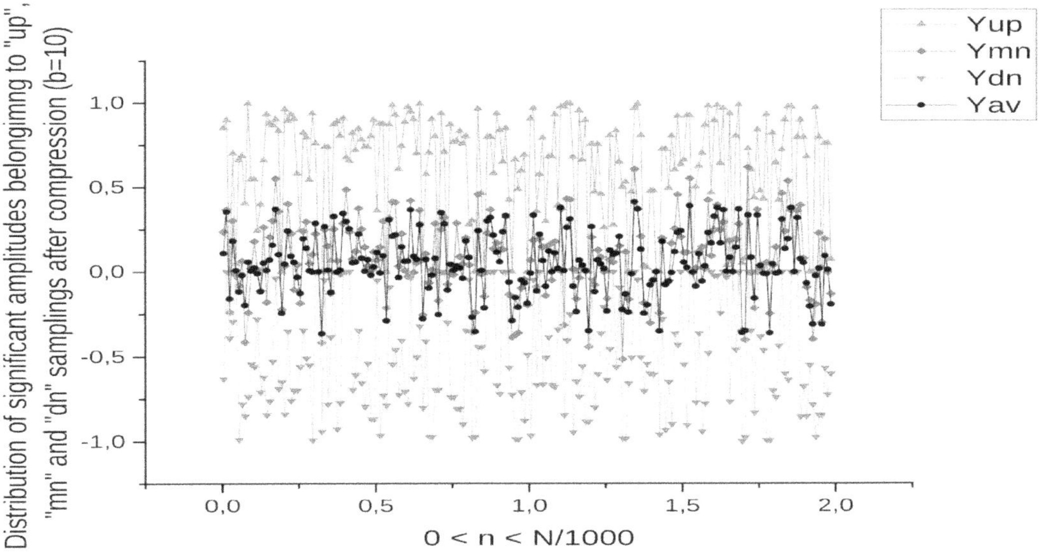

Figure 3.26 Here we demonstrate the results of reduction procedure to three incident points for significant amplitudes. The initial curve and the curve $Yav(t_n) = (Yup(t_n) + Ymn(t_n) + Ydn(t_n))/3$ is strongly correlated with the initial one, depicted in Fig. 3.23. The Pearson correlation coefficient equals 0.95–0.99. Other definitions are the same as in the previous figure.

Namely, these curves obtained from the initial TLS will be used for fitting purposes. Then we obtain the total F-spectra for the integrated curves $JYav_m(t_n)$ belonging to the matrices $M_{1,2}$, correspondingly. We use expression [2], where we put $\sigma = \pi$ in order to see all amplitudes distribution. Then, for the fitting purposes, we select only the right-part of these spectra shown in Figs. 3.33 and 3.34. Namely, these averaged parts of the F-spectra, presented in the right-hand sides of Fig. 3.33 will be used in further calculations. For fitting purposes, it is sufficient to choose m (number of modes) from the interval $[0, Nc]$ (Nc=150, see expression 3.7) and put the value of =0.75. (See expression (3.8)).

In the result of these simple manipulations, one can calculate the desired set of the reduced frequencies Ω_k They are shown in Fig. 3.35 for two matrices $M_{1,2}$, accordingly.

Actually, these two segments of the frequencies with the almost equal minimal values ($\min(\Omega) = 2\pi/2 \cong 3.17$) and $\max(\Omega)$=130.04 (for a set of the integrated curves originated from the matrix M_1) and 193.48 (for matrix M_2)) allow to fit a wide set of the integrated curves $JYup_m(t_n)$, $JYdn_m(t_n)$, $JYmn_m(t_n)$, $JYav_m(t_n)$ forming 4 matrices with number of columns M=148 (for M_1) and M=211(M_2). The authors cannot show the fit of all sets of these curves; therefore, one can show the fit of three selected curves $JYup_m(t_n)$, $JYdn_m(t_n)$, $JYav_m(t_n)$ corresponding only to the limiting cases (m=0 and M=148 (M_1), 211(M_2). They are shown in Figs. 3.36 and 3.37 as well as Figs. 3.38 and 3.39, correspondingly.

In order to put a final point in consideration of this example, it is necessary to show also the amplitude-frequency response (AFR) for these selected three curves,

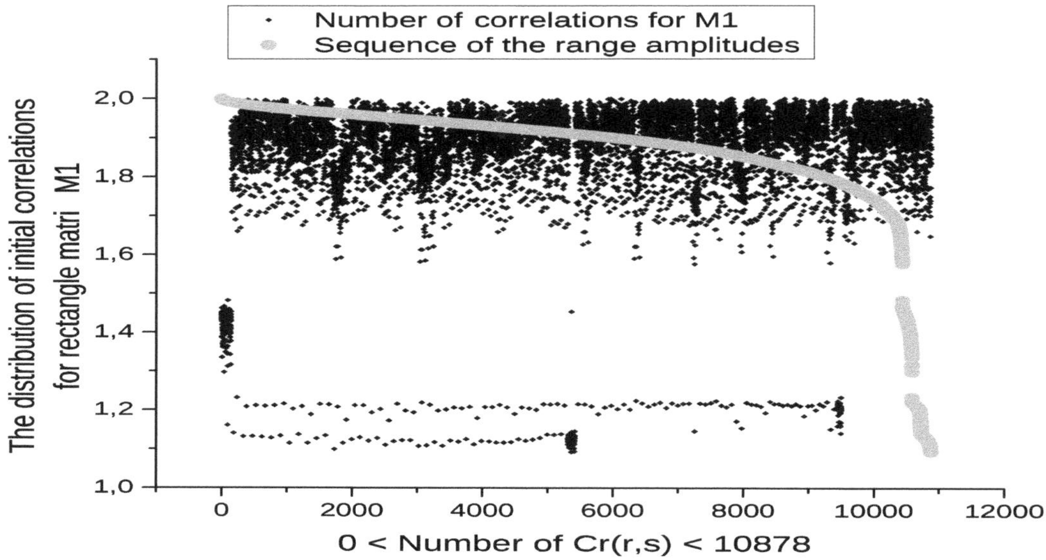

Figure 3.27 The distribution of initial correlations calculated with respect of expression (16) for rectangle matrix M_1. As it follows from this figure all correlations are located in the interval (1.1–2.0). The thick downwards trend line shows the distribution of the SRA (sequence of the ranged amplitudes). Three groups of correlations are seen clearly.

including two limiting cases. As it is well known, the AFR includes the dependence of the amplitude modulus $Amd_k(\Omega_k) = \sqrt{Ac_k^2 + As_k^2}$, $\Phi_k(\Omega_k) = \tan^{-1}(As_k/Ac_k)$ with respect to the reduced spectrum $\Omega_k = \frac{\max(\Omega)-\min(\Omega)}{\max(k)}k + \min(\Omega)$. These dependencies are shown in Figs. 3.40 and 3.41 as well as Figs. 3.42 and 3.43, correspondingly.

3.7 RESULTS AND DISCUSSION

- In this chapter, we found the answer for the problem formulated above: how to obtain the reduced/shortcut F-spectrum from the complete one that helps to fit the desired random signal with high accuracy (the value of the relative error does not exceed 5%). This simple approach can be realized easily in microcontrollers and other portable devices for analysis of different beatings and multi-periodic signals.

- In this study, we deliberately do *not* use the orthogonal and symmetrical properties of the F-transformation because the main problem (let us say main challenge/ advantage) of the proposed approach, defined as NOCFASS-method, is to find the shortcut/reduced spectrum that provides the fit of the given complex and multi-periodic signal.

- Nontrivial examples associated with real data show that AFR cannot contain the leading amplitudes that are typical for the mixed harmonic signals. Instead

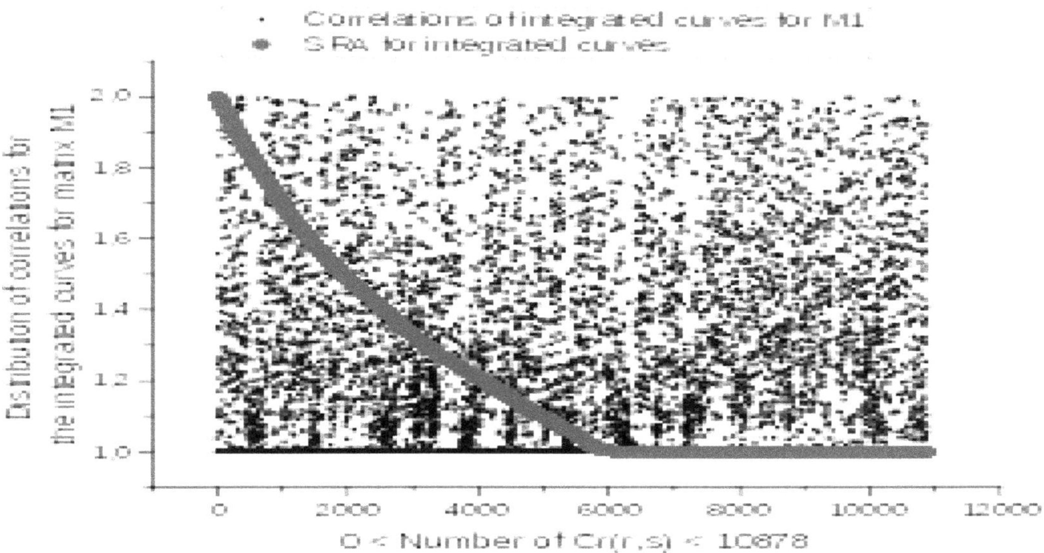

Figure 3.28 The distribution of initial correlations calculated for integrated curves with respect of expression (3.12) for matrix M_1. As it follows from this figure, the integrated correlations become weaker and part of them equals one. As before, the thick downward trend line shows the distribution of the SRA (sequence of the ranged amplitudes).

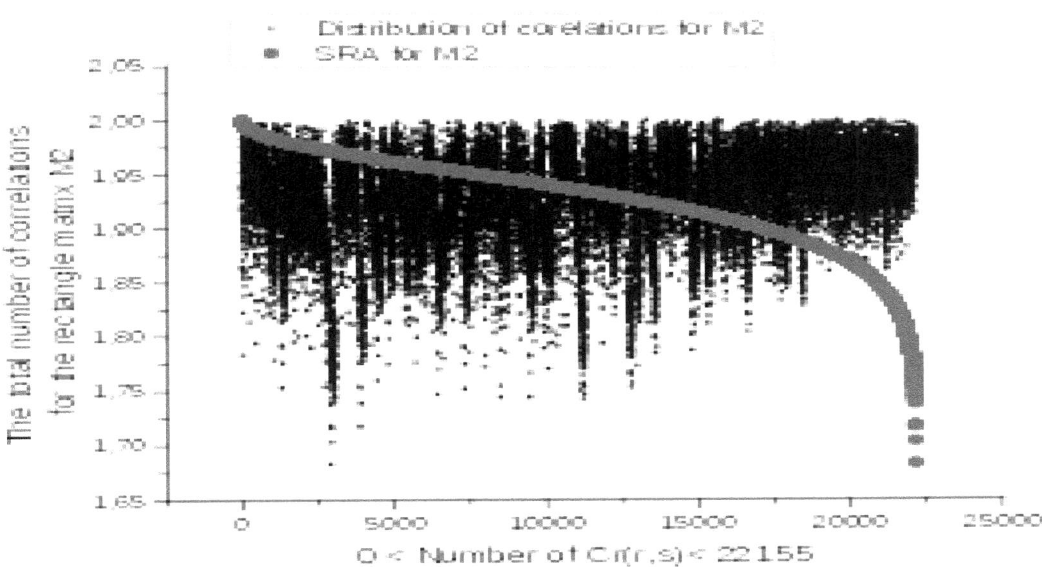

Figure 3.29 The distribution of initial correlations calculated with respect of expression (12 for rectangle matrix M_2. As it follows from this figure all correlations are located in the interval (1.1–2.0). The thick downward trend line show the distribution of the SRA (sequence of the ranged amplitudes).

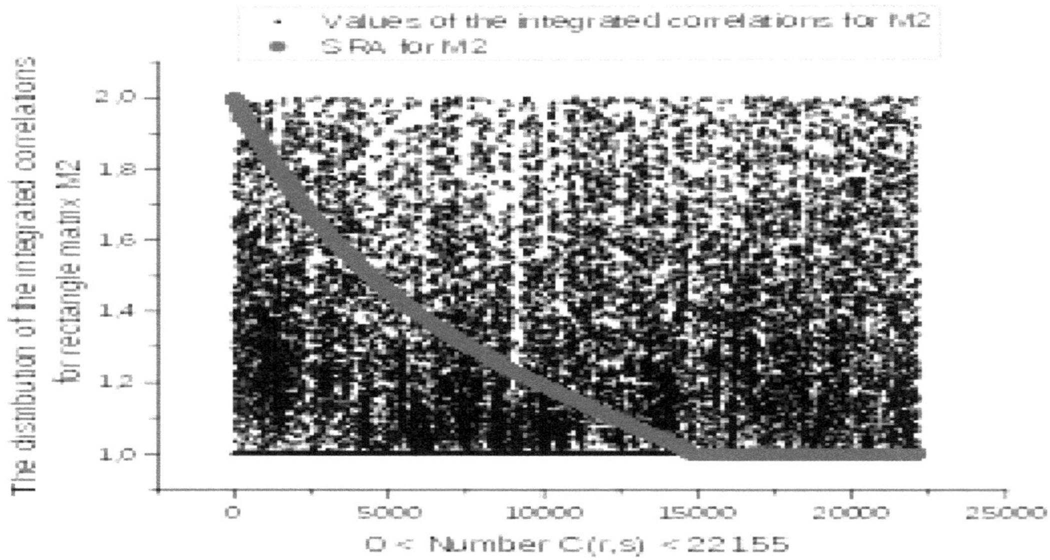

Figure 3.30 The distribution of initial correlations calculated for integrated curves with respect of expression (3.12) for matrix M_2. As it follows from this figure, the integrated correlations become weaker and part of them equals one. As before, the thick downwards trend line show the distribution of the SRA (sequence of the ranged amplitudes).

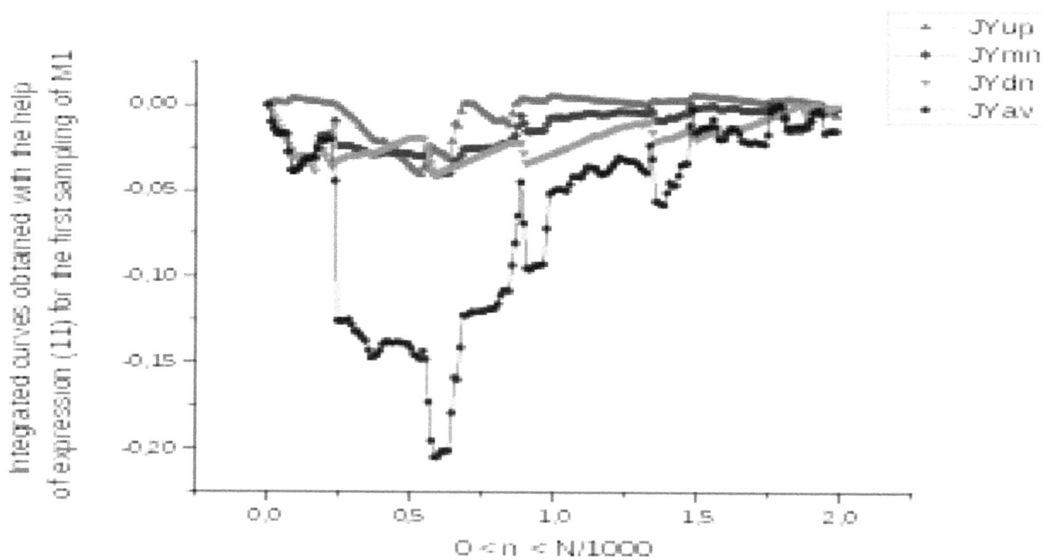

Figure 3.31 Here we show the integrated curves obtained with the help of expression (3.11) from the TLS depicted in Fig. 3.23 for the first sampling of M_1. Namely, these four integrated curves obtained from the initial matrix M_1 (M=148) will be fitted with the help of NOCFASS method.

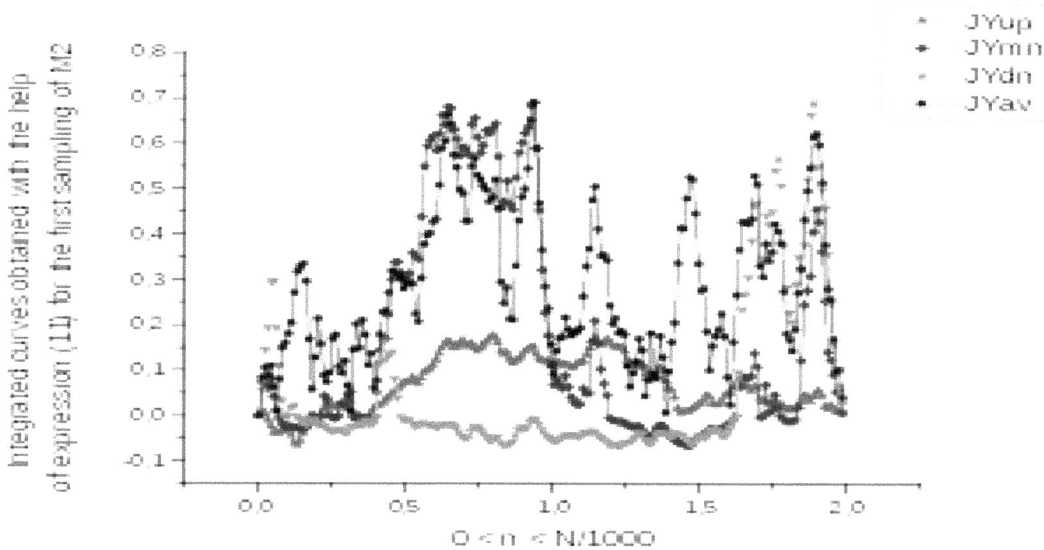

Figure 3.32 Here we show the integrated curves obtained with the help of expression (3.15) from the TLS depicted in Fig. 3.24 for the first sampling of the matrix M_2. Namely, these four integrated curves obtained from the initial matrix M_2 (M=211) will be fitted with the help of NOCFASS method.

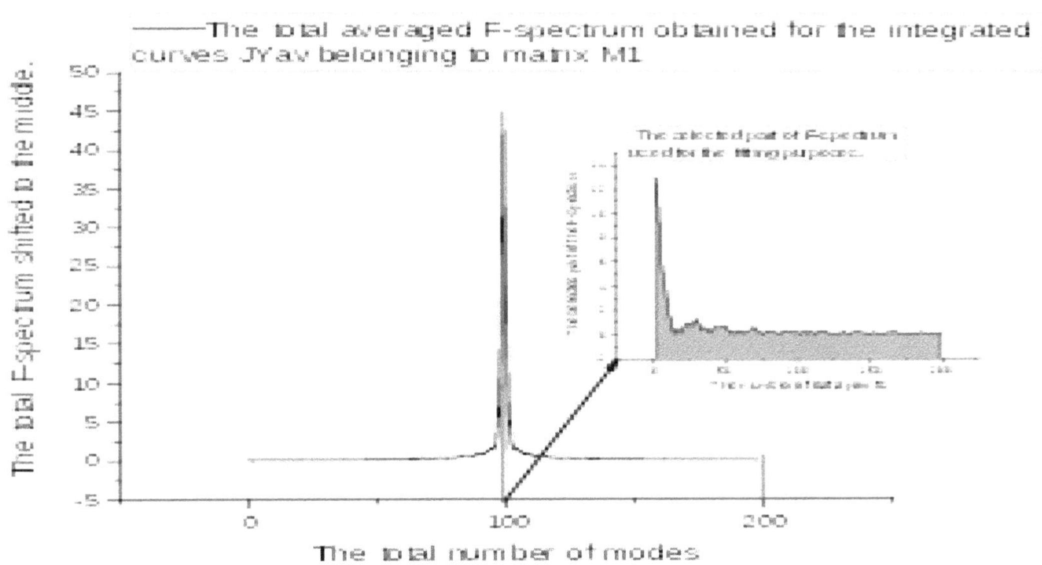

Figure 3.33 Here we show the averaged part of the F-spectrum shifted to the middle part of the total data points. This spectrum is obtained for all integrated curves $JYav_m(t_n)$ from matrix M_1. Only the right-hand side (shown on the small figure) will be used for the fitting purposes.

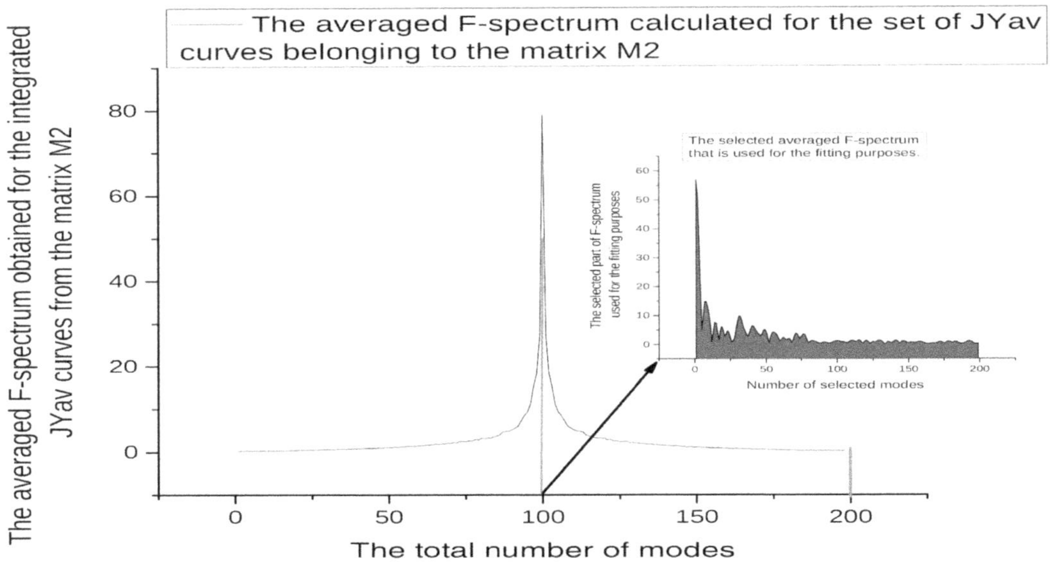

Figure 3.34 Here we show the averaged part of the F-spectrum shifted to the middle part of the total data points. This spectrum is obtained for all integrated curves $\mathrm{JYav}_m(t_n)$ from matrix M_2. Only the right-hand side (shown on the small figure) will be used for the fitting purposes. This spectrum is obtained for all integrated curves $\mathrm{JYav}_m(t_n)$ from matrix M_2.

of AFR containing frequencies with values of amplitudes, we obtain the AFR with high-frequency bands where amplitudes are almost equal to each other (economic data). This specific plateau formed from the high-frequency set of the almost equal to each other amplitudes can be used for analysis of the different signals as well.

- Specific location of the amplitudes in certain intervals and their distribution in the corresponding AFRs obtained for transcendental (π, **E**) and irrational number ($2^{0.5}$) can be used as crypto-keys in differentiation of an "ideal" noise.

- It is necessary to add some important remarks related to analysis of geophysical data. The preliminary results of the first stage of random fluctuations (RF) are based on the recordings of the seismic gradient set. This new type of RF modeless analysis allows us to consider the TLS as the reduced curve for the necessary extraction of the trends presenting themselves the "hidden" information. Without the application of filters of different kinds and without loss of the initial data in the low-frequency region, it becomes possible to describe *quantitatively* a massive of data covering some days. It is interesting to notice that, from the first site, a quasi-stationary background initial noise is located near-zero values. At the same time, the reduced curves (*Yup*, *Ymn* and *Ydn* obtained from the three-incident points procedure) have another character that is differed from the zero values. Application of the NOCFASS-method allows us to "see" in the chaotic trend distributions the desired dispersion law

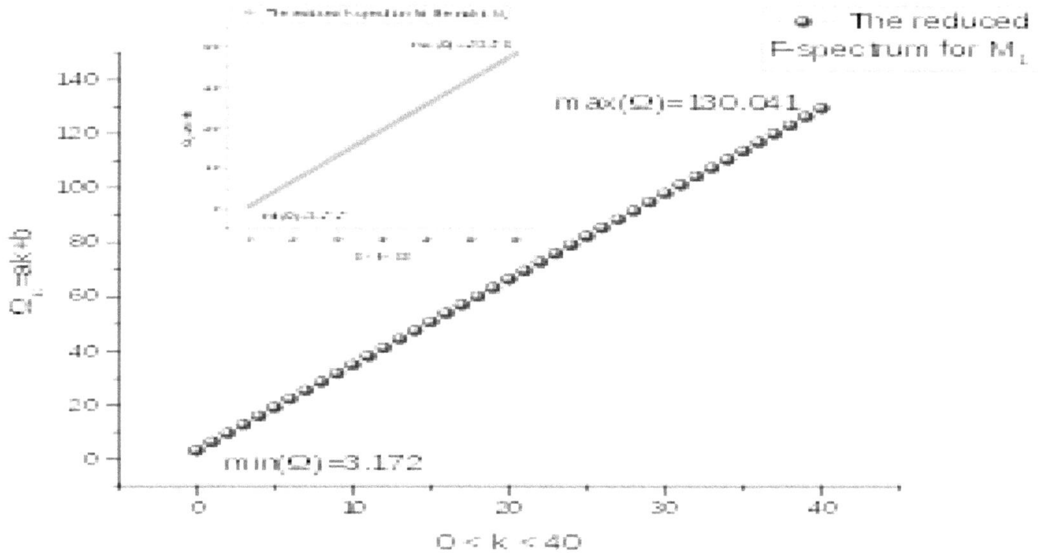

Figure 3.35 This figure demonstrates the reduced F-spectra obtained for the both cases $M_{1,2}$. As one can see from comparison of these spectra the basic difference between them lies in their maximal values max()=130.041 (M_1) and max()=193.476 (M_2). Actually, these reduced spectra can be considered as the specific invariant spectra that allow to fit a wide set of the integrated curves obtained from initial matrices $M_{1,2}$.

applied for description of the calculated frequencies. The found dispersion law is characterized by the segment of the straight line that is rather unusual for RF distribution (noise).

After a detailed analysis of the AFR for matrices M_1 (the small-amplitude signal fluctuations) and M_2 (the amplitude signal fluctuations), one can see that for the given three days recordings (from 14 July 2018 (00:00:00 GMT+6) to 16 July 2018 (23:59:59 GMT+6)) at the monitoring point Ukok-2 the beginnings of the recordings are characterized by values of the AFRs in comparison with the amplitudes corresponding to the record endings. At the same time, the amplitude values for the low and mean curves for M_1 are located linearly near-zero values, while the upper AFR curve is falling down like a saw to zero values in the given frequency segment. The phase characteristics AFR of these three curves are presented by random uniform distributions. The sequences of the ranged amplitudes (SRA) are kept their closeness to the segments, and here the SRA are close to the straight lines. For the matrix M_2, the AFRs distributions for the down and mean curves differ from the distributions corresponding to the matrix M_1. This peculiarity serves as an additional argument to the separate consideration of the M_1 and M_2 recordings. However, for both matrices, the AFRs demonstrate the saw-like falling down of amplitudes starting from small frequencies Ω to the ones. This fact serves as an indicator of

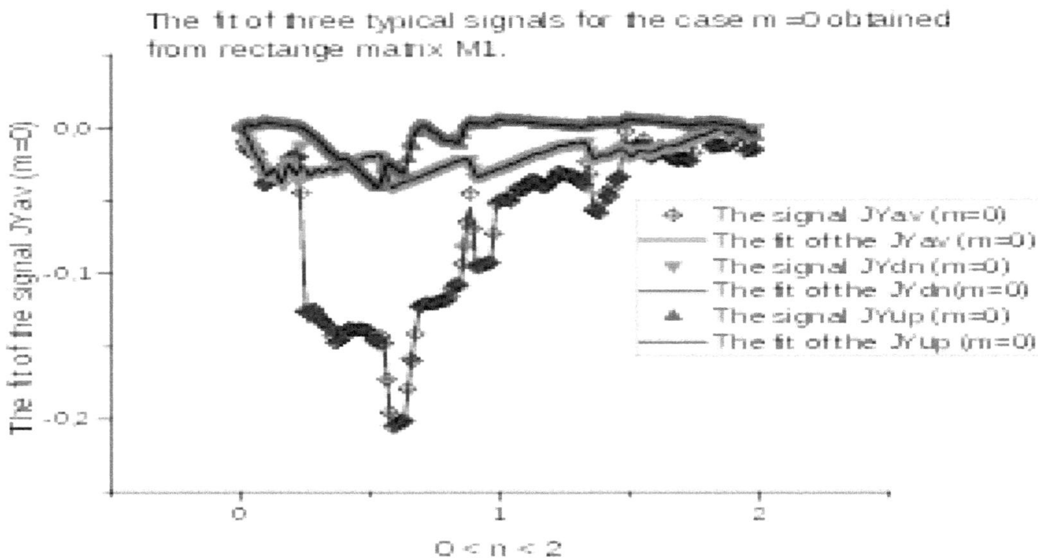

Figure 3.36 Here we demonstrate the fit of three typical curves $JYup(t_n)$, $JYdn(t_n)$, $JYav(t_n)$ for m=0 obtained from the matrix M_1. We are not able to show all fitting curves (4148=592), obtained with the help of the reduced spectrum $\Omega_k = \frac{\max(\Omega)-\min(\Omega)}{\max(k)}k + \min(\Omega)$ depicted in the previous figure.

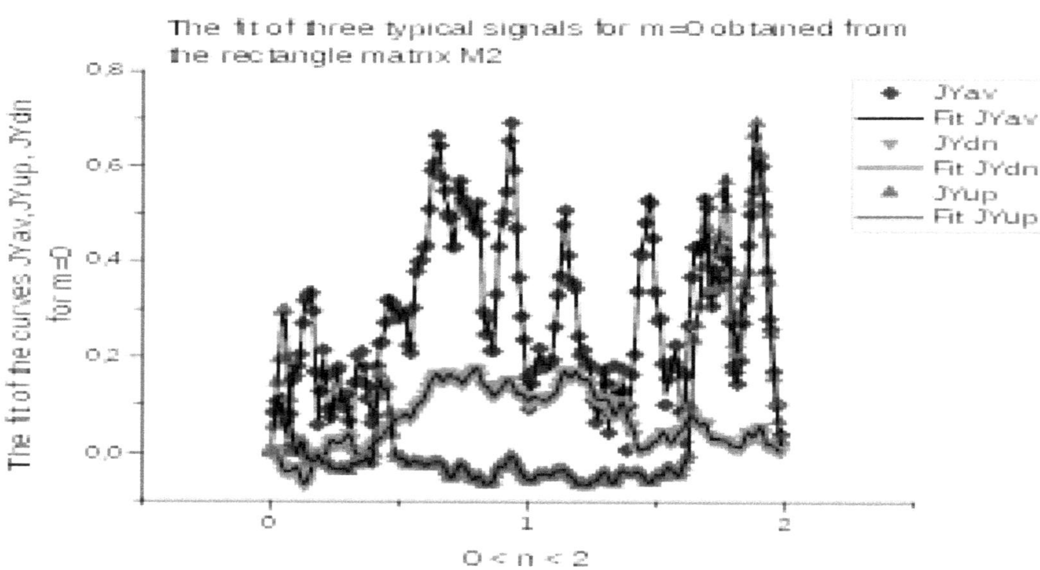

Figure 3.37 The fit of three typical curves $JYup(t_n)$, $JYdn(t_n)$, $JYav(t_n)$ for m=0 obtained from the matrix M_2. We are not able to show all fitting curves (4211=844), obtained with the help of the reduced spectrum $\Omega_k = \frac{\max(\Omega)-\min(\Omega)}{\max(k)}k + \min(\Omega)$ depicted in the previous figure. We are limited by demonstration of the limiting cases.

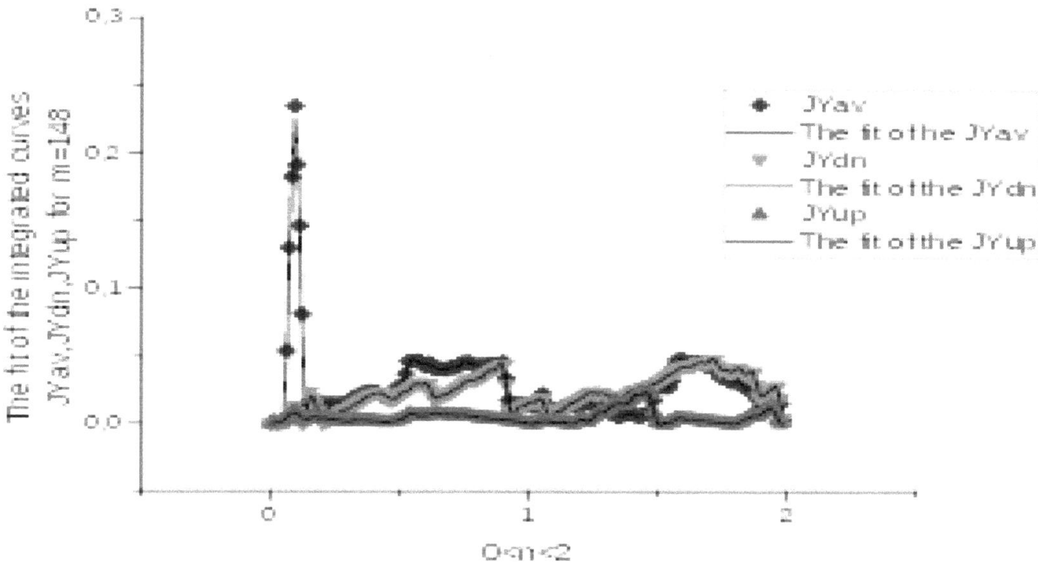

Figure 3.38 Here we demonstrate the fit of three integrated curves $JYup(t_n)$, $JYdn(t_n)$, $JYav(t_n)$ for the final sampling M=148 obtained from the matrix M_1. The relative error does not exceed the value 1.5%.

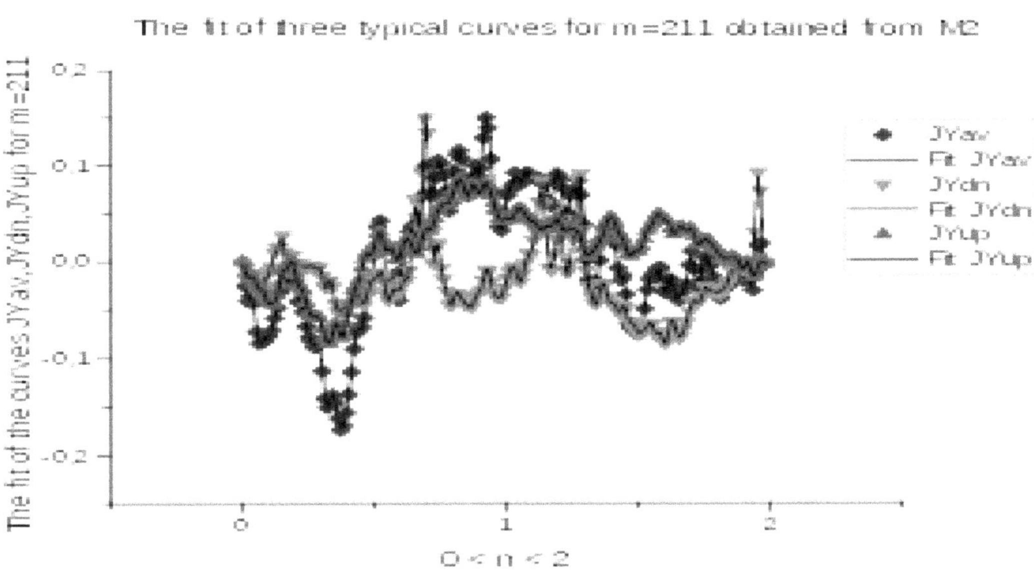

Figure 3.39 Here we demonstrate the fit of three integrated curves $JYup(t_n)$, $JYdn(t_n)$, $JYav(t_n)$ for the final sampling M=211 obtained from the matrix M_2. The mean relative error value equals 1.5% and lies in the interval (1–2%).

the tense-deformed medium state. Really, in the beginning of the recordings, the medium has potential energy. During three observation days, three local earthquakes took place. After that, all system loses its potential energy and

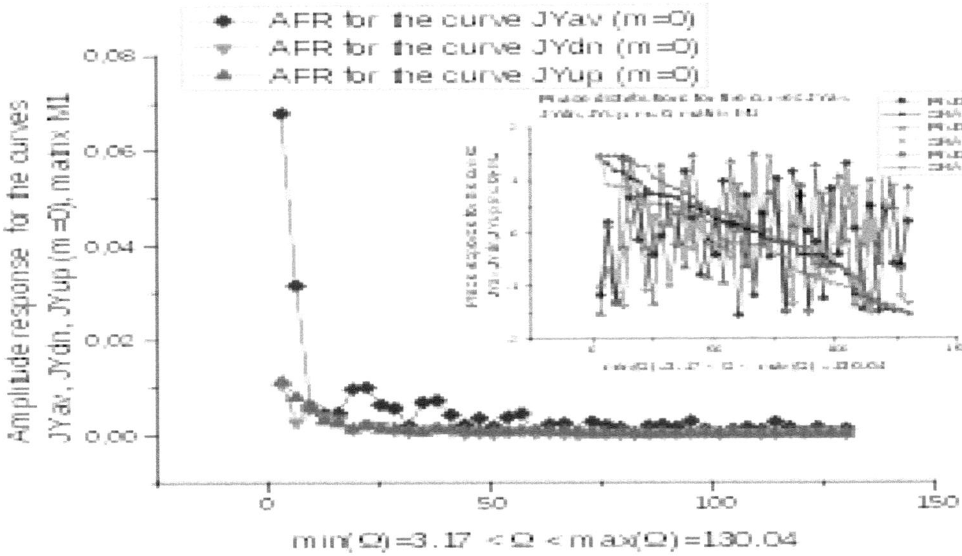

Figure 3.40 AFR for three curves depicted in Fig. 3.36 for m=0 (matrix M_1). As one can notice from these curves, the behavior of the amplitude modulus presented presumably by decaying curves, while the phase characteristics of these curves are presented by random uniform distributions. The sequences of the ranged amplitudes (SRA) are close to the segments of the straight lines.

amplitudes became small and close to zero values. Therefore, starting from the values $\Omega \approx 100$, the significant amplitude and phase outliers are not observed.

These results can help in providing an additional useful information for characterization of the "hidden" geodynamic processes of the Tien Shan. The complex process of pre-quake deformation is formed due to fracturing of the geological medium under the influence of ionospheric disturbances, either temperature variations, in-rock fluid injections and other factors.

Our results can serve as a rather strong argument for the further decomposition of noise-like seismic fluctuations. A hypothesis of the mechanism of appearance of noise-like sequence with somehow regrouping of noise is explained in favor of periodicity for earthquake preparation. Therefore, the data obtained from the system are promising for further study in conjunction with statistical signal processing and vector field analysis.

The authors hope that the usage of the reduced fractal model on the base of NOCFASS-method, will help avoid extensive calculations using signal seeking of the standard methods and directly construct a quantitative description of short ambient seismic noise-like recordings in terms of a common set of calculated fitting parameters without application of a model for noise/RF distribution.

- 6. We should repeat that two powerful tools as NAFASS/NOCFASS will be useful for the fitting of arbitrary random sequences with trend in cases, where

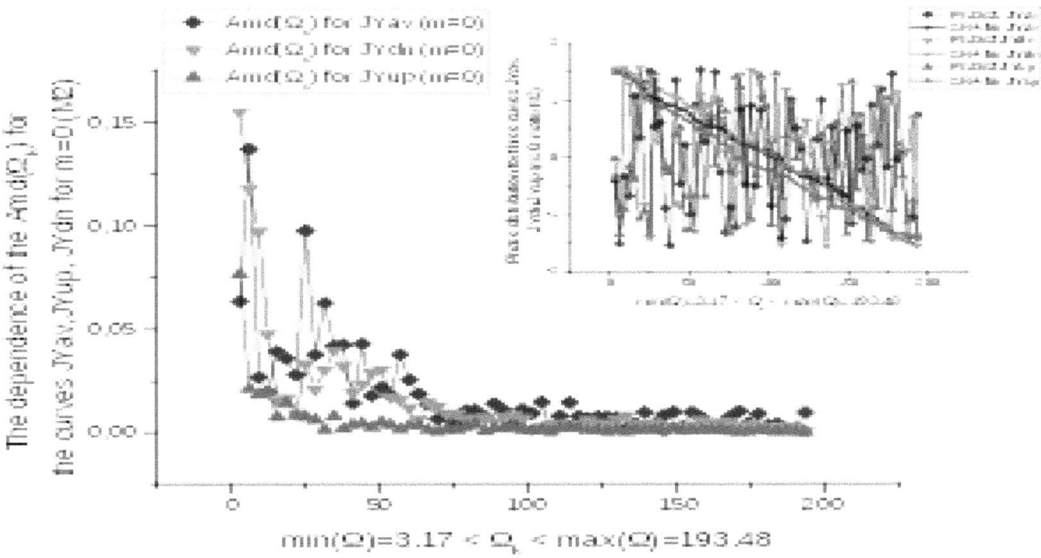

Figure 3.41 AFR for three curves depicted in Fig. 3.37 for m=0, (matrix M_2). As one can notice from these curves the behavior of the amplitude modulus presented presumably by decaying curves with some clearly distinguished peaks, while the phase characteristics of these curves are presented by random uniform distributions. The sequences of the ranged amplitudes (SRA) are close to the segments of the straight lines.

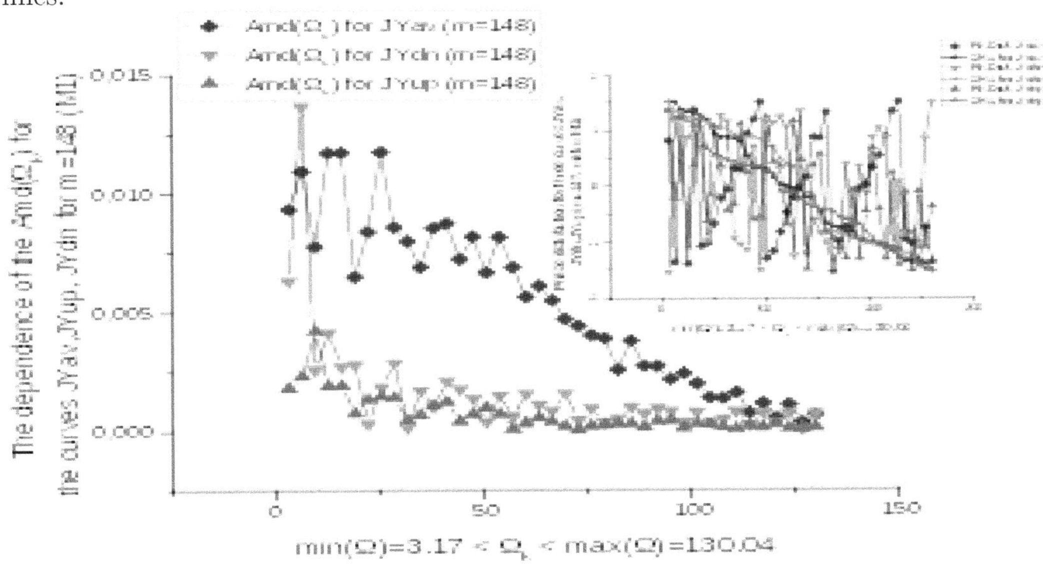

Figure 3.42 AFR for three curves depicted in Fig. 3.38 for M=147 (matrix M_1). As one can notice from these curves the behavior of the amplitude modulus presented presumably by decaying curves, while the phase characteristics of these curves are presented by random uniform distributions. The sequences of the ranged amplitudes (SRA) are close to the segments of the straight lines.

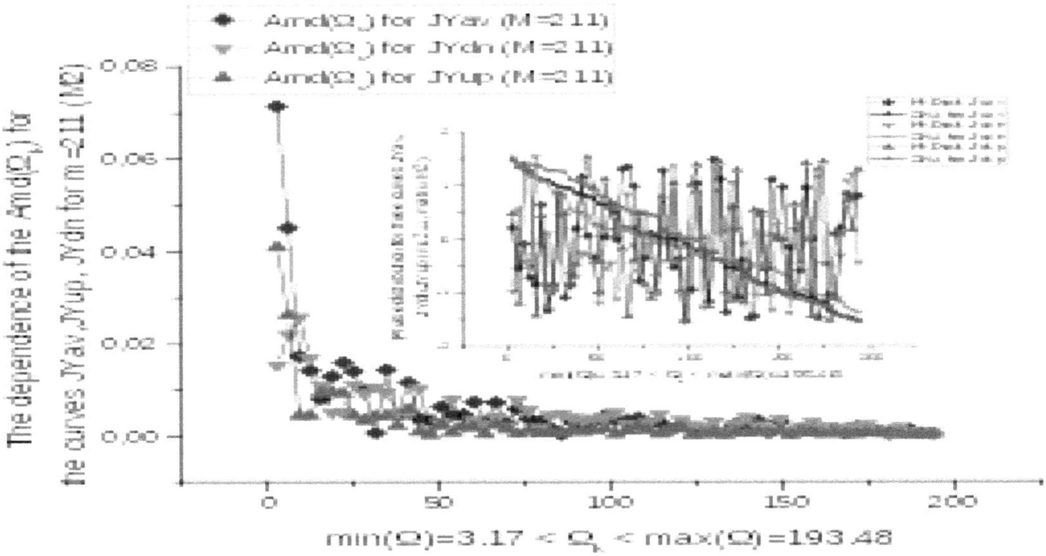

Figure 3.43 AFR for three curves depicted in Fig. 3.39 for M=211 (matrix M_2). As one can notice from these curves the behavior of the amplitude modulus presented presumably by decaying curves, while the phase characteristics of these curves are presented by random uniform distributions. As one can notice earlier and here the sequences of the ranged amplitudes (SRA) maintain their closeness to the segments of the straight lines.

the "best-fit" model is *absent.* It is important, especially, in cases when the receiving of reliable information expressed in the form of amplitude/frequency distributions that are contained in multi-periodic signals is necessary.

For potential readers one can recommend to reproduce the NOCFASS approach based on their own data. This reproduction should convince the skeptically tuned readers that this tool enables to fit any random function with clearly expressed trend.

BIBLIOGRAPHY

[1] R. R. Nigmatullin and I. M. Gubaidullin, "NAFASS: Fluctuation spectroscopy and the Prony spectrum for description of multi-frequency signals in complex systems," *Communications in Nonlinear Science and Numerical Simulation*, vol. 18, pp. 153–171, Jan. 2013.

[2] R. R. Nigmatullin and V. A. Toboev, "Non-Orthogonal Amplitude-Frequency Analysis of the Smoothed Signals (NAFASS): Dynamics and the Fine Structure of the Sunspots," *Journal of Applied Nonlinear Dynamics*, vol. 4, no. 1, pp. 67–80, 2015.

[3] R. R. Nigmatullin and W. Zhang, "NAFASS in action: How to control randomness?," *Communications in Nonlinear Science and Numerical Simulation*, vol. 18, no. 3, pp. 547–558, 2013.

[4] E. M. Stein and G. Weiss, *Introduction to Fourier Analysis on Euclidean Spaces*, vol. 32 of *Princeton Mathematical Series*. Princeton, NJ: Princeton University Press, 1971.

[5] L. Cohen, *Time-Frequency Analysis*. Upper Saddle River, NJ: Prentice Hall PTR, 1995.

[6] A. Sović and D. Seršić, "Signal decomposition methods for reducing drawbacks of the DWT," *Engineering Review*, vol. 32, no. 2, pp. 70–77, 2012.

[7] P. A. Delgado-Arredondo, D. Morinigo-Sotelo, R. A. Osornio-Rios, J. G. Avina-Cervantes, H. Rostro-Gonzalez, and R. d. J. Romero-Troncoso, "Methodology for fault detection in induction motors via sound and vibration signals," *Mechanical Systems and Signal Processing*, vol. 83, pp. 568–589, 2017.

[8] N. E. Huang, Z. Shen, S. R. Long, M. C. Wu, H. H. Shih, Q. Zheng, N.-C. Yen, C.-C. Tung, and H. H. Liu, "The empirical mode decomposition and the Hilbert spectrum for nonlinear and non-stationary time series analysis," *Proceedings of the Royal Society of London. Series A: Mathematical, Physical and Engineering Sciences*, vol. 454, no. 1971, pp. 903–995, 1998.

[9] M. E. Torres, M. A. Colominas, G. Schlotthauer, and P. Flandrin, "A complete ensemble empirical mode decomposition with adaptive noise," in *2011 IEEE International Conference on Acoustics, Speech and Signal Processing (ICASSP)*, pp. 4144–4147, May 2011.

[10] P. Flandrin, G. Rilling, and P. Goncalves, "Empirical mode decomposition as a filter bank," *IEEE Signal Processing Letters*, vol. 11, no. 2, pp. 112–114, 2004.

[11] Z. Wu and N. E. Huang, "Ensemble empirical mode decomposition: a noise-assisted data analysis method," *Advances in Adaptive Data Analysis*, vol. 1, no. 01, pp. 1–41, 2009.

[12] A. Garcia-Perez, O. G. Ibarra-Manzano, and R. d. J. Romero-Troncoso, "Analysis of partially broken rotor bar by using a novel empirical mode decomposition method," in *IECON 2014 – 40th Annual Conference of the IEEE Industrial Electronics Society*, pp. 3403–3408, 2014.

[13] E. H. El Bouchikhi, V. Choqueuse, M. Benbouzid, and J.-F. Charpentier, "Induction machine fault detection enhancement using a stator current high resolution spectrum," in *IECON 2012 – 38th Annual Conference on IEEE Industrial Electronics Society*, pp. 3913–3918, 2012.

[14] F. Cupertino, E. De Vanna, L. Salvatore, and S. Stasi, "Comparison of spectral estimation techniques applied to induction motor broken bars detection," in *4th IEEE International Symposium on Diagnostics for Electric Machines, Power Electronics and Drives (SDEMPED 2003)*, pp. 129–134, 2003.

[15] F. Cupertino, G. Martorana, L. Salvatore, and S. Stasi, "Diagnostic startup test to detect induction motor broken bars via short-time MUSIC algorithm applied to current signals," in *Proceedings of the 2003 IEEE International Symposium on Industrial Electronics (ISIE 2003)*, vol. 2, pp. 833–838, 2003.

[16] F. Cupertino, G. Martorana, L. Salvatore, and S. Stasi, "Detection of IM broken rotor bars using MUSIC pseudo-spectrum and pattern recognition," in *IECON 2003. 29th Annual Conference of the IEEE Industrial Electronics Society*, vol. 3, pp. 2829–2834, 2003.

[17] R. G. Lyons, *Understanding Digital Signal Processing.* Upper Saddle River, NJ, USA: Prentice Hall PTR, 2001. First published in 1997 by Addison-Wesley.

[18] R. R. Nigmatullin, P. Lino, and G. Maione, *New Digital Signal Processing Methods: Applications to Measurement and Diagnostics.* Cham, Switzerland: Springer Nature Switzerland AG, 2020.

[19] D. Striccoli, G. Boggia, C. Ceglie, and R. R. Nigmatullin, "A novel approach for characterizing multimedia 3D video streams by means of quasiperiodic processes," *Multimedia Tools and Applications*, vol. 75, no. 22, pp. 14201–14234, 2016.

[20] R. R. Nigmatullin, V. A. Toboev, P. Lino, and G. Maione, "Reduced fractal model for quantitative analysis of averaged micromotions in mesoscale: Characterization of blow-like signals," *Chaos, Solitons & Fractals*, vol. 76, pp. 166–181, 2015.

[21] L. de Arcangelis, C. Godano, J. R. Grasso, and E. Lippiello, "Statistical physics approach to earthquake occurrence and forecasting," *Physics Reports*, vol. 628, pp. 1–91, 2016.

[22] A. Atef, K. H. Liu, and S. S. Gao, "Apparent Weekly and Daily Earthquake Periodicities in the Western United States," *Bulletin of the Seismological Society of America*, vol. 99, no. 4, pp. 2273–2279, 2009.

[23] A. V. Desherevskii, V. I. Zhuravlev, A. N. Nikolsky, and A. Y. Sidorin, "Analysis of rhythms in experimental signals," *Izvestiya, Atmospheric and Oceanic Physics*, vol. 53, pp. 847–858, 2017.

[24] H. Zhang, X.-H. Chen, and X.-M. Wu, "Seismic data reconstruction based on CS and Fourier theory," *Applied Geophysics*, vol. 10, no. 2, pp. 170–180, 2013.

[25] D.-Y. Wang, J.-P. Huang, X. Kong, Z.-C. Li, and J. Wang, "Improving the resolution of seismic traces based on the secondary time," *Applied Geophysics*, vol. 14, no. 2, pp. 273–285, 2017.

[26] Y.-Y. Ma, G.-F. Li, Y.-J. Wang, H. Zhou, and B.-J. Zhang, "Random noise attenuation by f-x spatial projection-based complex empirical mode decomposition predictive filtering," *Applied Geophysics*, vol. 12, no. 1, pp. 47–54, 2015.

[27] A. Lyubushin, "Prognostic properties of low-frequency seismic noise," *Natural Science*, vol. 4, pp. 659–666, 2012.

[28] Z. Duputel, V. Ferrazzini, F. Brenguier, N. Shapiro, M. Campillo, and A. Nercessian, "Real time monitoring of relative velocity changes using ambient seismic noise at the Piton de la Fournaise volcano (La Réunion) from January 2006 to June 2007," *Journal of Volcanology and Geothermal Research*, vol. 184, no. 1–2, pp. 164–173, 2009.

[29] I. O. Kitov, I. A. Sanina, K. N. Nepeina, and N. L. Konstantinovskaya, "Using a matched-filter technique at the Mikhnevo small-aperture seismic array," *Seismic Instruments*, vol. 51, no. 3, pp. 191–200, 2015.

[30] K. Löer, N. Riahi, and E. H. Saenger, "Three-component ambient noise beamforming in the Parkfield area," *Geophysical Journal International*, vol. 213, no. 3, pp. 1478–1491, 2018.

[31] L. Stehly, M. Campillo, and N. M. Shapiro, "A study of the seismic noise from its long-range correlation properties," *Journal of Geophysical Research: Solid Earth*, vol. 111, no. B10, p. B10306, 2006.

[32] V. V. Adushkin, I. O. Kitov, N. L. Konstantinovskaya, K. N. Nepeina, M. V. Nesterkina, and I. A. Sanina, "Detection of ultraweak signals on the Mikhnevo small-aperture seismic array by using cross-correlation of waveforms," *Doklady Earth Sciences*, vol. 460, no. 1, pp. 189–191, 2015.

[33] J. Groos, *Broadband Seismic Noise: Classification and Green's Function Estimation*. PhD thesis, Karlsruhe Institute of Technology (KIT), Karlsruhe, Germany, 2010. Dr. rer. nat. dissertation.

[34] S. M. Mousavi, S. P. Horton, C. A. Langston, and B. Samei, "Seismic features and automatic discrimination of deep and shallow induced-microearthquakes using neural network and logistic regression," *Geophysical Journal International*, vol. 207, no. 1, pp. 29–46, 2016.

[35] Y. Chen, "Non-stationary least-squares complex decomposition for microseismic noise attenuation," *Geophysical Journal International*, vol. 213, no. 3, pp. 1572–1585, 2018.

[36] X. Zhang, Z. Jia, Z. E. Ross, and R. W. Clayton, "Extracting dispersion curves from ambient noise correlations using deep learning," *IEEE Transactions on Geoscience and Remote Sensing*, vol. 58, no. 12, pp. 8932–8939, 2020.

[37] S. E. Popov and R. Y. Zamaraev, "A fast algorithm for classifying seismic events using distributed computations in Apache Spark framework," *Programming and Computer Software*, vol. 46, no. 1, pp. 35–48, 2020.

[38] R. E. Learned and A. S. Willsky, "A wavelet packet approach to transient signal classification," *Applied and Computational Harmonic Analysis*, vol. 2, no. 3, pp. 265–278, 1995.

[39] C. Riggelsen, M. Ohrnberger, and F. Scherbaum, "Dynamic Bayesian networks for real-time classification of seismic signals," in *Knowledge Discovery in Databases: PKDD 2007* (J. N. Kok, J. Koronacki, R. Lopez de Mantaras, S. Matwin, D. Mladenic, and A. Skowron, eds.), vol. 4702 of *Lecture Notes in Computer Science*, pp. 565–572, Berlin, Heidelberg: Springer, 2007.

[40] K. Plenkers, S. Husen, and T. Kraft, "A multi-step assessment scheme for seismic network site selection in densely populated areas," *Journal of Seismology*, vol. 19, no. 4, pp. 861–879, 2015.

[41] G. Ekström, "Time domain analysis of Earth's long-period background seismic radiation," *Journal of Geophysical Research: Solid Earth*, vol. 106, no. B11, pp. 26483–26493, 2001.

[42] A. A. Lyubushin, "Synchronization trends and rhythms of multifractal parameters of the field of low-frequency microseisms," *Izvestiya, Physics of the Solid Earth*, vol. 45, no. 5, pp. 381–394, 2009.

[43] K. S. Nepeina, P. A. Kaznacheev, and P. N. Aleksandrov, "Seismic monitoring of modern geodynamic processes using a gradient system," *Bulletin of Kamchatka Regional Association "Educational-Scientific Center" Earth Sciences*, vol. 44, no. 4, pp. 84–92, 2019. [In Russian].

[44] A. N. Kamshilin and P. A. Kaznacheev, "Local current gauge: Instrument for geoelectric measurements," *Seismic Instruments*, vol. 54, no. 5, pp. 573–578, 2018.

[45] A. K. Rybin, E. A. Bataleva, K. S. Nepeina, P. A. Kaznacheev, P. A. Matiukov, and P. N. Aleksandrov, "Definition of the Seismic Field of the Underground Sources in the Ambient Seismic Noise in the Tien Shan Region Using a Three-Component Gradient System," *Journal of Earth Science*, vol. 31, no. 5, pp. 988–992, 2020.

CHAPTER 4

Generalized Correlations-Based Methods

We collect some original formulas in order to stress one important feature – the internal correlations have relative character and the conventional definitions are needed in some modifications. In this chapter we are going to give some generalizations that help to specify this feature.

4.1 INTERNAL AND EXTERNAL CORRELATIONS (THEORETICAL PART)

As we know, the modern science of the 21-st century is trying to find and understand the properties of the complex systems. What do we understand under the definition "complex system"? A reader can find many definitions in "Google" search, however, for us it is important to stress and repeat again two key points:

1. For description of complex systems it is necessary to take into account the fact that a simple fitting function having minimal number of the fitting parameters and reflecting the basic relationship (as a simple physical law) in many cases is *absent*.

2. The most complex systems have *diffusive* nature. What does it mean? In ideal case it has many relationships that are *not* stable in time and the "solid/rigid" relationship that is remained stable during long period of time is *absent* also.

We have to take into account these two basic features in definition of the most complex systems. For quantitative description of complex systems, there are many different methods. In this chapter, we want to concentrate presumably on the *correlation analysis* and add some new features to this important branch of the mathematical statistics. In solution of applied problems, a researcher deals with necessity to establish some functional or other dependencies between different variables. These variables can be deterministic or random. Therefore, we should establish some relationships between these variables. The problem of correlation analysis can be understood better on the example of the "black box". Input of the black box is characterized by the set of variables or vector $\boldsymbol{X} = (X_1, X_2 \ldots, X_p)$. The vector of output variables

DOI: 10.1201/9781003794691-4

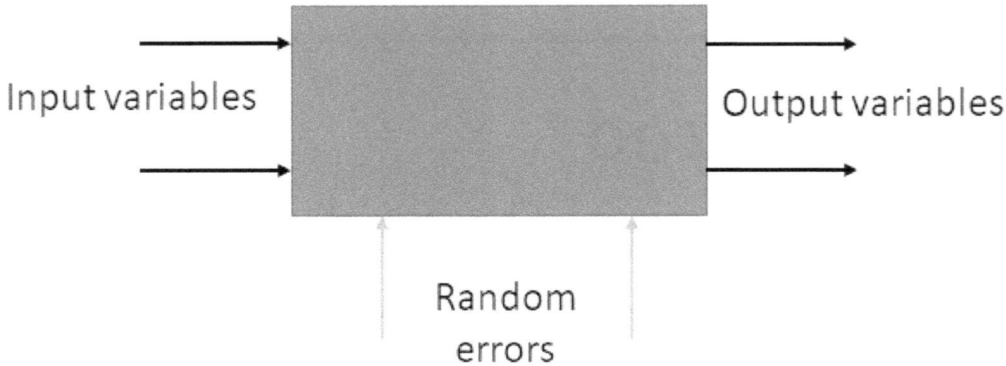

Figure 4.1 The schematic model of the black box that shows the "hidden" relationship between input/output variables.

is characterized by the vector $\boldsymbol{Y} = (Y_1, Y_2..., Y_m)$. Other variables are shown in Fig. 4.1.

If the input vector is random then we have also the vector of random errors, in the form of different uncontrollable factors. Quantitatively, they can be written approximately as a function of random fluctuations $\epsilon = \boldsymbol{Y} - f(\boldsymbol{X})$. It is necessary to establish the relationship between the input/output variables taking into account the influence of the errors/uncontrollable factors. These errors are appeared in the result of measurements, they can be contained also because of the chosen model and treatment errors. Usually, the correlation analysis problems are divided on three classes.

1. The problem of the *correlation analysis* is reduced to finding possible correlations between the above-mentioned groups of variables $\boldsymbol{X} = (X_1, X_2..., X_p)$ and $\boldsymbol{Y} = (Y_1, Y_2..., Y_m)$.

2. The problem of *regression analysis* is reduced to establishing of possible analytical dependencies (function $f\ (\boldsymbol{X})$ is supposed to be unknown) between these variables.

3. The problem of dispersion analysis is reduced to establishing a possible influence of the selected (we want to underline that in the most cases they are *not* complete) variables $\boldsymbol{X}$ on the variable $\boldsymbol{Y}$.

We are not going to consider these problems in detail, because there is a huge set of books (manuals and monographs) related to correlation analysis.

However, it is necessary to specify these problems on concrete examples.

Problem 1. Let variable Y specifies the value of the tire tread rubbing wear (in mm) after a certain mileage, X_1 is the load value for the fixed car's wheel (in kg(s).), X_2 – the protector type used. It is necessary to establish the influence of variables $X_{1,2}$ $(Y = F(X_1, X_2))$ on the variable Y and present its influence in the form of some experimental plot.

Problem 2. Let Y_1 – defines the productivity of a chemical producer plant (tons per hour), Y_2 is a spoiled goods (in %). Experimentally was established that the following technological parameters exert influence on $Y_{1,2}$: X_1 – the row material humidity, X_2 – temperature of the reactor plant, X_3 – the mixture content. If one could find a possible relationship between the input parameters $X_{1,2,3}$ and output parameters $Y_{1,2}$ (i.e., $Y_1 = F_1(X_1, X_2, X_3)$ and $Y_2 = F_2(X_1, X_2, X_3)$) then it becomes possible to find the optimal technological regime (minimization of the value of Y_2 at the given technological regime). In other words, it is necessary to find the fitting expression $Y_1 = F(Y_2)$ and express this dependence in terms of some fitting parameters.

Problem 3. Let Y be a progress in the students results (as a quantitative parameter one can take the mean exams ball during a semester). The authorities decide to realize the students screening study and find the significant factors that can exert on this parameter. In the results of the preliminary analysis, it was established that these factors are: X_1 - time that was spent of a student for self-instruction; X_2 – number of the missed lessons; X_3 – the size of the living allowance/stipend. Is there relationship between these chosen factors $X_{1,2,3}$ and the output value Y?

An attentive reader can classify easily these randomly taken problems and refer them to one of the classes (1, 2, or 3) mentioned above. A "classical" solution of the correlation analysis problem is reduced to the usage of the conventional Pearson correlation formula

$$\cos(\theta) = \frac{(y_1 \cdot y_2)}{\sqrt{y_1{}^2}\sqrt{y_2{}^2}}, \ (y_1 \cdot y_2) = \sum_{j=1}^{N} y_1(j) \cdot y_2(j) \tag{4.1}$$

and in receiving the value of the angle located in the interval (−1,1) in the significant space having a size P. In many cases researches add a supposition about the distribution of the vectors $\boldsymbol{y}_1$, $\boldsymbol{y}_2$ if they are random. How to increase the possibilities of expression (4.1) and find its new applications? In Chapter 3 of the book [1] published in the year 2020 and in papers [2, 3], one of us introduced the definition of the generalized Pearson correlation function (GPCF) for a pair of the chosen random sequences $y_1(j)$, $y_2(j)$ (j=1,2,...,N) based on the generalized mean value function

$$GPCF_p = \frac{G_p(y_1, y_2)}{\sqrt{G_p(y_1, y_1) \cdot G_p(y_2, y_2)}}, \tag{4.2}$$

$$G_p(y_1, y_2, ..., y_k) = \left(\frac{1}{N}\sum_{j=1}^{N} |yn_1(j) \cdot yn_2(j) \cdot ... \cdot yn_k(j)|^{mom_p}\right)^{\frac{1}{mom_p}}. \tag{4.3}$$

Here, $yn_k(j)$, k=1,2,...,K the normalized values of the compared sequences reduced to the interval [0,1]. The normalized sequence $yn(j)$, with $0 < yn(j) < 1$, j = 1, 2,..., N, is defined as:

$$yn(j) = \begin{cases} \frac{y_j+|y_j|}{\max(y_j+|y_j|)} - \frac{y_j-|y_j|}{\min(y_j-|y_j|)} \\ \frac{y_j-\min(y_j)}{\max(y_j-\min(y_j))} \text{ (for initial positive sequence)} \end{cases} \tag{4.4}$$

where y_j denotes the initial random sequence that can contain a trend, or that is to be compared with another trendless sequence. The minimum of the GMV function at mom_p =0 coincides with its geometric mean, i.e.,

$$\underset{mom_p \to 0}{GMV_p}(|y|) = \exp\left(\frac{1}{N}\sum_{j=1}^{N}\ln(|y_j|)\right) = \prod_{j=1}^{N}|y_j|^{1/N} \tag{4.5}$$

while the maximum coincides with the maximum of the normalized sequence. This function is monotone and always increasing for all positive values of the mom_p and the values of the moments that are taken from interval:

$$mom_p = \exp\left(-r+2\frac{p}{P}r\right),\ e^{-r} \leq mom_p \leq e^r,\ p=0,1,...,P. \tag{4.6}$$

This chosen interval allows to avoid the limiting case (4.5) (when $r \to \infty$) and covers practically all positive values of the real moments and, therefore, can be considered as an optimal one. Usually, for practical purposes the limiting values of r shown in (4.6) equaled in the interval (10–15) is sufficient for presenting all possible correlations that are given by the function (4.2). We should notice also that expression (4.2) being located in the interval [0,1] has "universal" behavior. At negative values of the argument mom_p it tends to the unit value, at positive values of the mom_p it achieves the limiting value L at p=P. The value L, in turn, can accept three values: (a) L=1 corresponds to the case of strong correlations; (b) the interval $M < L < 1$ corresponds to the case of intermediate correlations; (c) the case $L \approx M$ (M is minimal value of (4.2)) corresponds to the weak correlations. We should notice also that weak correlations *always* exist and the case of the total absence of correlations between two random sequences can be considered as a theoretical *supposition*. This supposition can be realized only in cases, if preliminary the random sequences compared can be orthonormalized with the help of the Gram-Schmidt procedure. This "universal" behavior of the function (4.2) allows to determine the complete correlation factor as the product

$$CF = L \cdot M,\ M^2 \leq CF \leq M. \tag{4.7}$$

The complete correlation factor (CF) can characterize the degree of correlation between a couple of the chosen random sequences. The analysis based on expression (4.2) was applied to available real data [2, 3].

Let us define the complex moments as follows.

$$GPFCM_p = \frac{GMCM_p(y_1,y_2)}{\sqrt{GMCM_p(y_1,y_1)\cdot GMCM_p(y_2,y_2)}}, \tag{4.8}$$

$$GMCM_{Z_p}(y_1,y_2,...,y_k) = \left(\frac{1}{N}\sum_{j=1}^{N}|yn_1(j)\,yn_2(j)...yn_k(j)|^{Z_p}\right)^{\frac{1}{Z_p}}. \tag{4.9}$$

In expression (4.8), random sequences $yn_1(j)$, $yn_2(j)$,..., $yn_k(j)$ are normalized to the interval [0,1] in accordance with relationship (4.4), the value of Z_p accept *complex* values compared from two parts and located in the following intervals:

$$Z_p = R_p + i\Omega_p,\ R_p = \exp\left(-r + 2r\frac{p}{P}\right),\ \Omega_p = \Omega_0 + \frac{p}{P}\left(\Omega_L - \Omega_0\right), \tag{4.10}$$

$$e^{-r} \le R_p \le e^{r},\ \Omega_0 \le \Omega_p \le \Omega_L. \tag{4.11}$$

In order to separate the real and imaginary parts from numerator of expression (4.6), we introduce the following sums

$$Sr_p(1,2) = \frac{1}{N}\sum_{j=1}^{N}(yn_1(j)yn_2(j))^{R_p}\cos\left[\Omega_p\ln(yn_1(j)yn_2(j))\right], \tag{4.12}$$

$$Sm_p(1,2) = \frac{1}{N}\sum_{j=1}^{N}(yn_1(j)yn_2(j))^{R_p}\sin\left[\Omega_p\ln(yn_1(j)yn_2(j))\right] \tag{4.13}$$

With the help of expressions (4.8), the numerator of expression (4.6) can be presented in the form

$$G_{Z_p}(1,2) = \left(\frac{1}{N}\sum_{j=1}^{N}(yn_1(j)yn_2(j))^{Z_p}\right)^{\frac{1}{Z_p}} = \left|G_{Z_p}(1,2)\right|\exp(i\Phi_p(1,2)) \tag{4.14}$$

$$\left|G_{Z_p}(1,2)\right| = \frac{\exp\left[\ln\left|S_p(1,2)\right|\right]}{\left(R_p^2+\Omega_p^2\right)^{1/2}},\ \Phi_p(1,2) = \varphi_p(1,2) - \tan^{-1}\left(\frac{\Omega_p}{R_p}\right). \tag{4.15}$$

The values $|S_p(1,2)|, \varphi_p(1,2)$ are determined with the help of expressions (4.8) as

$$|S_p(1,2)| = \sqrt{Sr_p^2(1,2) + Sm_p^2(1,2)},\ \varphi_p(1,2) = \tan^{-1}\left(\frac{Sm_p(1,2)}{Sr_p(1,2)}\right). \tag{4.16}$$

The expressions in the dominator in (4.6) are obtained easily with the help of last expressions (4.8)–(4.12), where it is necessary to replace the corresponding sequences $(1 \leftrightarrow 2)$. Finally, expression (4.6) after separation of the real and imaginary parts can be presented in the form

$$GPFCM_p = \frac{\left|G_{Z_p}(1,2)\right|}{\sqrt{\left|G_{Z_p}(1,1)\right|\cdot\left|G_{Z_p}(2,2)\right|}}\exp\left(i\Phi_p(1,2) - \frac{i}{2}\left(\Phi_p(1,1)+\Phi_p(2,2)\right)\right). \tag{4.17}$$

Other expressions, entering into (4.14) are defined by expressions from (4.10). Expression (4.14) represents itself the final formula that generalizes the previous results obtained earlier in the book [1] and papers [2, 3, 4] for the case of real fractional moments.

With the help of new expressions (4.8)–4.14) we want to show that possible generalization of the concept of correlations attaches to the conventional definition (4.1) has a *relative* character. It means that correlations from the fixed number is transformed to a distribution; they can be arbitrary and occupy all admissible interval [0,1].

Attentive analysis of the results obtained above for pair correlations allows to propagate them for consideration of multi-correlations that include a set of data expressed in the form of an array. Really, one can introduce the generalized scalar product between a set of vectors having the same dimension

$$(y_1 \cdot y_2 \cdot \ldots \cdot y_k) = \sum_{j=1}^{N} (y_1(j) \cdot y_2(j) \cdot \ldots \cdot y_k(j)). \tag{4.18}$$

Here $\{y_s\}$ (s=1,2,...,k) is a set of vectors forming an array, defined in the N-th dimensional space of discrete variable j=1,2,..., N. Definition (4.18) allows to generalize the conventional Pearson correlation parameter defined by (4.1) and propagate it for a set of vectors

$$K_1(y_1, y_2, \ldots, y_k) = \frac{\prod_{s=1}^{k}(y_s)}{\prod_{s=1}^{k}|y_s|} \equiv \frac{\sum_{j=1}^{N}(y_1(j) \cdot y_2(j) \cdot \ldots \cdot y_k(j))}{\prod_{s=1}^{k}(y_s \cdot y_s)^{1/2}}. \tag{4.19}$$

Expression (4.19) defines the correlation of an array normalized to the values of the single vectors. These correlations have relative character and one can define the correlation to respect of pair correlations. For practical applications we write down expression for the multi-correlations of an array normalized to the values of pair correlations

$$K_2(y_1, y_2, \ldots, y_k) = \frac{\sum_{j=1}^{N}(y_1(j) \cdot y_2(j) \cdot \ldots \cdot y_k(j))}{\left((y_1 \cdot y_2)^{1/2} \cdot (y_1 \cdot y_k)^{1/2} \cdot \ldots \cdot (y_{k-1} \cdot y_k)^{1/2}\right)}. \tag{4.20}$$

The square root operation is used for normalization purposes that makes expression (4.20) dimensionless. In a similar manner, one can normalize the scalar product (4.18) for triple correlations and etc. Expressions (4.18–4.20) allow to generalize them for the fractional moments, including the real and complex values. It is easy to write down expression (4.2) for the multi-correlation case including the values of the real moments

$$K_{p,1}(y_1, y_2, \ldots, y_k) = \frac{G_p(y_1, y_2, \ldots, y_k)}{\prod_{s=1}^{k}\left[G_p(y_s, y_s)\right]^{1/2}}. \tag{4.21}$$

Here $G_p(y_1,y_2,\ldots,y_k)$ is defined by the second row in expression (4.2). Similarly, expression (4.19) can be normalized to the pair-correlation in the space of the fractional moments

$$K_{p,1}(y_1,y_2,...,y_k) = \frac{G_p(y_1,y_2,...,y_k)}{[G_p(y_1,y_2)]^{1/2}\cdot[G_p(y_1,y_k)]^{1/2}\cdot...\cdot[G_p(y_{k-1},y_k)]^{1/2}}. \tag{4.22}$$

It is easy to notice that multi-correlation arrays are normalized and located in the interval [0,1]. The final generalization can be realized if we replace mom_p in definition (4.6) by its complex values as Z_p defined by expression (4.10). Expressions (4.21) and (4.22) conserve their form, however, we take into account their real and imaginary part for their numerical evaluation.

We deliberately reproduce these important expressions from our previous book [1] in order to stress two important points:

1. The generalized mean value (GMV) function (4.4) generalizes the concept of the mean value and covers all possible means (*harmonic* mean with mom_p =-1, *geometric* mean (with the limiting value equaled $mom_p = 0$) and arithmetic mean when $mom_p = 1$. The GMV – function is monotone function and, therefore, it defines other set of mean values at mom_p >1. It can be easily generalized for multi-correlation cases.

2. The second generalization is related to propagation of the scalar product to an arbitrary set of the compared vectors $(s_1,s_2,\ldots,s_K)$ as it is highlighted by expressions (4.21) and (4.22). As one can notice these expressions enable to compare the sequences having the *same* length and, therefore, these expressions can be applied for evaluation of the *internal* correlations only.

However, in many cases there is a necessity to compare the sequences having different lengths. How to do it? This problem can be solved also in the frame of the SFMs (Statistics of the Fractional Moments) if one uses the GMV-functions written for a single vector

$$GMV_p(y_k) = \left(\frac{1}{N_k}\sum_{j=1}^{N_k}|yn_k(j)|^{mom_p}\right)^{\frac{1}{mom_p}}, \tag{4.23}$$

$$k = 1,2,...,K,\ N_1 \neq N_2 \neq ... \neq N_K. \tag{4.24}$$

Expression (4.23) allows to evaluate the *external* correlations if to compare for the chosen pair (l,m) of random sequences their slopes and intercepts.

$$GMV_p(y_l) = S_{l,m}GMV_p(y_m) + B_{l,m}, \tag{4.25}$$

$$S_{l.l} = 1,\ B_{l,l} = 0. \tag{4.26}$$

The application of external correlations was proved to be useful and considered also in Chapter 3 of our previous book [1]. Another simple expression for evaluation of external correlations is associated with expression

$$Cr_{l,m} = \frac{Range(s_l) + Range(s_m)}{Range(s_l, s_m)}, \; 0 < Cr_{l,m} \leq 2, \tag{4.27}$$

$$Range(s_l, s_m) = \max(s_l, s_m) - \min(s_l, s_m). \tag{4.28}$$

Expressions (4.25) and (4.27) allow to compare the random sequences having different lengths also and the last one equals two in the case of complete coincidence for the pair sequences compared. The interval (0,1) corresponds to the absence of *external* correlations, while the internal correlations calculated with the help of the basic expression (4.2) and its possible generalizations shown above can be applied also.

Is that all? Are there other interesting relationships associated with correlations possible or not? Concluding this section related to the theoretical part we want to show the relationship of the generalized mean value function with well-known Prony decomposition.

Let us write down the GMV_p-function for the complex moment

$$G_p(y) = \frac{1}{N}\sum_{j=1}^{N}\left(|yn(j)|^{Z_p}\right) \equiv \frac{1}{N}\sum_{j=1}^{N}\exp\left(Z_p \ln|yn(j)|\right), \tag{4.29}$$

$$0 < |yn(j)| < 1, \; Z_p = R_p + i\Omega_p. \tag{4.30}$$

The limits of the values R_p, Ω_p are defined by expression (4.10). Here we deliberately normalized an arbitrarily sequence $yn(j)$ to the positive interval ($[yn(j)/\max(yn(j))] - [yn(j)/\min(yn(j))]$) and then reduced it to the given interval (0,1). Substitution of the intervals R_p, Ω_p to definition (4.29) yields to the following expressions:

$$G_p(y) = \mathrm{Re}\left(G_p(y)\right) + i\,\mathrm{Im}\left(G_p(y)\right), \tag{4.31}$$

$$\mathrm{Re}\left(G_p(y)\right) = \frac{1}{N}\sum_{j=1}^{N}\exp\left(\left(e^{-r} + 2rx_p\right)\ln|yn(j)|\right)\cos\left(\left(\Omega_0 + x_p\Delta\Omega\right)\ln|yn(j)|\right), \tag{4.32}$$

$$\mathrm{Im}\left(G_p(y)\right) = \frac{1}{N}\sum_{j=1}^{N}\exp\left(\left(e^{-r} + 2rx_p\right)\ln|yn(j)|\right)\sin\left(\left(\Omega_0 + x_p\Delta\Omega\right)\ln|yn(j)|\right), \tag{4.33}$$

$$x_p = \frac{p}{P}, \; 0 \leq x_p \leq 1. \tag{4.34}$$

Two last lines in (4.31) remind the well-known Prony decomposition

$$F(x_p) = A_0 + \sum_{j=1}^{N} \exp(-\lambda_j x_p) \left[Ac_j \cdot \cos(\omega_j x_p) + As_j \cdot \sin(\omega_j x_p)\right]. \tag{4.35}$$

For identifying of expressions (4.35) to (4.31) it is necessary to identify the value λ_j and open the cos(...) and sin(...) arguments. Therefore, we put

$$\lambda_j = -2r \ln|yn(j)|, \quad \omega_j = \Delta\Omega \ln|yn(j)| \equiv (\Omega_L - \Omega_0) \ln|yn(j)|, \tag{4.36}$$

$$Ac_j^{(r)} = \exp\left(e^{-r} \ln|yn(j)|\right) \cos(\Omega_0 \ln|yn(j)|), \tag{4.37}$$

$$As_j^{(r)} = -\exp\left(e^{-r} \ln|yn(j)|\right) \sin(\Omega_0 \ln|yn(j)|), \tag{4.38}$$

$$Ac_j^{(im)} = -As_j^{(r)}, \; As_j^{(im)} = Ac_j^{(r)}, \tag{4.39}$$

and with the help of (4.36) one can obtain relationships similar to expression (4.35)

$$G_p(y) = \mathrm{Re}(G_p(y)) + i\,\mathrm{Im}(G_p(y)), \tag{4.40}$$

$$\mathrm{Re}(G_p(y)) = \frac{1}{N} \sum_{j=1}^{N} \exp(-x_p \lambda_j) \left[Ac_j^{(r)} \cos(\omega_j x_p) + As_j^{(r)} \sin(\omega_j x_p)\right], \tag{4.41}$$

$$\mathrm{Im}(G_p(y)) = \frac{1}{N} \sum_{j=1}^{N} \exp(-x_p \lambda_j) \left[Ac_j^{(im)} \cos(\omega_j x_p) + As_j^{(im)} \sin(\omega_j x_p)\right]. \tag{4.42}$$

Comparing expression (4.40) with (4.35), we find their similarity at A_0=0. Therefore, this relationship will be very helpful for the finding of other useful relationships between complex GMV-functions and conventional relationships associated with Prony's decompositions and their useful generalizations. It is easy to notice that the previous expressions keep their definitions, however, in the complex plane it is necessary to replace the real value mom_p from (4.6) by its complex value Z_p from expression (4.10). Having an additional degree of freedom expressed in the form of the imaginary part, one can add new features in analysis of correlations between two and more compared sequences.

4.2 VERIFICATION ON AVAILABLE DATA

In order to "percept" the usefulness of the newly proposed expressions we demonstrate them on real data without trend. Initially we consider a trendless noise obtained from the given transmitter that was fixed by the antenna of some receiver. Here it is not important to describe other details. For us it is important only that the recorded signals are presented in the form of two matrices. Each rectangle matrix has N=10000 data points and M=50 columns. The first matrix represents a real part of the accepted signal, while the second matrix contains its imaginary part. Their plots

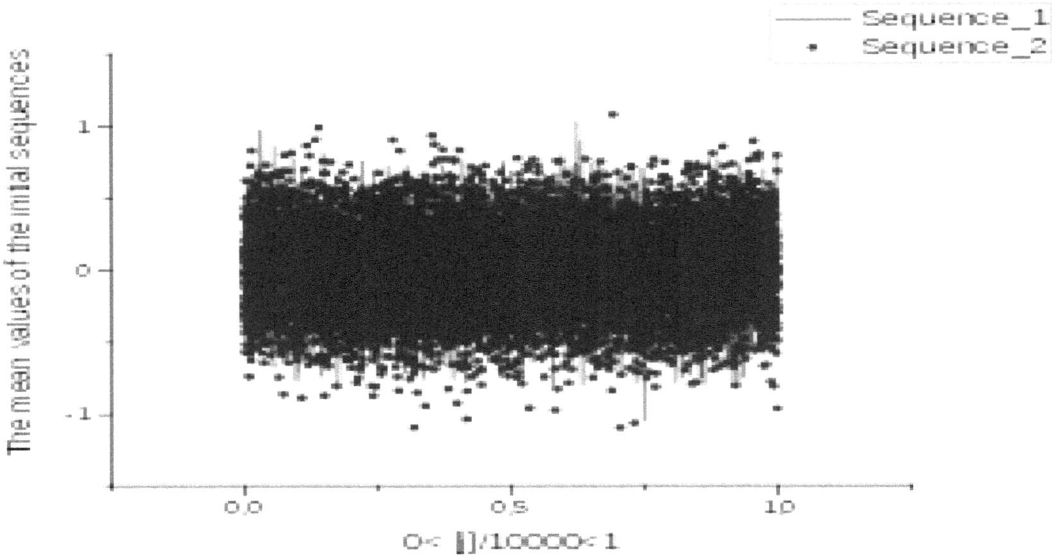

Figure 4.2 This is initial trendless sequences averaged over all 50 columns. They describe the real (black lines) and imaginary (black points) part of the transmitter that is accepted by the fixed receiver.

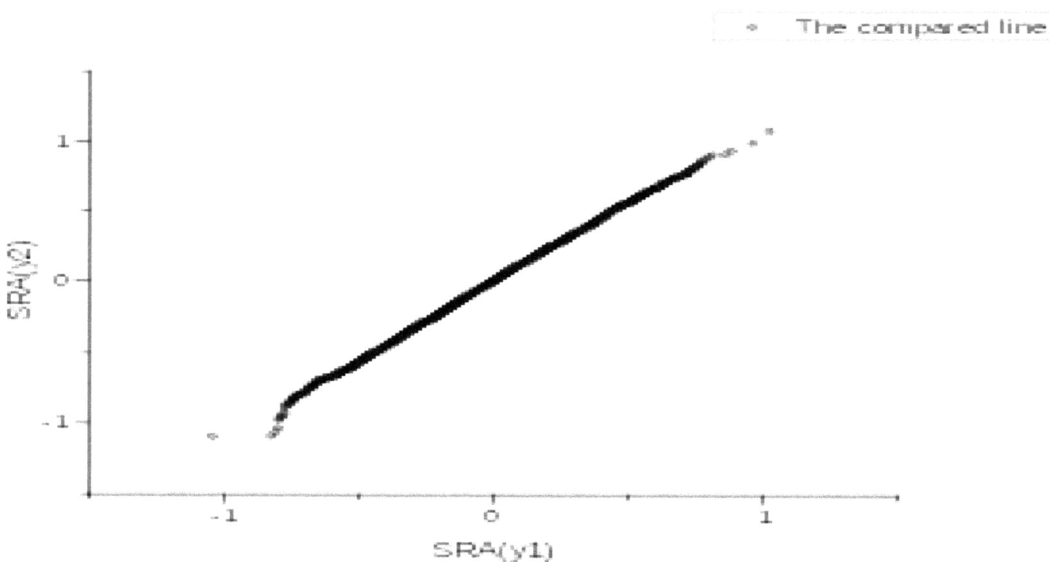

Figure 4.3 If one plots the sequence of the ranged amplitudes (SRA) SRA(y1,y2) (when the amplitudes are located in the descended order $y_1 > y_2 > \ldots > y_N$) with respect to each other, then the combined plot demonstrates their strong correlations.

are shown below. As one can see from these plots, these trendless sequences (TLS(s)) are strongly correlated. The plot on the right-hand side demonstrates clearly this correlation. These correlations are shown in Fig. 4.2 and 4.3, correspondingly.

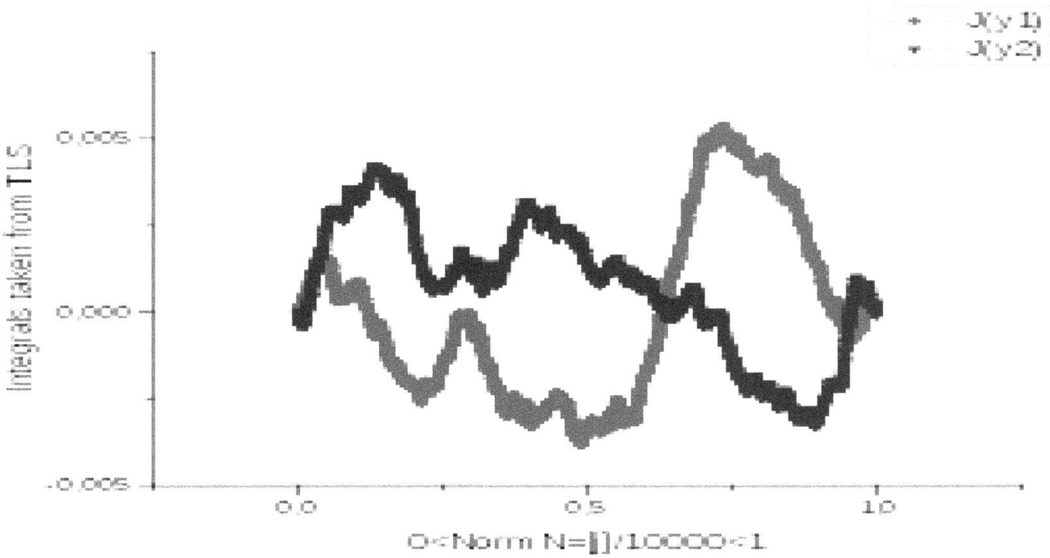

Figure 4.4 If one calculates the integrals from the initial TLS by the trapezoid method (when their mean values are preliminary subtracted), one can see that their trends are strongly different.

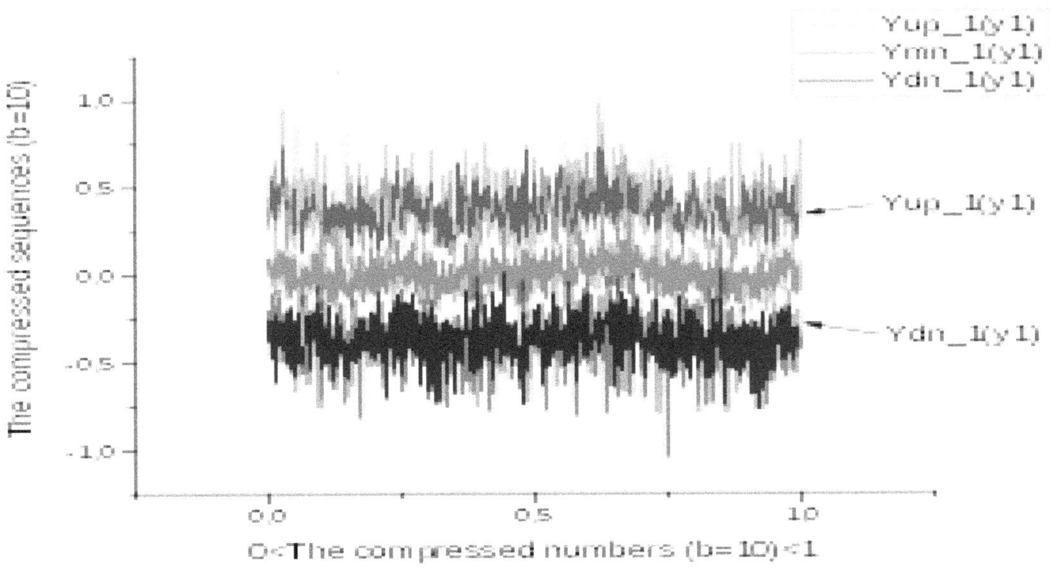

Figure 4.5 Here, we demonstrate the procedure of the reduction to three incident points (max, min and mean points that are invariant relatively permutations inside fixed b points; compression parameter b=10) helps essentially simplify the treatment procedure and to keep their initial similarity. As calculations show that the invariant sequence is associated with the central Ymn_1 sequence (marked by the middle curve), corresponding to distribution of mean values.

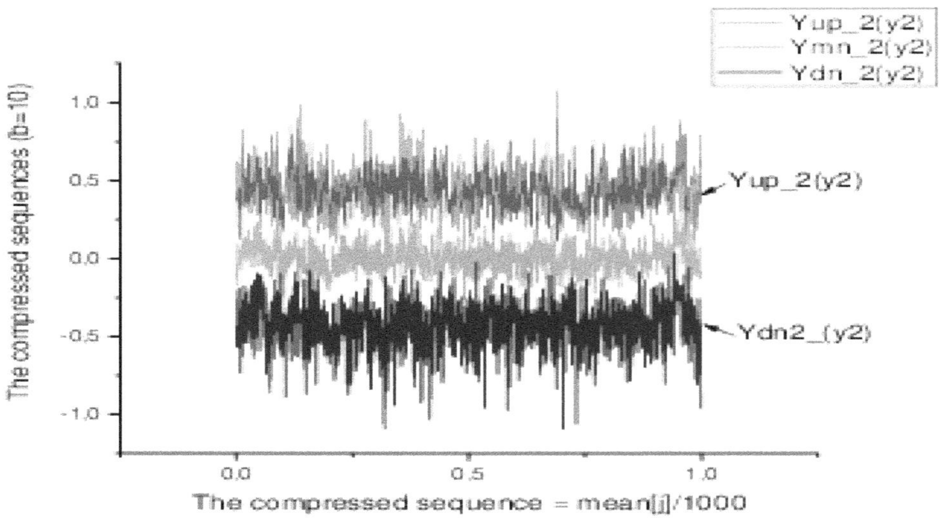

Figure 4.6 A similar reduction procedure to three incident points is applied to the second random sequence. In Figs. 4.5 and 4.6, we counted the reduced sequences relatively the mean values of the compressed sequences.

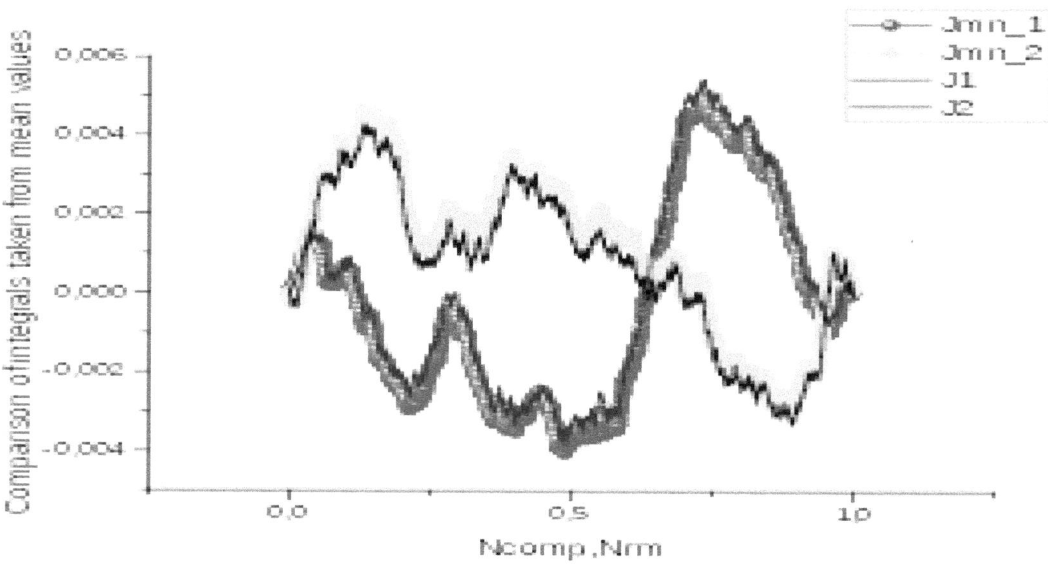

Figure 4.7 Here we compare the initial integrals shown in Fig. 4.4 with integrals taken from the reduced mean sequences as Ymn_1,2. One can notice that the initial and reduced integrals are very close to each other.

We would like to stress here that after reduction procedure to three incidents points, we discovered very interesting phenomenon. The integrals taken from the reduced curves corresponding to their mean values Ymn_1,2 and initial integrals

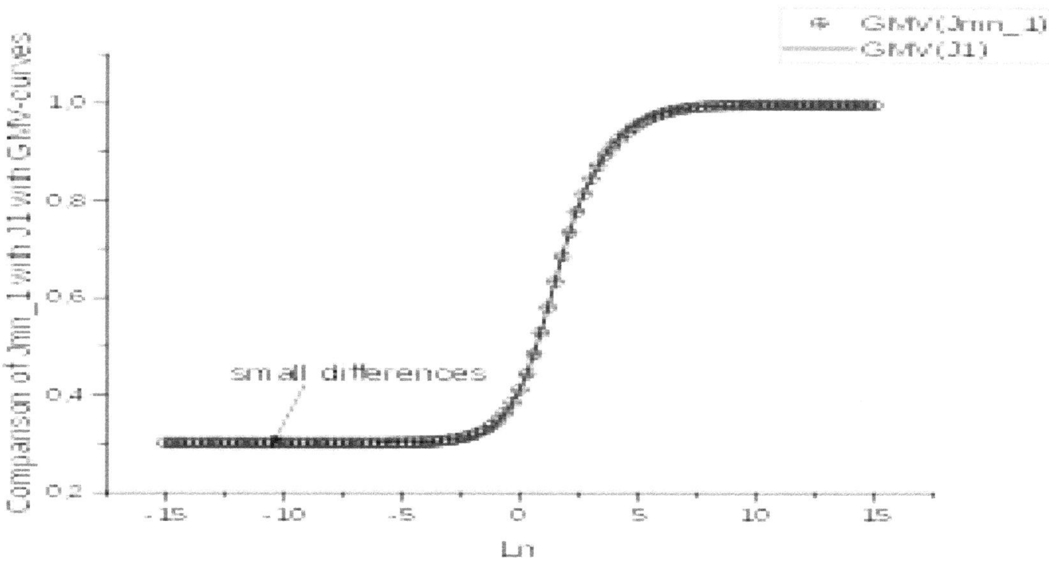

Figure 4.8 The statistics of the fractional moments allows to compare the sequences having different number of data points. With the help of the generalized mean values functions one can see that the integral curves shown on the previous figure are very close to each other. Practically, compression in 10–50 times keeps the scaling invariance that is expressed in the form of integrals shown in Fig. 4.7.

Jy(1,2) are similar to each other. One can define these curves as *specific invariants* obtained in the result of compression procedure. Fig. 4.8 demonstrates the statistics of the FM.

Analysis of the complex moments shows that the phase behavior is more sensitive in comparison with modulus part. They are defined by expressions (4.14). Comparison of the initial TLS Ymn_1 with Ytst shows that influence of a small signal is almost noticeable in Fig. 4.15. The comparison of modulus parts depicted in Fig. 4.14 leads to identical coincidence for two compared curves. How to extract these differences? In order to extract these differences one can form SRAs for the phase curves and find their difference. The difference is defined as Diff = SRA(Ph(Ytst))-SRA(Ph(Ymn_1). This difference is shown in Fig. 4.16. The correlations calculated with respect to expressions (4.17) reveal also some peculiarities in comparison with real case. The modulus ratio can exceed the unit value. It is happened because of the behavior of the modulus in denominator can be close to zero values. See, for example, Fig. 4.14, when the curve started from small values, then for the positive values of the moments it tends to zero values again. The behavior of the complex correlation function is shown in Fig. 4.17.

It is instructive also to compare the initial curves Ymn_1 and Ymn_2 based on the statistics of complex moments. In Fig. 4.9, we show the comparison of the real part calculated with respect to expression (4.2). The comparison of the phases and detection of their differences are shown in figures below.

Finishing this section, it is instructive to formulate some basic conclusions and recommendations especially for young researchers that can have own data for analysis by these proposed methods.

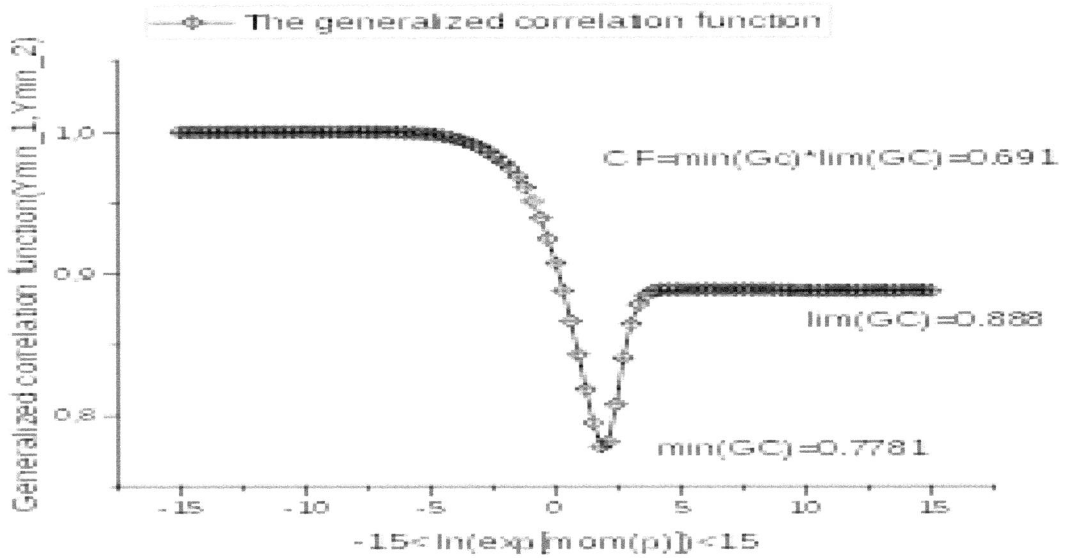

Figure 4.9 If one compares the reduced sequences associated with mean values having the same number of points (N=1000) with the help of the generalized Pearson correlation function, then one can notice that the value of the complete correlation factor CF=0.691. The conventional Person correlation gives the value 0.0295. This small analysis shows that the values of the Pearson correlations have *relative* character and do not coincide with correlation that is given by the complete correlation factor CF.

1. The conventional Pearson formula (4.1) can be essentially generalized. The belief to the value of the Pearson correlation coefficient becomes *questionable.* It can be generalized for the set of real moments (expression (4.2)) and this generalization can be continued for complex case (expression (4.17)). Any modifications of initial sequences (TLS, integrals, SRAs, etc.) can change the value of correlations. These correlations are changed also if for their evaluations we use different expressions (4.1), (4.2), and 4.17) also.

2. The test of new proposed tool definitely shows that the phase distribution (see expression (4.17) in comparison of two random sequences is extremely sensitive to the presence of a "small" perturbation/signal. This observation is needed in further confirmations on real data. The Fig. 4.16 demonstrate clearly this exceptional sensitivity.

3. It is interesting to note that the SRA curve depicted in Fig. 4.11 can be fitted with high accuracy (Rel. Err < 2%) by the generalized Gaussian distribution of the type

$$y(x) = A \cdot \exp\left[-\left(\frac{x-x_0}{\sigma_1}\right)^{\theta_1} - \left(\frac{x-x_0}{\sigma_2}\right)^{\theta_2}\right]. \tag{4.43}$$

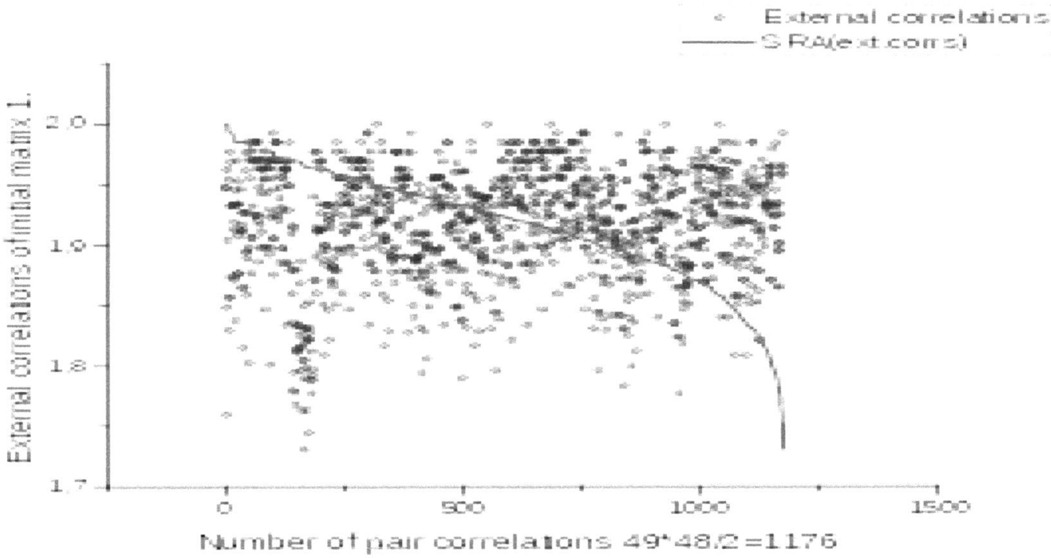

Figure 4.10 Simple formula (4.27) associated with comparison of the ranges allows to compare the external correlations. If this correlation parameter lies in the interval [1,2] then one can say that this rectangle matrix represents a *good* experiment. In the opposite case, when the ratio enters into the interval [0,1], then the columns of the compared matrix are uncorrelated and this experiment needs further analysis. Black solid line demonstrates the SRA of these external correlations. The rectangular matrix associated with TLS (y_2) looks similar and, therefore, it is not shown. The further analysis shows that the SRA-curve can be fitted by the generalized Gauss distribution. See the next figure.

The Generalized Gaussian distribution having 6 fitting parameters (A, x_0, $\sigma_{1,2}$, $\theta_{1,2}$) can be used for reduction purposes and quantitative description of similar histograms depicted in Fig. 4.11.

4.3 ANOTHER APPROACH TO THE CONSIDERATION OF POSSIBLE CORRELATIONS

In this section, a new method for evaluation of the desired correlations is proposed. It allows to evaluate the "content" of the external factors (l=1,2,...,L) setting in the form of data arrays $y_m{}^{(l)}(x)$ (m=1,2,...,M) inside the given $Y_m(x)$ function that is supposed to be subjected by the influence of these factors. As contrasted to the conventional correlation analysis, the proposed method allows finding the "influence" functions $b_l(x)$ (l=1,2,...,L) and evaluating the "remnant" array $G_m(x)$ that is remained as a "quasi-independent" part from the influence of the factors $y_m{}^{(l)}(x)$. The general expression works as a specific "balance" and reproduces the well-knows cases, when b_l(x) $=C_l$ (it is reduced to the linear least squares method with $G_m(x) \approx 0$) and coincides with the remnant function $Y_m(x) \approx G_m(x)$, when the influence functions become negligible ($b_l(x) \approx 0$). The available data show that the proposed method

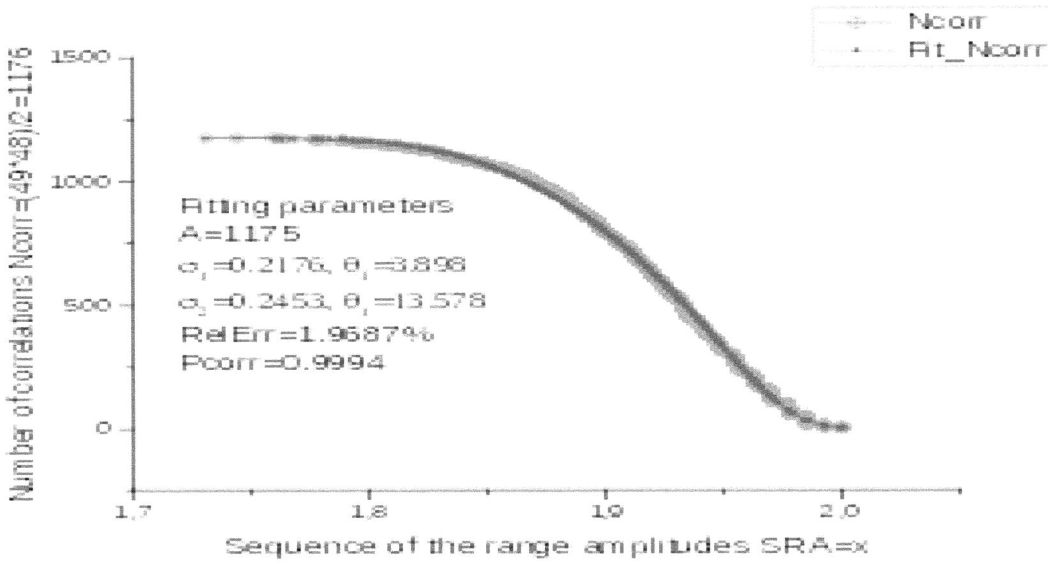

Figure 4.11 If we invert the axis of the SRA shown on the previous figure, then this inverted curve can be fitted by the generalized Gaussian distribution function $y(x) = A \cdot \exp\left[-\left(\frac{x-x_0}{\sigma_1}\right)^{\theta_1} - \left(\frac{x-x_0}{\sigma_2}\right)^{\theta_2}\right]$. The values of the fitting parameters are shown inside the figure.

allows to extract a small signal $S(x)$ from the "pattern" background and it conserves its stability/robustness in the presence of a random fluctuations/noise. The method is rather flexible and allows to consider the cases of strong correlations, when the external factors act *successively*, forming the cause-and-effect chains. It can be generalized for expressions containing the bonds expressed in the form of memory functions. The proposed method adds new quantitative ties expressed in the form of the desired functions to the conventional correlation relationships expressed in the form of the correlation coefficients forming, in turn, the correlation matrices. New relationships allow to understand deeper the existing correlations and make them more informative, especially in detection of the desired deterministic and stable bonds/laws that can be hidden inside. This short preambular taken from paper [5] explains clearly the approach that will be outlined below.

If our potential reader has a look at the wide set of books and papers meant for practitioners [6, 7, 8, 9, 10, 11, 12, 13, 14] related to correlation analysis then he has a right to pose a reasonable question: what kind of new elements the ambitious authors of the book are trying to introduce in order to convince a skeptically tuned expert/reader that these innovative elements are really *new* and will be *useful* for practical applications? If somebody will print in his computer "correlation and regression analysis" then he can find about 20 million results (!) associated with this subject. This huge "information wall" can serve as a real obstacle in attempts to suggest some new and original elements that "go out" from the conventional Pearson correlation paradigm and its existing generalizations. Nevertheless, the analysis of the

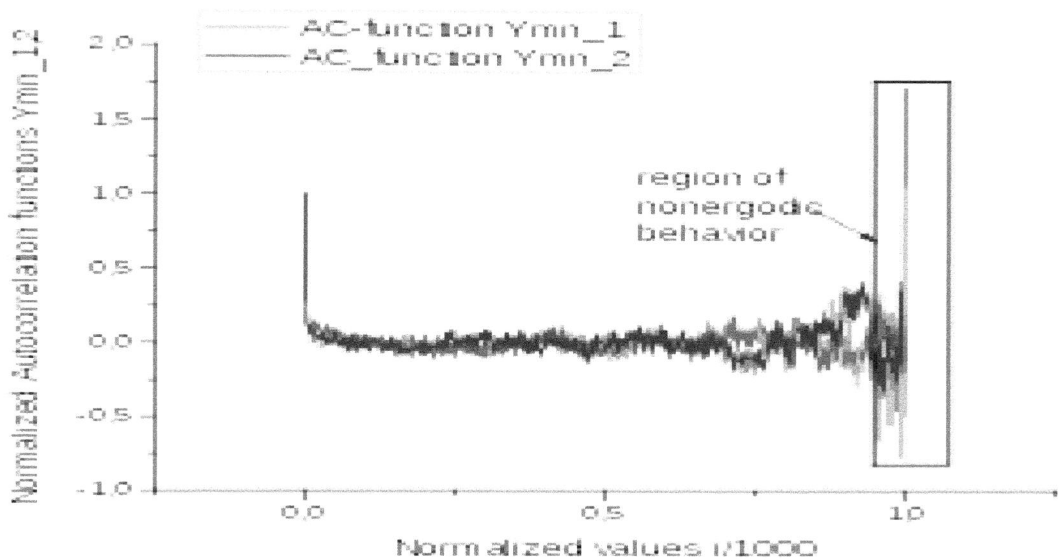

Figure 4.12 Finishing this small and instructive analysis, we decided to calculate the autocorrelation function (ACF). Usually, calculation of the ACFs is needed for verification of the correlation weakening principle (CWP). If these correlations become weaker with time, then these correlations are defined as normal or ergodic. In the opposite case, when they are increasing with time then are classified as *nonergodic*. Comparison of the sequences Ymn_1 and Ymn_2 depicted in Figs. 4.5 and 4.6 clearly shows that these sequences are nonergodic.

basic literature [6, 7, 8, 9, 10, 11, 12, 13, 14] related to the establishing of the desired relationships between two correlated random sets of data allows finding the desired *new* and important elements that can be helpful and useful in many applications. In order to be more specific, let us formulate the problem in details, including some necessary definitions.

Let the set $\{x_j\}$ (j=1,2,...,N) coincides with the input variable determined in N data points and the discrete function $Y_m(x_j)$ determines the rectangle matrix of the corresponding responses (outputs), where the parameter m=(1,...,$M \leq N$, $M > 1$) determines the number of successive measurements related to the given input set.

In accordance with the general theory related to the quasi-reproducible experiment (GTQRE) [15, 16] one can derive a general functional dependence for the set of $Y_m(x_j)$ and this dependence can be expressed in the form of the *generalized* fitting function related to the Prony function. In many practical applications, it is necessary to establish the correlation relationships between measurements that were obtained for the *same* set of input variables and corresponding to similar experimental conditions. We define these output variables as: $y_m^{(l)}(x_j), l = 1, 2, ..., L$, where parameter l determines the number of the external and controllable factors that can be related with $Y_m(x_j)$. Mathematically, the problem that we are going to solve in this chapter can be formulated as:

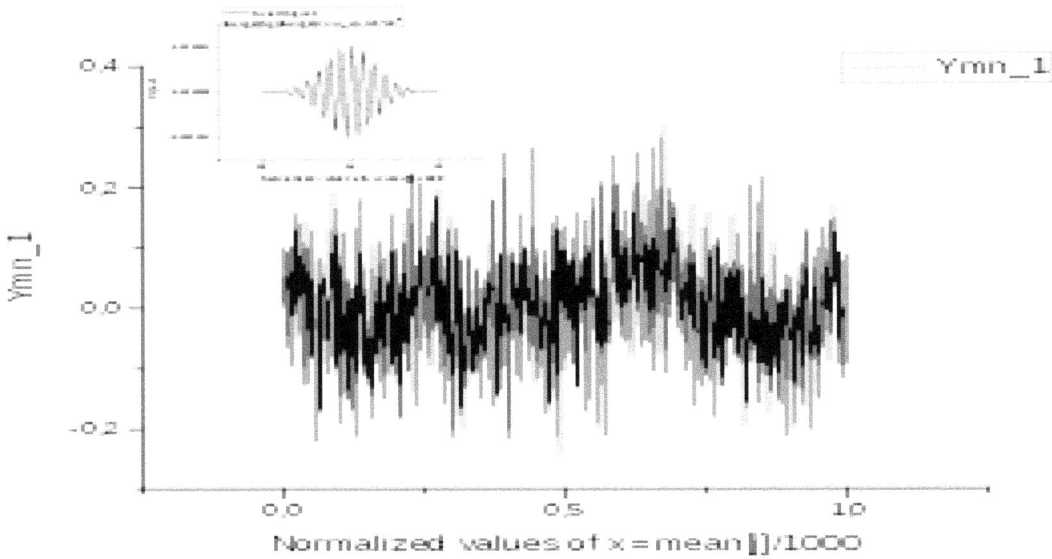

Figure 4.13 In order to demonstrate the exceptional sensitivity of the imaginary part of the complex moments to the presence of a small signal we simply added to the selected initial signal Ymn_1 a small signal (Sg) with ratio Range(Sg)/Range(Ymn_1) = 3.6$*10^{-6}$! The form of this signal is shown on the left-hand side above.

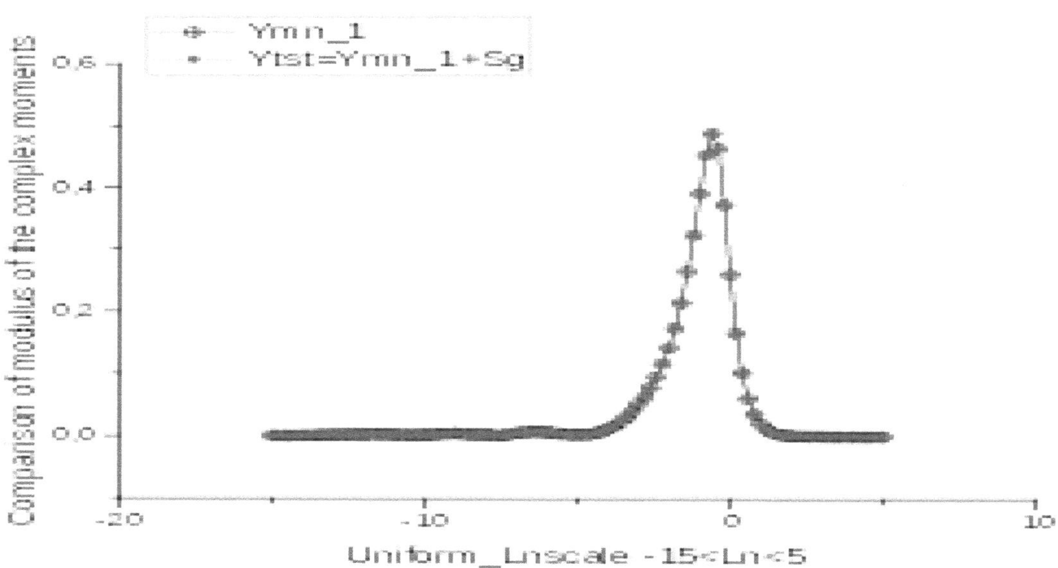

Figure 4.14 The influence of a weak signal in behavior of the modulus is *unnoticeable.* Two curves one is Ymn_1 and with added signal Ytst=Ymn_1+Sg look identical.

Is it possible to relate at least approximately the set of the functions $Y_m(x_j)$ with other functions $y_m{}^{(l)}(x_j)$ for removing their correlation dependence and separate its “remnant” or almost independent part?

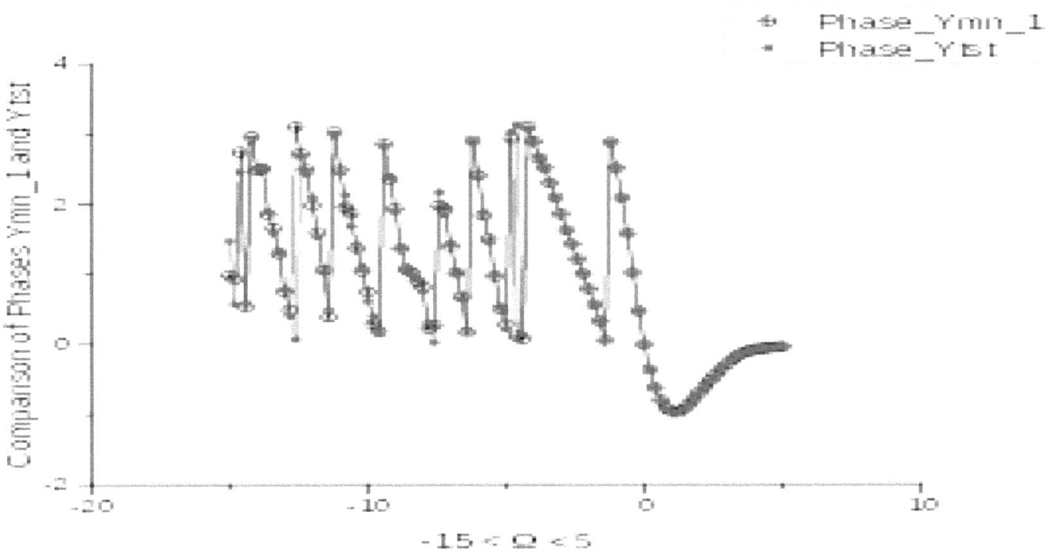

Figure 4.15 We want to compare also the phase behavior corresponding for two curves compared. As one notice from the visual comparison small differences between two curves become noticeable. The positive part is not informative while the negative part demonstrates clearly the desired difference.

We try to find the original solution of this problem that will have large applications related to different data.

4.4 DESCRIPTION OF THE PROPOSED ALGORITHM

In many cases, it is impossible to express the single-valued variable x_j through the chosen function ${y_m}^{(l)}(x_j)$ (except the cases of the simplest dependence), because the output ${y_m}^{(l)}(x_j)$ for any fixed m (m=1,2,...,M; M is a number of columns of the matrix; j =1,2,...N as before is the number of data points) expressed in the form of the multi-valued functions (coinciding with rectangle matrices) and the analytical transformation $x_j \to F(y_m^{(l)})$ becomes *impossible*. Usually, in this case one can use the extended statistics of the fractional moments [17, 18] and calculate the complete correlation factor that generalizes the conventional Pearson correlation coefficient over all space of the fractional moments. However, in many cases, it is necessary to know (and evaluate, if necessary) the "share" of correlations of one random variable with respect to another one in the form of an "influence" function and the general problem in this formulation, as far as we know, is *not* solved. Mathematically, this correlation relationship (in the case, when different external factors are considered independently from each other) can be expressed approximately in the form of the following linear combination:

$$Y_m(x_j) \cong \sum_{l=1}^{L} \langle b_l(x_j)\rangle\, y_m^{(l)}(x_j) + G_m(x_j), \tag{4.44}$$

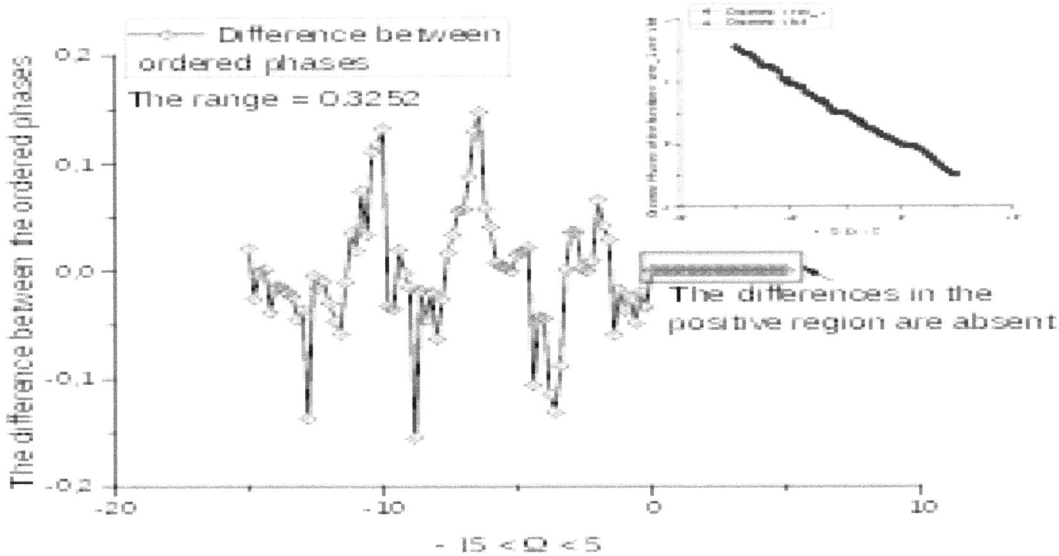

Figure 4.16 In order to notice the difference between the phases shown on the previous figure we ordered these curves (it is shown above on the right-hand side) and find the differences between them. These differences are shown on the central figure. They can be characterized by the value of the range (in our case this value equals 0.3252). Compare this value with the initial value of the S/N ratio equaled $3.6*10^{-6}$!

where the sets $b_l(x_j)$ and $G_m(x_j)$ determine the *unknown* functions that are needed to be evaluated. In expression (4.44), we assume that the "influence" function is expressed in the *averaged sense* $b_{ml}(x) \to <b_l(x)>$ and the "remnant" function $G_m(x) \to <G\ (x)>$ also (for simplicity we omit the data points index j) that implies that they do *not* depend on index m. We should make this replacement having in mind that the functional least squares method suggested below enables to calculate only the functions $<b_l(x)>$ and $<G\ (x)>$. However, after calculation of the mean value function $<G\ (x)>$, expression (4.44) allows to restore the complete set of the functions $G_m(x)$ forming the desired array.

We suppose also that variables $Y_m(x_j)$ and ${y_m}^{(l)}(x_j)$ can be close in the statistical sense and, therefore, they can be correlated to each other in some extend. For their evaluation, one can use the functional linear least squares (FLLS) method, proposed earlier in papers [15, 16], requiring that the *functional* dispersion between the functions in (1) achieve a minimal value:

$$\sigma(x) = \min\left[\frac{1}{M}\sum_{m=1}^{M}\left(Y_m(x) - \sum_{s=1}^{L}\langle b_s(x)\rangle\, y_m^{(s)}(x) - \langle G(x)\rangle\right)\right]^2. \tag{4.45}$$

Taking the functional derivatives with respect to unknown functions $< b_l(x) >$ and $<G(x)>$, we obtain:

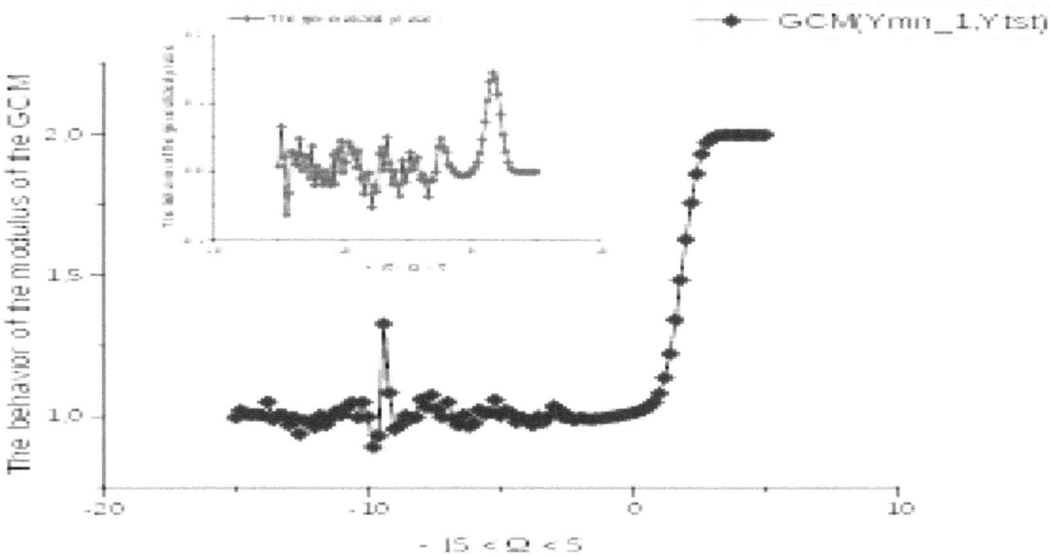

Figure 4.17 These curves demonstrate the behavior of the complex moments. The modulus is given in the central figure, while on the small figure above we demonstrate the behavior of the phases. One concludes that the complex moments are more sensitive in comparison with the real parts.

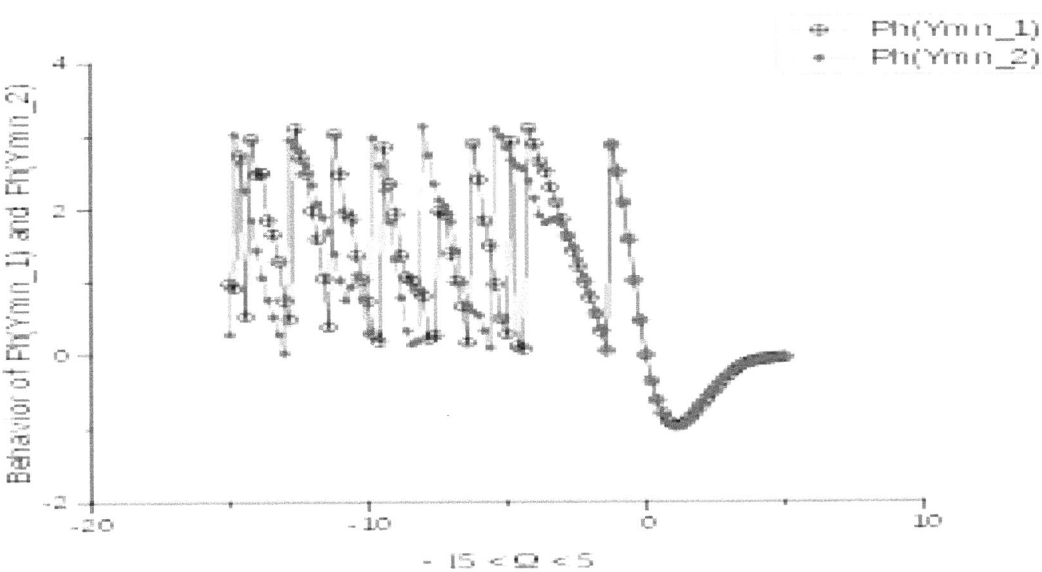

Figure 4.18 The phase behavior corresponding for two curves Ph(Ymn_1) and Ph(Ym_2). As one notice from the visual comparison the difference between two curves is essential. The positive part as before is not so informative, while the negative part demonstrates clearly the differences.

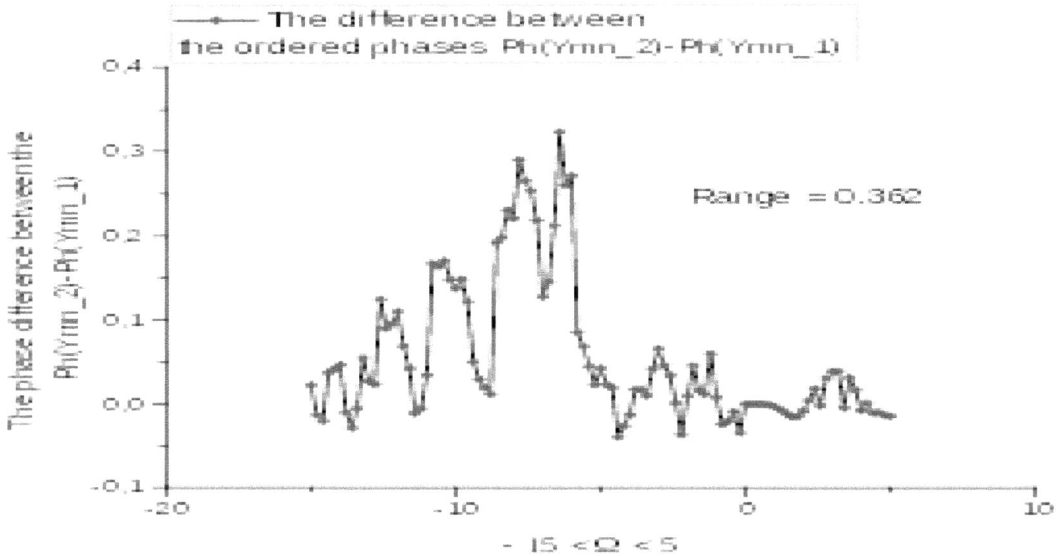

Figure 4.19 The difference between the phases depicted on the previous figure is shown here. They can be characterized by the value of the range (in this case this value equals 0.36205).

$$-\frac{\delta\sigma(x)}{\delta\langle b_l(x)\rangle}=\frac{1}{M}\sum_{m=1}^{M}\left[y_m^{(l)}(x)\left(Y_m(x)-\sum_{s=1}^{L}\langle b_s(x)\rangle y_m^{(s)}(x)-\langle G(x)\rangle\right)\right]=0 \quad (4.46)$$

$$-\frac{\delta\sigma(x)}{\delta\langle G(x)\rangle}=\frac{1}{M}\sum_{m=1}^{M}\left[\left(Y_m(x)-\sum_{s=1}^{L}\langle b_s(x)\rangle y_m(x)-\langle G(x)\rangle\right)\right]=0. \quad (4.47)$$

Here, we apply the averaging procedure over all admissible measurements (m=1,2,...,M) supposing that the influence functions $<b_l(x)>$ do *not* depend on the current index m. Defining the pair correlation functions

$$Q_l(x)=\frac{1}{M}\sum_{m=1}^{M}Y_m(x)y_m^{(l)}(x),\ q_{s,l}(x)=\frac{1}{M}\sum_{m=1}^{M}y_m^{(s)}(x)y_m^{(l)}(x), \quad (4.48)$$

$$\langle Y(x)\rangle=\frac{1}{M}\sum_{m=1}^{M}Y_m(x),\ \left\langle y^{(l)}(x)\right\rangle=\frac{1}{M}\sum_{m=1}^{M}y_m^{(l)}(x),\ s,l=1,2,...,L, \quad (4.49)$$

one can obtain the system of linear equations with respect to unknown functions $<b_s(x)>$ in (4.46)

$$\sum_{s=1}^{L}q_{s,l}(x)\langle b_s(x)\rangle+\langle G(x)\rangle\left\langle y^{(l)}(x)\right\rangle=Q_l(x), \quad (4.50)$$

$$\sum_{s=1}^{L}\langle b_s(x)\rangle\left\langle y^{(s)}(x)\right\rangle+\langle G(x)\rangle=\langle Y(x)\rangle. \quad (4.51)$$

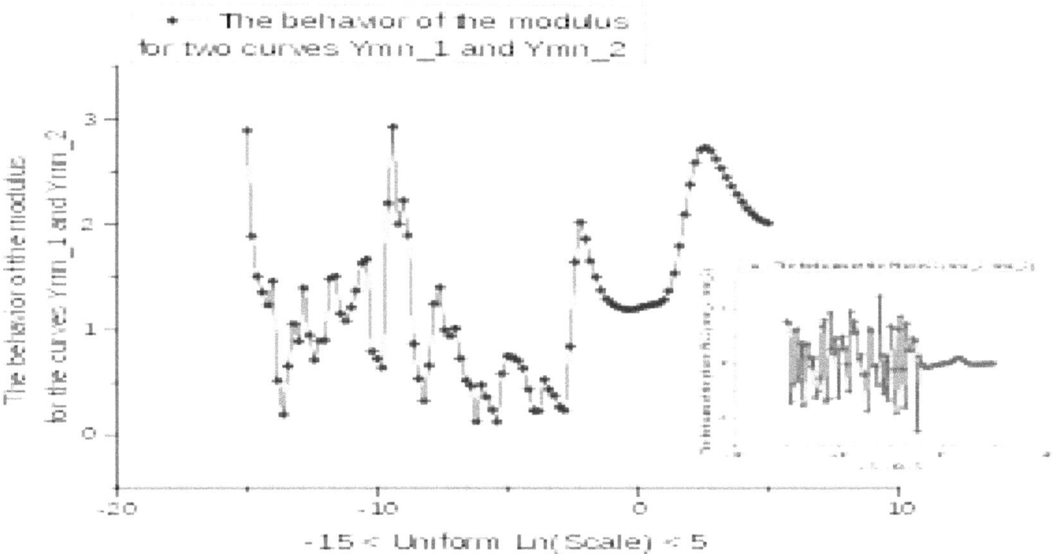

Figure 4.20 These curves demonstrate the behavior of the complex moments calculated for comparison of two curves Ymn_1 and Ymn_2. For evaluation we use again relationship (4.17). The modulus is given in the central figure, while on the small figure on the right-hand side we demonstrate the behavior of the phases. One concludes that the complex moments are more sensitive in comparison with the real parts. We should notice also that the values of correlations can exceed the unit value. We should notice that in the case of comparison of two identical curves the modulus coincides with unit value, while the phase curve coincides with zero.

From linear system (4.50), one can calculate the desired functions $<b_s(x)>$ and $<G(x)>$ and, finally, find the solution of equation (4.44). Here we want to notice that for calculation of the unknown functions $<b_s(x)>$ we need to know only the value of $<G(x)>$, however from (4.44), one can restore directly the set of desired functions $G_m(x)$ for any fixed m. One can simplify equation (4.50) also, if one can express the unknown function $<G(x)>$ from the second line and insert this expression into the first line of (4.50). After simple algebraic manipulations, one can obtain the following expression:

$$\sum_{s=1}^{L} \tilde{q}_{s,l}(x) \langle b_s(x) \rangle = \tilde{Q}_l(x), \tag{4.52}$$

$$\tilde{q}_{s,l}(x) = q_{s,l}(x) - \left\langle y^{(s)}(x) \right\rangle \left\langle y^{(l)}(x) \right\rangle, \tag{4.53}$$

$$\tilde{Q}_l(x) = Q_l(x) - \langle Y(x) \rangle \left\langle y^{(l)}(x) \right\rangle. \tag{4.54}$$

$$\langle G(x) \rangle = \langle Y(x) \rangle - \sum_{s=1}^{L} \langle b_s(x) \rangle \left\langle y^{(s)}(x) \right\rangle. \tag{4.55}$$

These expressions solve the problem of calculation of the set of *true* quasi-independent functions $G_m(x)$ from (4.44) based on the calculated set of the influence functions $<b_s(x)>$ from (4.52). It is instructive to write down some expressions for the simplest cases, having large practical importance.

1. The influence of one factor (L=1).

For this, case system (4.52) has the following solution:

$$\langle b_1(x)\rangle = \frac{\tilde{Q}_1(x)}{\tilde{q}_{11}(x)}, \tag{4.56}$$

$$\langle G(x)\rangle = \langle Y(x)\rangle - \langle b_1(x)\rangle \left\langle y^{(1)}(x)\right\rangle, \tag{4.57}$$

$$\tilde{q}_{11}(x) = \frac{1}{M}\sum_{m=1}^{M}\left(\Delta y_m^{(1)}(x)\right)^2, \ \Delta y_m^{(1)}(x) = y_m^{(1)}(x) - \left\langle y^{(1)}(x)\right\rangle \tag{4.58}$$

$$\tilde{Q}_1(x) = \frac{1}{M}\sum_{m=1}^{M}\Delta Y_m(x)\Delta y_m^{(1)}, \ \Delta Y_m(x) = Y_m(x) - \langle Y(x)\rangle. \tag{4.59}$$

It is useful to consider the limiting cases.

(a) Let us suppose that

$$\Delta Y_m(x) \cong \lambda \Delta y_m^{(1)}(x). \tag{4.60}$$

In this case, (after averaging (4.60) over all successive measurements m=1,2,...M) we have

$$\langle \Delta Y(x)\rangle = \lambda\left\langle \Delta y^{(1)}(x)\right\rangle + \varepsilon(x) \tag{4.61}$$

and the calculation of the unknown parameter λ is reduced to the linear least squares method (LLSM).

In this expression, we want to stress the closeness of the proposed method with the conventional LLSM. However, the functional least squares method allows to consider also the dependence of a small function $\epsilon_m(x)$ against the array index m. In this case, it is necessary to modify relationship (4.60) adding small function $\epsilon_m(x)$ in the right-hand side and replace the constant λ by a function $<\lambda(x)>$. This modification reproduces the relationship (4.44) again, when the function $\epsilon_m(x)$ is supposed to be small. The mean value in (4.61) is replaced by numbers that are calculated over the input variable x. If $\lambda = 1$ (it implies the complete coincidence of the function $Y_m(x)$ with ${y_m}^{(1)}(x)$), then $G_m(x) = 0$. For $\lambda \approx 1 =$ const, the influence of the function $G(x)$ becomes negligible and it is replaced in (4.61) by a fitting error function $\epsilon(x)$.

(b) Let us suppose that number of the measurements is small $Y_m(x) \approx <Y(x)>$ (it corresponds the case of an "ideal" experiment considered in [15, 16]). In this case, the influence of correlation with the function $<y_1(x)>$ becomes negligible ($<b_1(x)> \approx 0$). In practice, it is easy to verify another condition $<Y(x)> \approx <G(x)>$ with the help of the Pearson correlation coefficient. If this coefficient lies in the interval [0.9–0.99] then one concludes that the compared sequences $Y_m(x)$ and ${y_m}^{(1)}(x)$ are *uncorrelated*, or, in other words, *independent* from each other.

Therefore, expression (4.56) covers the cases known earlier.

2. The influence of two factors (L=2).

For this case from system (4.52) one can receive the following expressions for $<b_l(x)>$, l=1,2.

$$\langle b_1(x)\rangle = \frac{\tilde{Q}_1(x)\cdot\tilde{q}_{2,2}(x)-\tilde{Q}_2(x)\cdot\tilde{q}_{1,2}(x)}{\tilde{q}_{1,1}(x)\cdot\tilde{q}_{2,2}(x)-\left(\tilde{q}_{1,2}(x)\right)^2}, \tag{4.62}$$

$$\langle b_2(x)\rangle = \frac{\tilde{Q}_2(x)\cdot\tilde{q}_{1,1}(x)-\tilde{Q}_1(x)\cdot\tilde{q}_{1,2}(x)}{\tilde{q}_{1,1}(x)\cdot\tilde{q}_{2,2}(x)-\left(\tilde{q}_{1,2}(x)\right)^2}. \tag{4.63}$$

Here the matrix components are defined by the following expressions:

$$\tilde{q}_{s,l}(x) = \frac{1}{M}\sum_{m=1}^{M}\Delta y_m^{(s)}(x)\Delta y_m^{(l)}(x), \tag{4.64}$$

$$\tilde{Q}_l(x) = \frac{1}{M}\sum_{m=1}^{M}\Delta Y_m(x)\Delta y_m^{(l)}(x), \tag{4.65}$$

$$\Delta f_m(x) = f_m(x) - \langle f(x)\rangle, \;\; s,l = 1,2. \tag{4.66}$$

The "remnant" array (defined as the weakly correlated part of $Y_m(x)$) is given by expression:

$$G_m(x) \cong Y_m(x) - \sum_{l=1}^{2}\langle b_l(x)\rangle\, y_m^{(l)}(x). \tag{4.67}$$

Expressions (4.62)–(4.67) are considered as the final expressions for the case of influence of two additive factors. In any case considered above, it is supposed that the number of similar measurements M exceeds the number of external factors *L(M>L)*. The previous calculations were associated with the case when all external factors act independently/additively from each other. This case is expressed mathematically by expression (4.44). Is it possible to consider the case when the factors are strongly correlated and act *successively*, i.e., when one dominant factor (cause) evokes the appearance of another factor (effect). In this case, locating the cause-and-effect relations in the right order one can write the following relationship:

$$Y_m(x) = G_m(x)\prod_{s=1}^{L}\left(y_m^{(s)}(x)\right)^{\langle b_s(x)\rangle}. \tag{4.68}$$

Taking the natural logarithm from the both parts, we obtain

$$\ln(Y_m(x_j)) \cong \sum_{l=1}^{L}\langle b_l(x_j)\rangle \ln(y_m^{(l)}(x_j)) + \ln(G_m(x_j)). \tag{4.69}$$

Formally, equation (4.69) is similar to expression (4.44), however, in addition one can determine the values of the logarithmic functions, when the function $y_m^{(l)}(x_j)$

becomes negative and equals zero. Therefore, all calculations considered above can be applicable for this case too. The proposed method allows also combining these two cases and considering more complex problem. In order to show its generality, we present expression (4.68) in the following form

$$F_m^{(l)}(x) = G_m^{(l)}(x) \prod_{s=1}^{l} \left(y_m^{(s)}(x)\right)^{\langle w_s(x)\rangle}, \; s = 1, 2, ..., l. \tag{4.70}$$

Equation (4.70) allows finding the set of the unknown functions $G_m^{(l)}(x)$ and the weighting factors $<w_s(x)>$ with the help of equation (4.69). After that, one can apply *again* equation (4.44) representing it in the form:

$$Y_m(x) \cong \sum_{l=1}^{L} \langle b_l(x)\rangle \, G_m^{(l)}(x) + R_m(x). \tag{4.71}$$

From this equation, it is possible to find the set of influence functions $<b_l(x)>$ (l=1,2,...,L) and the remnant array $R_m(x)$ that is defined again as the weakly-correlated part of $Y_m(x)$.

In the case of cause-and-effect relations, a special attention should be paid for the *negative* values of functions in expressions (4.68) and (4.70). The negative values of a power-law function $f(x)$ in the space of real numbers can be defined as

$$\mathrm{Re}\left(V(x)\right) = \mathrm{Re}\left(\left[(-f(x))^{\pm\alpha(x)}\right]\right) = |f(x)|^{\pm\alpha(x)} \cos(\alpha(x)\pi). \tag{4.72}$$

The region when *f(x)≡0 requires some regularization procedure in any partial case, when α(x) becomes negative or close to zero value ($\alpha(x) \cong 0$). In conclusion of this section, one can show also another possibility, where the proposed algorithm could be applied also. Let us suppose that the measured function $F_m(t)$ is formed by a linear combination of the correlated functions related to the initial one by means of the convolution operation, i.e.,:*

$$F_m(t) = \sum_{l=1}^{L} \int_0^t \langle \mu_l(t-\tau)\rangle \, f_m^{(l)}(\tau)d\tau + \Phi_m(t). \tag{4.73}$$

In expression (4.73) the set of the functions $<\mu_l(t)>$ determines the so-called correlation memory that arises between the correlation function $f_m^{(l)}(t)$ evoked by the factor l (l=1,2,...,L) and the initial function $F_m(t)$ that can be affected by this factor. These correlations can be understood in the sense of "heritagetory" principle [19] that was introduced initially by Volterra and later by Prigogine [20], and Zwanzig [21] in understanding of irreversible kinetic equations. If one replaces the memory kernel by the power-law function proportional to $(t-\tau)^{\gamma-1}$ then one can obtain the Riemann-Liouville fractional integral [22, 23, 24] playing an important role in description of many natural processes with *remnant* memory. Applying the Laplace transform to both parts of (4.73), we obtain:

$$F_m(p) = \sum_{l=1}^{L} \langle \mu_l(p) \rangle f_m^{(l)}(p) + \Phi_m(p), \tag{4.74}$$

$$J(t) := J(p) = \int_0^{\infty} J(t) \exp(-pt) dt. \tag{4.75}$$

Here, we should stress also that any function having the finite Laplace transform and satisfying to three principles as (a) linearity, (b) fading behavior, (c) invariance principle marked in [19] can be used as the memory function. (4.74) is similar to expression (4.44) and, therefore, one can apply the algorithm developed above for this nontrivial case. From our point of view, expression (4.74) can be useful in analysis of different dielectric "mixtures", when the total complex permittivity can be presented approximately as an additive combination of different complex permittivities associated with different dielectric materials.

In order to outline the limits of the proposed algorithm one can find the answer to the following question: how many minimal measurements (that is given by the value of M) are necessary for its application? The analysis shows that for calculation of the mean values given by expressions (4.48) only *two* repetitions of the *same* measurement (M=2) are necessary. However, these two successive measurements *cannot* create the statistically significant sampling and, therefore, it is necessary to collect a "rich" sampling with number of measurements exceeding $M > 10$, as minimal. How to do it *artificially* if a researcher because of different reasons cannot realize the representative sampling? In this case, one can suggest an approach that is described below in Sec. 4.5.

4.5 CREATION OF A REPRESENTATIVE SAMPLING

Let us suppose that we definitely know only two measurements coinciding with the limits of a possible sampling $Ydn(x)$ and $Yup(x)$. These functions define the "down" and "up" limits of the desired sampling, respectively. Let us define the variable from the interval [0,1].

$$v_m = \frac{m-1}{M-1}, \; m = 1, ..., M. \tag{4.76}$$

Then we define the function from the same interval $F(v_m) = 3(v_m)^2$-$2(v_m)^3$. This specific choice is stipulated by the fact that at $F(0)$=0, $F(1)$=1, $F(1/2)$=1/2 and, therefore, the desired array can be obtained as:

$$Y_m(x_j) = Ydn(x_j) + F(v_m) \cdot (Yup(x_j) - Ydn(x_j)), \tag{4.77}$$

$$Ydn(x) \le Y_m(x) \le Yup(x), \; j = 1, 2, ..., N, \; m = 1, ..., M. \tag{4.78}$$

Let us suppose that *a priori* three functions are known: $Ydn(x)$, Yup(x) and some intermediate function $f_1(x)$. As it has been mentioned above, we choose the function $F(v)$ in the form of the polynomial of the second order:

$$F(v,x) = a(x) \cdot v + b(x) \cdot v^2. \tag{4.79}$$

For the finding of the unknown *functions*, $a(x)$ and $b(x)$, we take into account the following conditions: $F(1)=1$, $F(1/2)=w_1(x)=(f_1(x) - Ydn(x))/(Yup(\mathrm{x}) - Ydn(x))$. After simple algebraic transformations we obtain again the structure (4.77) with the function $F(v_\mathrm{m}, w_1(x))$ from (4.79), where the functions $a(x)$ and $b(x)$ are defined as

$$a(x) = 4w_1(x) - 1, \tag{4.80}$$
$$b(x) = 2 - 4w_1(x). \tag{4.81}$$

By complete analogy with this result, one can obtain the desired expressions for *two* known intermediate functions: $f_k(x)$ (k=1,2). We subordinate the function $F(v)$ to the following conditions: $F(1)=1$, $F(1/3)=w_1(x)$ and $F(2/3)=w_2(x)$ with $w_k(x)=(f_k(x) - Ydn(x))/(Yup(\mathrm{x}) - Ydn(x))$, k=1,2 and consider the polynomial of the third order:

$$F(v,x) = a(x)v + b(x)v^2 + c(x)v^3. \tag{4.82}$$

After some algebraic manipulations, one can find the desired functions from the linear system of equations

$$a(x) + b(x) + c(x) = 1, \tag{4.83}$$
$$\frac{1}{3}a(x) + \frac{1}{9}b(x) + \frac{1}{27}c(x) = w_1(x), \tag{4.84}$$
$$\frac{2}{3}a(x) + \frac{4}{9}b(x) + \frac{8}{27}c(x) = w_2(x). \tag{4.85}$$

The system admits the solution:

$$a(x) = \frac{1}{2}(2 + 18w_1(x) - 9w_2(x)), \tag{4.86}$$
$$b(x) = -\frac{9}{2}(1 + 5w_1(x) - 4w_2(x)), \tag{4.87}$$
$$c(x) = \frac{9}{2}(1 + 3w_1(x) - 3w_2(x)). \tag{4.88}$$

Finally, we obtain the same mathematical expression (4.77), where the function $F(v)$ should be replaced by expression (4.82) with the functions from (4.86). More complex cases with intermediate functions $f_k(x)$ (k=1,2,3...) can be considered with complete analogy with this simple algorithm and, therefore, the problem of generation of an artificial massive in general case can be solved by a similar way.

Expression (4.77) generates a set of functions from the admissible interval $[Ydn(x), Yup(x)]$. If necessary, one can add the intermediate functions $f_k(x)$ (k=1,2,3...) and use more complicated expressions (4.79) and (4.82) shown above.

These expressions can be applied for generation of the desired data in cases, when the creation of the arrays becomes impossible. This situation is frequently met in economics, geophysics and biological/medical sciences, when the existing conditions do not allow to repeat similar experiments. Let us suppose that we fixed the random curve $Ydn(t_0)$ at the moment t_0 and to the moment t_M it accepts another form $Yup(t_M)$. Then, introducing the normalized temporal variable

$$v_m = \frac{t_m - t_0}{t_M - t_0}, \; t_m = t_0 + \frac{m}{M}(t_M - t_0), \; m = 0, 1, ..., M \tag{4.89}$$

one can create an artificial data array with the help of expression (4.77) and its generalizations given below. Actually, expression (4.77) can be considered as a transformation of one random function to another one and this transformation can be useful in many practical applications. For example, let us consider the following situation. The measurement cell of an equipment generates a random function that slightly varied each day under the influence of a set of uncontrollable external factors. The aim of any engineer/experimentalist is to receive the stable characteristics of the measurement cell. In this case one can choose the function $Yup(t_M)$ as the pattern background and the initial function should coincide with $Ydn(t_0)$. With the help of expression (4.77) one can obtain the desired stability of the measured cell during the interval $[t_0, t_M]$.

In conclusion of this section, we would like to show how to obtain the *quantitative/numerical* estimations for the influence functions $<b_l(x)>$ and the remnant function $G_m(x)$. For this purpose, we take the average values from the functions $Y_m(x)$, $y_m{}^{(l)}(x)$ and $G_m(x)$

$$\langle Y(x)\rangle = \frac{1}{M}\sum_{m=1}^{M} Y_m(x), \; \left\langle y^{(l)}(x)\right\rangle = \frac{1}{M}\sum_{m=1}^{M} y_m^{(l)}(x), \langle G(x)\rangle = \frac{1}{M}\sum_{m=1}^{M} G_m(x) \tag{4.90}$$

and then we replace in expression (4.44) the functions $<b_l(x)> \to b_l$ by the set of unknown constants and $<G(x)>$ by the constant g. In the result of this transformation, we obtain

$$\langle Y(x)\rangle \cong \sum_{l=1}^{L} b_l \left\langle y^{(l)}(x)\right\rangle + g. \tag{4.91}$$

The set of the desired constants b_l (l=1,2,...,L) and g can be evaluated from (4.91) with the help of LLSM-method. In particular, for L=1 the constants b_1 and g coincide with the value of the slope and intercept, accordingly and, therefore, can be easily evaluated.

Attentive reading of the subsections 4.3–4.5 should convince a thoughtful reader that the problem of correlations is *not* completely solved. The existing statistical methods give only an initial solution of the correlation problem. In order to say definitely about the correlation of the compared random sequences we need finer instrument, including different types of correlations, because any transformation of the initial function can change the value of correlations. Below we show couple of examples that demonstrate the possibilities of the proposed approach.

4.6 EXAMPLES

4.6.1 Verification of expressions (4.44) on mimic data.

Let us consider the following problem. From some measurements we have a "pattern" background that is described mathematically by an array ${y_m}^{(l)}(x)$. This background is distorted by a "small" signal $Sgn(x)$ that did not appear in the background earlier. Is it possible to "notice" the presence of the function $Sgn(x)$ in the function $Y_m(x)$ and extract this small signal in the presence of a noise? We should stress here that this common problem is solved in many branches of natural/technical sciences:

1. Detection of a small additive $x(c)$ in some "pattern" background – electrochemistry, microbiology, chemical technology etc.;

2. Detection of a small signal $Sgn(t)$ in the presence of a radio-interference background;

3. The general problem "friend-or-foe" identification in military science. For demonstration of the purposes, we choose the following functions:

   a. "Pattern" array will be identified with ${y_m}^{(l)}(x)$ and generated by expression (4.77) with the limiting functions: $Ydn(x) = sin(0.4x - \pi/8)$, $Yup(\mathrm{x}) = cos(2x - \pi/8) + 3$; the array size: M=100; number of discrete points: j=0,1,...,N=500; the interval of the input variable: $x_{\min} = 10^{-3}$, $x_{\max} = 10$, $x_j = x_{\min} + (j/N)(x_{\max} - x_{\min})$.

   b. The tested array $Y_m(x)$ is generated again with the help of (4.77) and located in the interval $Ysdn(x) = Ydn(x) + Sgn(x)$ ($Sgn(x) = a/[(x\text{-}x_0)^2 + \gamma^2]$, a= 0.01, x_0=8, γ^2 =0.05), $Ysup(x) = w_1 \cdot Yup(x) + w_2$ $(Yrnm(x) + 3)$, where, in turn, $w_1 + w_2 = 1$, $Yrnm(x)$ defines the generator of the uniform random numbers from the interval [-1,1].

Having these functions, it is naturally to solve the problem of extraction of the useful signal $Sgn(x)$ from the function $Y_m(x)$ at different values of the parameter w_1; it admits a "mixture" of a noise (w_1=1 corresponds to the absence of the noise, w_1 =0 corresponds to the maximal influence of a noise). Figs. 4.21–4.24 demonstrate the results of this simple analysis. In Fig. 4.21, we show the "pattern" array corresponding to the background.

In Fig. 4.22, we show the location of a weak signal that is clearly expressed by the function $Ydn(x)$. Then it becomes unnoticeable when the array achieves its upper function $Yup(x)$. In Fig. 4.23, we demonstrate the influence of a noise that can appear in this mimic "experiment". With increasing of an experimental cycling (m= 1 → 100) the influence of random factors becomes essential in comparison with initial stage. Fig. 4.24 presents a final stage of this model research. In the absence of any signal, the comparison of the arrays $Y_m(x)$ with $y_m(x)$ leads to values $<b_1(x)> \equiv 1$ and $<G(x)> \equiv 0$. If a "strange reality" (in our case a "weak" signal) on the given background appears then it becomes noticeable. *Therefore*, having a reliable and "pure" background that serves as a specific detector one can notice the presence of a small mixture comparable with the noise or it can be detected as a signal.

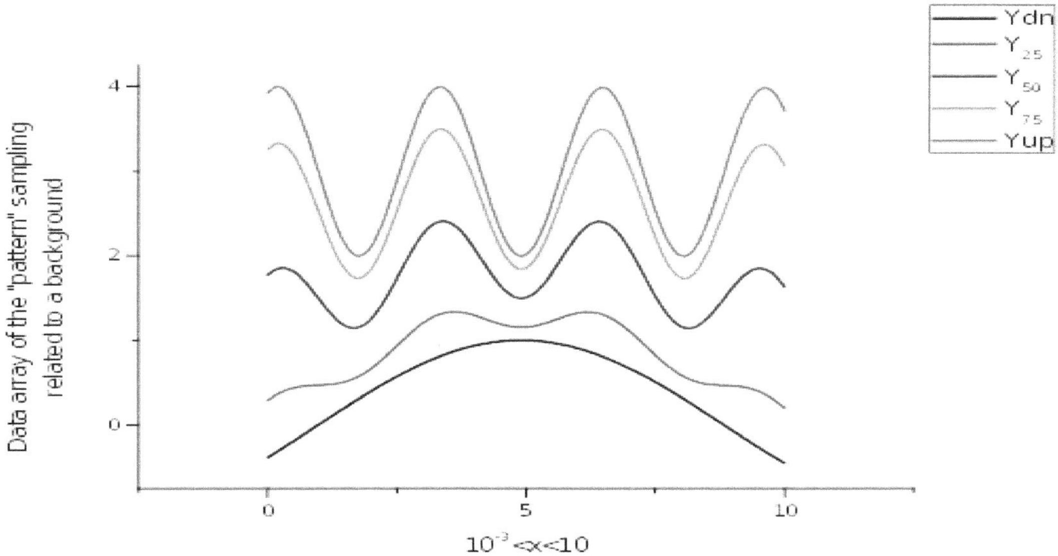

Figure 4.21 Data array corresponding to an "ideal"/pattern background. Only 3 intermediate functions $Y_{25,75,100}(x)$ from the total array (m=0,1,...,M=100) are shown.

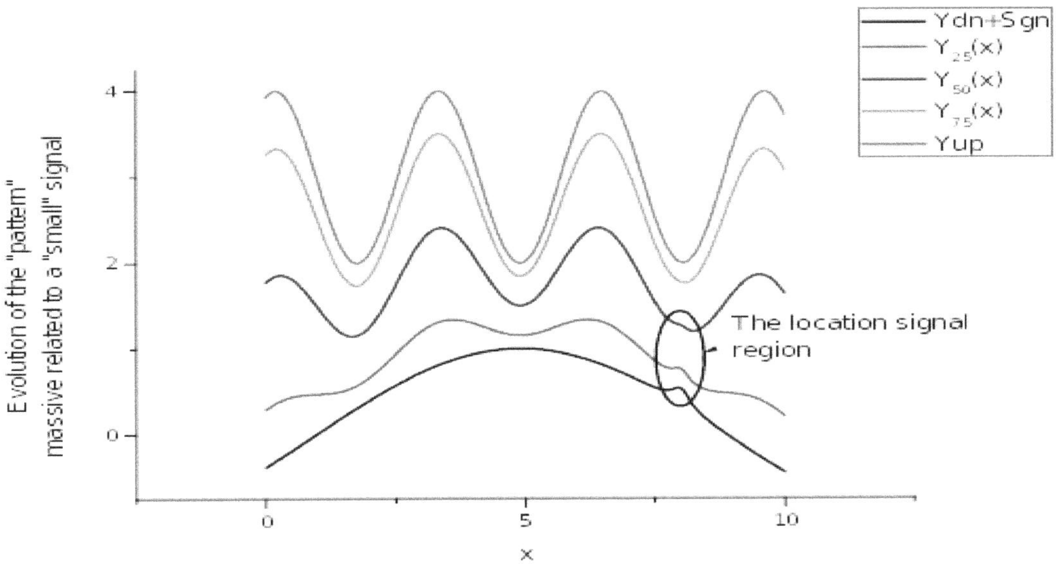

Figure 4.22 The creation of a small signal in another array without influence of the random component.

Figure 4.25 demonstrates the dependence of the influence remnant functions $b(w_2)$ and $g(w_2)$, respectively, obtained with the help of expression (4.91). In the beginning of the plots (where the influence of a noise is absent) the value of $b(0)$ close to the unit value while the value of $g(0)$ is minimal. The presence of a noise destroys this picture

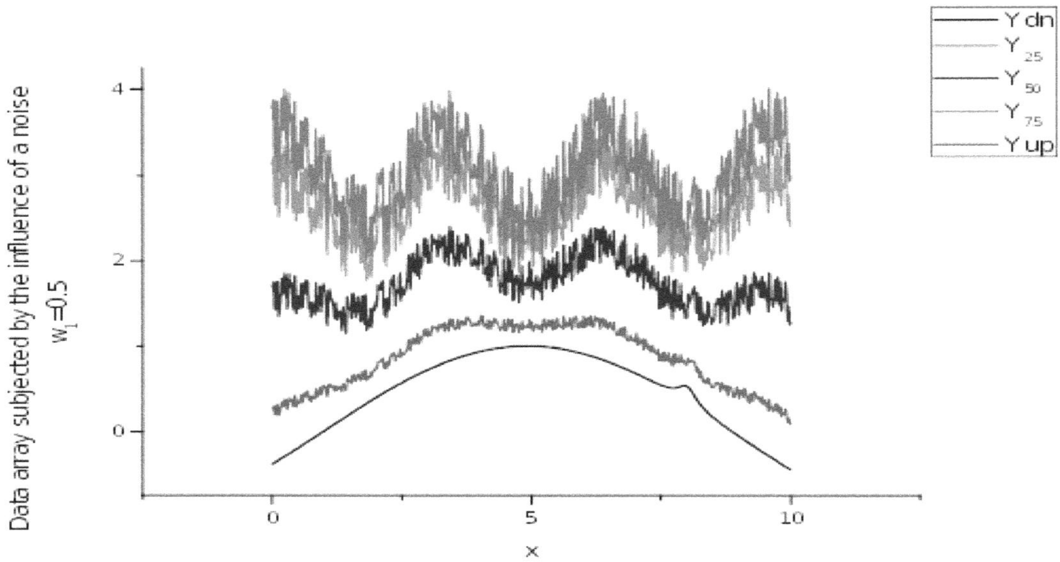

Figure 4.23 The sampling subjected by the influence of a noise (w_1=0.5). The signal is noticeable in the function $Ydn(x)$ and then it is "dissolved" into a noise.

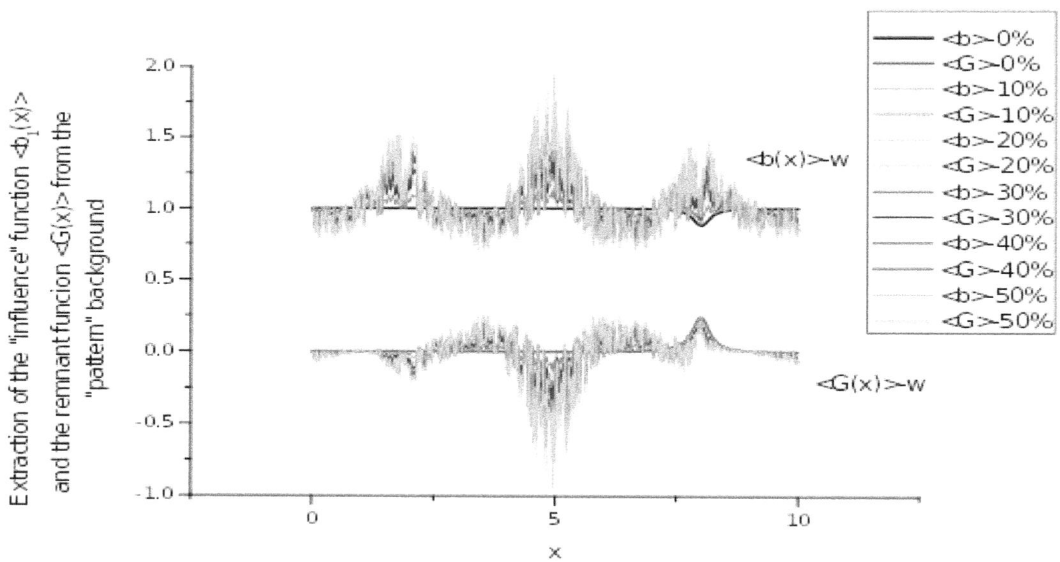

Figure 4.24 Here we present the results obtained with the help of expression (4.45) (two first lines) at different values of the parameter w_2=(0.0 -0.5). In "ideal" case (w_2=0) the functions $<b_1(x)>$ and $<G(x)>$ detect clearly the desired signal. With increasing of the parameter w_2 to the value 0.5 the influence of a noise becomes predominant; however, we keep deliberately the function $Ydn(x)$ with minimal influence of a noise.

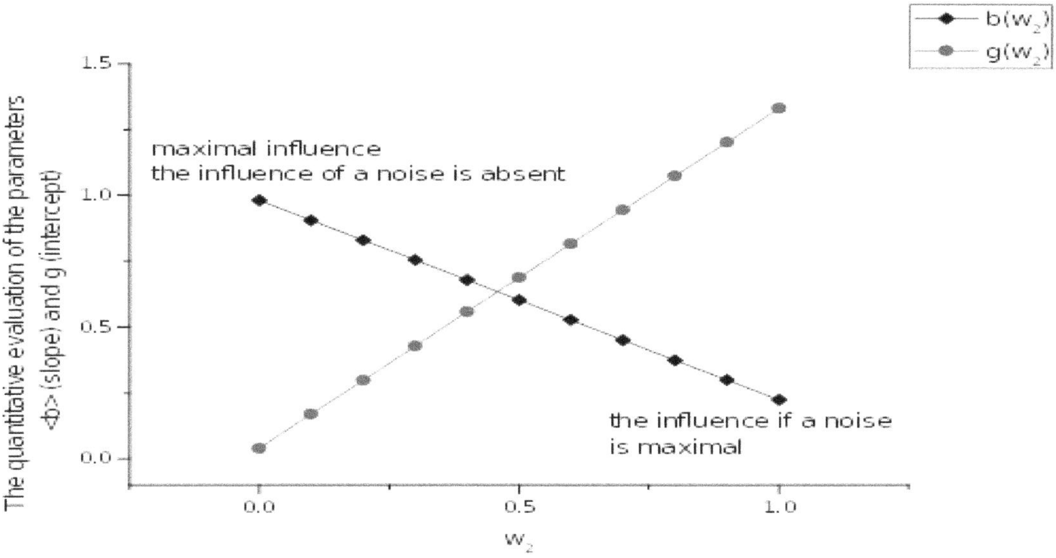

Figure 4.25 In this figure we demonstrate the quantitative evaluation of the parameters obtained with the usage of expression (4.91). In the beginning of the plots (where the influence of a noise is absent) the value of $b(0)$ close to the unit value while the value of $g(0)$ is minimal. The presence of a noise destroys this picture and with increasing of random fluctuations, the sensitivity is gradually lost.

and with increasing of the influence of a noise parameter $g(w_2)$ starts increasing (the corresponding array becomes more uncorrelated because of the strong influence of a noise) and its sensitivity associated with the influence function $b(w_2)$ is lost gradually.

4.6.2 Verification of expressions (4.56) on real data.

As one can notice from the previous analysis, the creation of a stable "pattern"/ideal background plays a *key* role in any original research. If a reliable and reproducible background can be created, then a researcher will obtain a sensitive instrument for detection of a "strange" signal or unexpected perturbation that should be located *out* of the boundaries of the stable/reproducible background. In order to dissipate a skepticism of a potential expert it would be desirable to demonstrate the proposed approach on real data. The author received the voltammograms (VAGs) from a typical electrochemical experiment: to the chosen background solute, some "unknown x" substance for its detection was added. Is it possible to notice its presence or not? The brief description includes the following cases:

Case 1. The background contained 100 measured voltammograms (VAGs) corresponding to 0.1 M(mole) of H_2SO_4 (the detecting electrode is located near the anode branch) in the range [0-1.3] V with the step 0.002V. See Fig. 4.26, where part of the measured VAGs is shown.

Case 2. Background + ("x"=Pd) (the detection electrode was replaced by an-

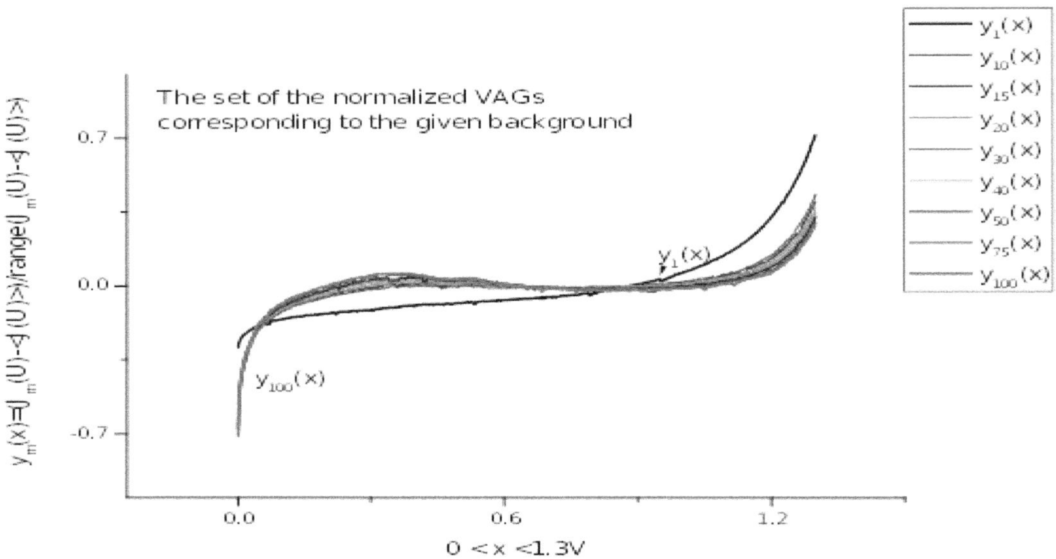

Figure 4.26 Here we demonstrate the array corresponding to the measured background. (See explanation Case 1 for details.)

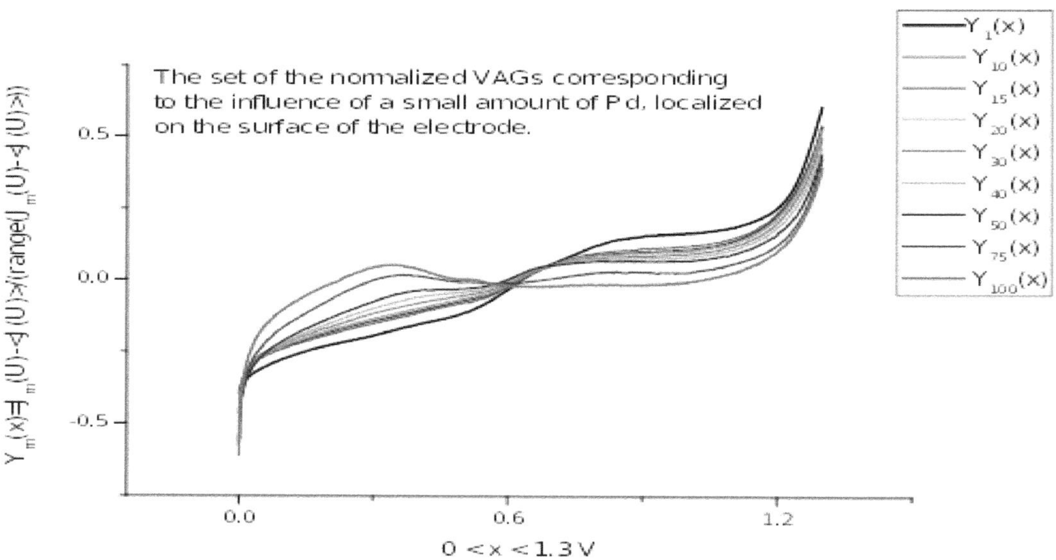

Figure 4.27 The set of the normalized VAGs corresponding to the influence of a small amount of Pd, localized on the surface of electrode (Case 2).

other microelectrode containing Pd metal particles covering its monolayer surface). See Fig. 4.27, where part of the VAGs is shown.

Case 3. The dopamine (DA) serving as "x" substance in concentration 10^{-5}M was added into the same background. The detected electrode in this case remains the same that was used in Case 1. These VAGs are depicted in Fig. 4.28.

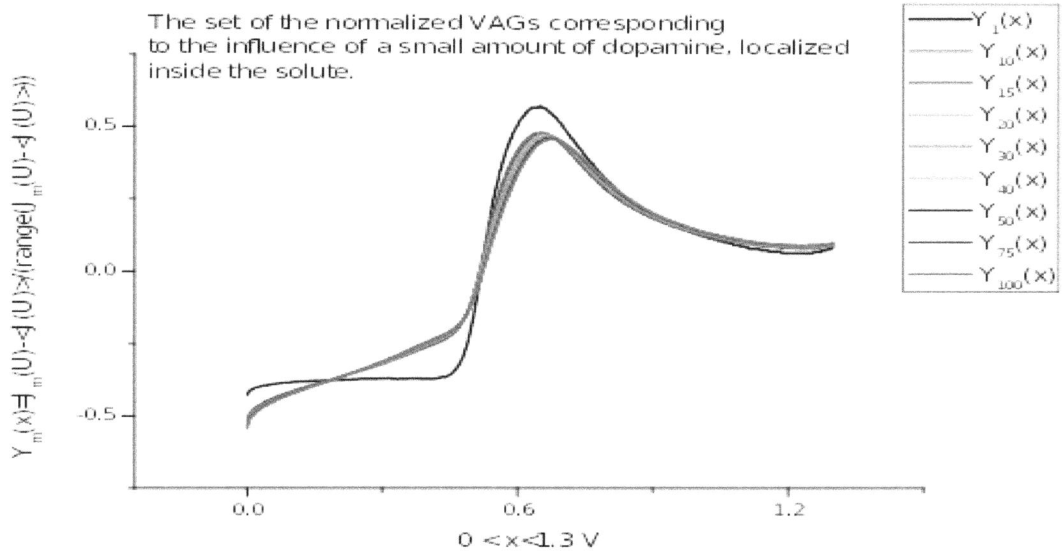

Figure 4.28 The set of the normalized VAGs related to addition of a small amount of dopamine (DA) inside the buffer background (Case 3).

All three experiments were performed in the *same* experimental conditions. It implies that any surrounding parameters as -room temperature (T), humidity (H), and normal pressure (P) are kept in the same intervals.

The problem can be formulated as follows: is it possible to "notice" a presence of "x" substance inside the given background, if each experiment was repeated M=100 times? This repetition is necessary for the creation of some representative and reproducible background array.

In this real experiment we associate the background array with the functions $y_m(x)$ (L=1), where y_m=J_m is the registered current, x= V (applied potential) and arrays $Y_m(x)$ for cases 2 and 3 as affected arrays with the presence of "x" substance. For this case, it is sufficient to use simple expressions (4.56) as in the previous mimic experiment explained above.

The Figs. 4.26–4.31 demonstrate the details that are explained also in the corresponding captions. The most important figures, from our point of view, are Figs. 4.30 and 4.31, where we demonstrate the principal differences between the functions $<G(x)>$ and their mean values $<Y(x)>$ averaged over the arrays containing "x"-factor. The comparison of these functions allows to detect small signals corresponding to the influence of small factor for both cases. The proposed approach forces to reconsider also the experimental results that are obtained for further analysis, when a theoretical hypothesis pretends for the fitting of experimental data is *absent*. We omit the figures demonstrating the corresponding arrays $G_m(x)$ because they are unimportant for this case.

We think that these two examples are sufficient for convincing of a young researcher in importance of the proposed approach. It can be useful for detection of a small "X-factor" that can be present (or unexpectedly appeared) in the array ana-

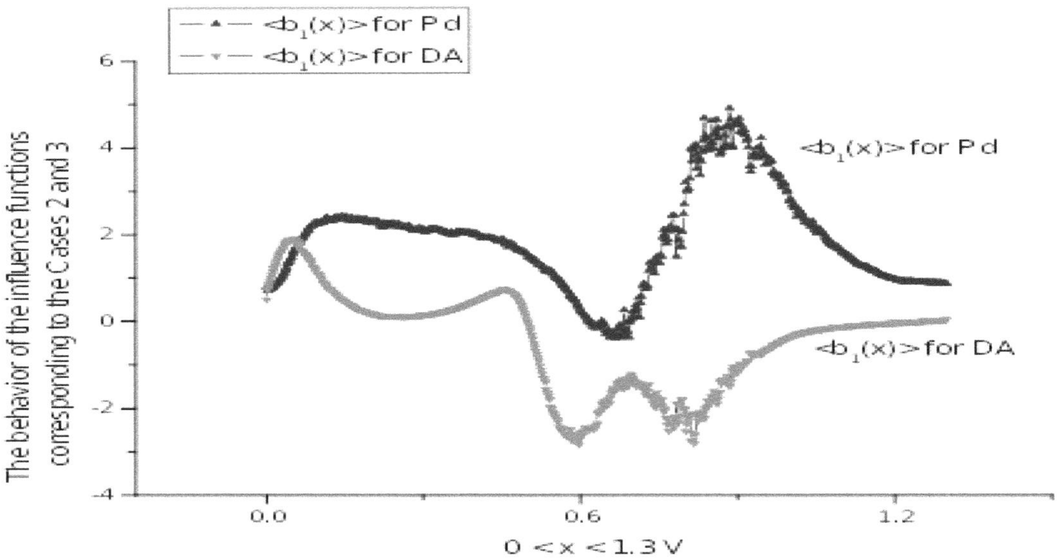

Figure 4.29 Here we show the behavior of the "influence" functions for the both cases. The behavior of these functions are different and these functions can serve as specific "markers" for identification of each case 1 and 2 described in the text.

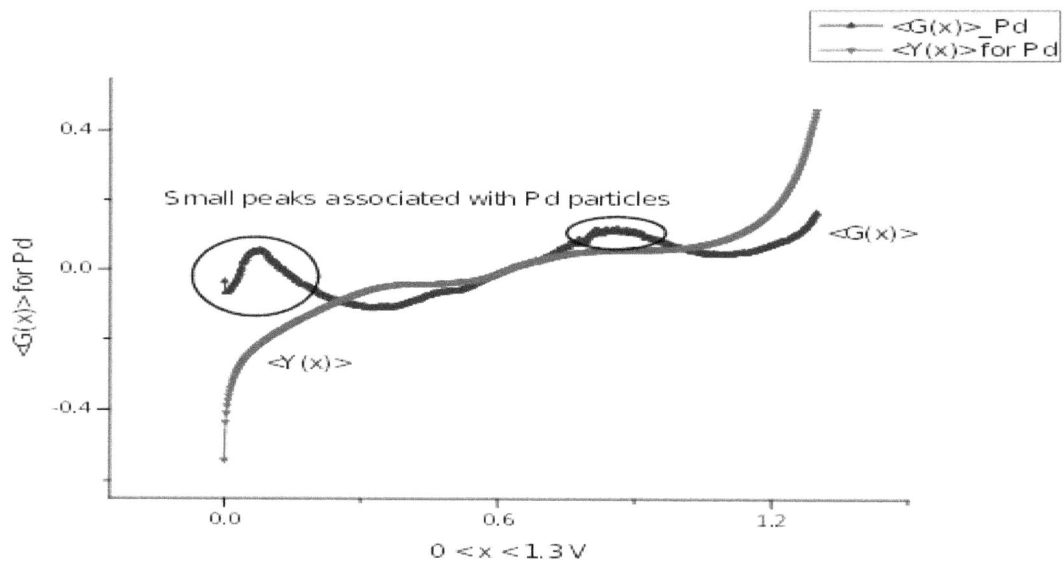

Figure 4.30 Comparison of the remnant function $<G(x)>$ with the mean function $<Y(x)>$ belonging to the VAGs associated with the influence of the Pd metal particles allows to differentiate small peaks corresponding to their influence. They are shown inside black circles.

lyzed. As one can see from these examples the successive measurements (the more the better) are necessary for solution of this problem.

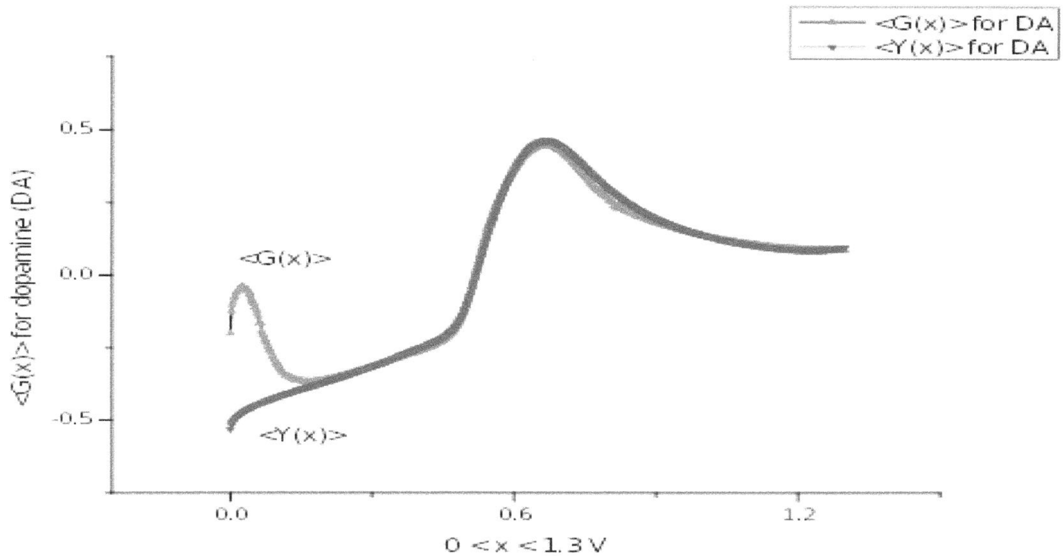

Figure 4.31 Similar comparison the function $<G(x)>$ with the mean function $<Y(x)>$ allows to identify a small peak at small values of the applied potential corresponding to the influence of dopamine.

4.7 PRELIMINARY ANALYSIS OF THE OBTAINED RESULTS

In conclusion of this section, we want to stress the advantages of the proposed approach:

1. We generalized the existing expressions related to analysis of correlations and, in the first time, we obtain the solution in the explicit form that allows to separate *directly* the correlated factor (taken in the form of background array) from a "mixture" of other factors and to obtain the desired uncorrelated/remnant part $G_m(x)$ in the form of an array.

2. Expression (4.44) can be generalized and allows considering the factors having a temporal memory (4.73) and nonlinear factors that are related by the cause-and-effects relations (expressions (4.70) and (4.71)).

3. In the limiting cases the general expression (4.44) is reduced to the linear least squares method (when $<b_l(x)>$ can be replaced by the constant values) and, in another case (when $Y_m(x) \approx < Y(x) >$), it restores directly the uncorrelated/remnant part of the array $G_m(x)$, when the degree of correlations becomes negligible $y_m^{(l)}(t) \cong 0$.

4. The examples based on mimic and real data show and confirm some important details that can be obtained in the result of the proposed approach. Special attention should be paid for the receiving of the stable/robust *background* serving a specific *detector* for registration of any small signal that could appear out of the limits of the background confidence "tube" because of unexpected "x" perturbation. This registered perturbation can help in detection of the desired "signal" that separates this new "quantity" from the random fluctuations containing inside the "pattern" background.

4.8 RELATIVE CHARACTER OF THE CORRELATIONS.

In this last section we want to present the mathematical evidences about the relative character of correlations that follows from the definition of the Pearson's correlations (see expression (4.1)). It becomes possible to transform well-known formulas and receive from the given set of the functions with arbitrary correlations the new set of the functions with the *given* correlations having the value λ. To be more specific we formulate the mathematical part as follows.

Initially, we start from a set of the normalized functions $[f_i(x), (f_i(x) \cdot f_i(x)) = Q_{ii} = 1,\ i = 0, 1, \cdots]$, where $x \equiv \{x_j;\ j = 1, 2, \cdots, N-1\}$ is a discrete variable having an arbitrary set of correlations $(f_i(x) \cdot f_j(x)) = Q_{ii}$. Here the symbol $(fg) = Q_{fg}$ defines the scalar product of a pair of arbitrary functions $f(x)$, $g(x)$ in the N-th dimensional discrete space, i.e.,

$$(f \cdot g) = \sum_{j=1}^{N} f_j g_j \equiv \sum_{j=1}^{N} f(x_j) g(x_j). \tag{4.92}$$

Using the well-known Gram-Schmidt expression

$$\varphi_j(x) = f_j(x) - \sum_{k=0}^{j-1} (f_j \cdot \varphi_k) \varphi_k(x), \tag{4.93}$$

$$\varphi_0(x) = f_0(x), j = 1, 2, ... \tag{4.94}$$

one can derive from initial set of arbitrary functions $f_j(x)$ having different correlations (in the conventional Pearson's sense) the orthonormalized set of functions that satisfy to conditions: $(\varphi_i(x).\varphi_j(x)) = \delta_{ij}$ (where δ_{ij} defines the well-known Kronecker's symbol).

The desired decomposition we present in the following form

$$V_j(x) = \mu_j \varphi_j(x) + \sum_{k=0}^{j-1} P_{jk} \varphi_k(x). \tag{4.95}$$

We should note that the argument $x \equiv \{x_j; j = 0, 1, \cdots, N-1\}$ accepts discrete values. The desired set of the vectors $V_j\ (x)$ should satisfy to condition

$$\{V_i(x) \cdot V_j(x)\} = \begin{cases} 1,\ i = j \\ \lambda,\ i \neq j,\ \lambda \in (-1, 1). \end{cases} \tag{4.96}$$

If one forms the scalar product and multiplies expression (4.95) by the set of orthonormalized vectors φ_j $(\varphi_i(x)\varphi_j(x) = \delta_{ij})$, then we obtain the obvious relationship

$$P_{j,k} = (V_j \cdot \varphi_k),\ k = 0, 1, ..., j, \tag{4.97}$$

$$(V_j \cdot \varphi_j) \equiv \mu_j. \tag{4.98}$$

From expression (4.95), it follows that for the index $k > j$ the scalar product

$$P_{j,k} = (V_j \cdot \varphi_k) = 0,\ k > j \tag{4.99}$$

should equal zero. Therefore, the problem is reduced to calculation of the unknown decomposition coefficients $P_{j,k}$. In order to understand better the structure of these calculated coefficients let us consider the partial cases

For j=0

$$V_0(x) \ = \mu_0\varphi_0(x). \tag{4.100}$$

Forming the scalar product with the function $V_0(x)$, one obtains the expression for the finding of unknown coefficient μ_0

$$\mu_0^2 = 1,\ or\ \mu_0 = \pm 1. \tag{4.101}$$

We keep the negative solution $\mu_0 = -1$ also because both values for μ_0 are acceptable.

For j=1 forming the scalar products with two functions $V_{0,1}(x)$ we obtain

$$V_1(x) = \mu_1\varphi_1(x) + P_{1,0}\varphi_0(x) \tag{4.102}$$

$$P_{1,0} = \lambda\mu_0 \tag{4.103}$$

$$1 = \mu_1^2 + P_{1,0}^2 \Rightarrow \mu_1 = \mu_0\sqrt{1 - P_{1,0}^2}. \tag{4.104}$$

In the same manner, we obtain solutions for j=2,3, forming the scalar products with $V_{0,1,2}(x)$, correspondingly

For j=2

$$V_2(x) = \mu_2\varphi_2(x) + P_{2,1}\varphi_1(x) + P_{2,0}\varphi_0(x), \tag{4.105}$$

$$P_{2,0} = P_{1,0} = \lambda\mu_0,\ P_{2,1} = \frac{\lambda - P_{1,0}^2}{\sqrt{1 - P_{1,0}^2}}, \tag{4.106}$$

$$\mu_2 = \sqrt{1 - P_{2,1}^2 - P_{2,0}^2}. \tag{4.107}$$

For j=3 (one forms the scalar product with 4 functions $V_{0,1,2,3}(x)$ with (4.52))

$$V_3(x) = \mu_3\varphi_3(x) + \sum_{k=0}^{2} P_{3,k}\varphi_k(x), \tag{4.108}$$

$$P_{3,0} = P_{2,0} = P_{1,0} = \lambda\mu_0, \tag{4.109}$$

$$P_{3,1} = P_{2,1} = \mu_0 \frac{\lambda - P_{3,0}^2}{\sqrt{1 - P_{3,0}^2}}, \; P_{3,2} = \frac{\lambda - P_{3,1}^2 - P_{3,0}^2}{\sqrt{1 - P_{3,0}^2 - P_{3,1}^2}}, \tag{4.110}$$

$$\mu_3 = \sqrt{1 - P_{3,0}^2 - P_{3,1}^2 - P_{3,2}^2}. \tag{4.111}$$

Attentive analysis of these partial solutions (4.102–4.108) allows to write down the desired decomposition coefficients for an arbitrary j.

$$V_j(x) = \mu_j\varphi_j(x) + \sum_{k=0}^{j-1} P_{j,k}\varphi_k(x), \tag{4.112}$$

$$P_{j,0} = P_{j-1,0} = ... = P_{1,0} = \lambda\mu_0, \tag{4.113}$$

$$P_{j,1} = P_{j-1,1} = ... = P_{2,1} = \mu_0 \frac{\lambda - P_{j,0}^2}{\sqrt{1 - P_{j,0}^2}}, \tag{4.114}$$

$$P_{j,k} = \mu_0 \frac{\lambda - \sum_{s=0}^{k-1} P_{j-1,s}^2}{\sqrt{1 - \sum_{s=0}^{k-1} P_{j-1,s}^2}}, \; k = 1, 2, ..., j-1, \tag{4.115}$$

$$P_{j,j-1} = \mu_0 \frac{\lambda - \sum_{s=1}^{j-1} P_{s,s-1}^2}{\sqrt{1 - \sum_{s=1}^{j-1} P_{s,s-1}^2}}, \tag{4.116}$$

$$\mu_j = \mu_0 \sqrt{1 - \sum_{s=1}^{j} P_{s,s-1}^2}. \tag{4.117}$$

These expressions solve the problem formulated above and allow to find the set of the functions $V_j(x)$ $(j = 0,1,..)$ for the given correlation factor λ. We should notice also that for the limiting values of λ equal to ± 1 the obtained expressions (4.102–4.112) become singular.

Finishing this chapter, one can conclude that the problem of correlations is not solved completely. The internal correlations have relative character and only comparison these correlations with expression (4.27) associated with external correlations allows to make more justified statements about the nature of correlations that arise between different responses corresponding to different measurements.

BIBLIOGRAPHY

[1] R. R. Nigmatullin, P. Lino, and G. Maione, *New Digital Signal Processing Methods: Applications to Measurement and Diagnostics.* Cham: Springer, 2020. Online ISBN 978-3-030-45359-6.

[2] R. R. Nigmatullin, A. Alekhi, D. Baleanu, E. Dinch, Z. Ustundag, H. Ekshi, and A. Solak, "Analysis of the effect of the potential cycles of the reflected infrared signals of nitro groups in nanofilms: Application of the fractional moments statistics," *Electroanalysis*, vol. 22, no. 4, pp. 419–426, 2010.

[3] R. R. Nigmatullin, S. Osokin, and S. Nelson, "Application of fractional-moments statistics to data for two-phase dielectric mixtures," *IEEE Transactions on Dielectrics and Electrical Insulation*, vol. 15, no. 5, pp. 1385–1392, 2008.

[4] R. R. Nigmatullin, "Strongly correlated variables and existence of the universal distribution function for relative fluctuations," *Physics of Wave Phenomena*, vol. 16, pp. 119–145, 2008.

[5] R. R. Nigmatullin and P. Agarwal, "Direct evaluation of the desired correlations: Verification on real data," *Physica A: Statistical Mechanics and its Applications*, vol. 534, no. 9, p. 121558, 2019.

[6] M. Haghighat, M. Abdel-Mottaleb, and W. Alhalabi, "Discriminant correlation analysis: Real-time feature level fusion for multimodal biometric recognition," *IEEE Transactions on Information Forensics and Security*, vol. 11, no. 9, pp. 1984–1996, 2016.

[7] T. W. Anderson, "Multivariate statistical analysis," *Statistical Science*, vol. 1, no. 3, pp. 334–365, 1986.

[8] M. J. Schervish, "A review of multivariate analysis," *Statistical Science*, vol. 2, no. 4, pp. 396–413, 1987.

[9] T. W. Anderson, *An Introduction to Multivariate Statistical Analysis.* Wiley, 1958.

[10] K. V. Mardia, J. T. Kent, and J. M. Bibby, *Multivariate Analysis.* Academic Press, 1979.

[11] A. R. Feinstein, *Multivariable Analysis.* Yale University Press, 1996.

[12] R. A. Johnson and D. W. Wichern, *Applied Multivariate Statistical Analysis.* Prentice Hall, 2007.

[13] J. L. Schafer, *Analysis of Incomplete Multivariate Data.* CRC Press, 1997.

[14] M. Haghighat, A. Ranjbar, and T. Gholami, "Discriminant correlation analysis: Real-time feature level fusion for multimodal biometric recognition," *IEEE Transactions on Information Forensics and Security*, vol. 11, no. 9, pp. 1984–1996, 2016.

[15] R. R. Nigmatullin, G. Maione, P. Lino, F. Saponaro, and W. Zhang, "The general theory of the quasi-reproducible experiments: How to describe the measured data of complex systems?," *Communications in Nonlinear Science and Numerical Simulation*, vol. 42, pp. 324–341, 2017.

[16] R. R. Nigmatullin, "Detection of quasi-periodic processes in experimental measurements by approximation of correlation function with the fractional exponential function," in *Complex Motions and Chaos in Nonlinear Systems*, vol. 15 of *Nonlinear Systems and Complexity*, pp. 1–37, Cham: Springer, 2016.

[17] R. R. Nigmatullin, V. A. Toboev, P. Lino, and G. Maione, "Reduced fractal model for quantitative analysis of averaged micromotions in mesoscale: Characterization of blow-like signals," *Chaos, Solitons & Fractals*, vol. 76, pp. 166–181, 2015.

[18] R. R. Nigmatullin, C. Ceglie, G. Maione, and D. Striccoli, "Reduced fractional modeling of 3D video streams: the FERMA approach," *Nonlinear Dynamics*, vol. 80, no. 4, pp. 1869–1882, 2015.

[19] V. V. Uchaikin, *The method of the fractional derivatives.* Artishok" Publishing house, 2008.

[20] B. D. Coleman and D. R. Owen, "On the thermodynamics of materials with memory," *Archive for Rational Mechanics and Analysis*, vol. 36, no. 4, pp. 245–269, 1970.

[21] I. Prigogine and P. Resibois, "On the kinetics of the approach to equilibrium," *Physica*, vol. 27, pp. 629–636, 1961.

[22] R. Zwanzig, "Statistical mechanics of irreversibility," in *Lectures in Theoretical Physics, Volume III*, pp. 106–141, New York: Interscience Publishers, 1961.

[23] K. B. Oldham and J. Spanier, *The Fractional Calculus.* Academic Press, 1974.

[24] K. S. Miller and B. Ross, *An Introduction to the Fractional Calculus and Fractional Differential Equation NY.* John Wiley & Sons, Inc, 1993.

CHAPTER 5

Generalized Methods Using Fractional Calculi

In this chapter, based on the "fuzzy" calculus covering the continuous range of operations between two couple of arithmetic operations (+, −) and (**x**, **:**), it becomes possible to propose the new form of the fractional integral occupying an intermediate position between the integral and derivative of the first order. This new form of the fractional integral satisfies the C_1 criterion according to the Ross classification [1]. The new calculus tightly related with the *continuous* values of S = 1 and can generalize the shifting expression for the fractional values of the shifting discrete index. This calculus can be interpreted as the appearance of the hidden states corresponding to *unobservable* values of the continuous spin S=1. Many well-known formulas can be generalized and receive new extended interpretation. In particular, one can factorize any rectangle matrix and receive the "perfect" filtering formula allowing transforming *any* function (deterministic or random) to another arbitrary function and vice versa. This transformation can find unexpected applications in data transmission, cryptography and calibration of different gadgets and devices. One can receive also the hybrid ("centaur") formula for the Fourier transformation unifying the both expressions for the direct in inverse F-transformations in one mathematical unit. The generalized Dirichlet formula, obtained in the frame of new calculus allowing selecting the desired resonance frequencies, will be useful also in discrete signals processing. The basic formulas are tested numerically on mimic data.

We would like to underline also that the modified formulas are free from treatment errors and model suppositions. They extend any arithmetic specific expression and make it *uncertain* that in some cases this uncertainty becomes useful and profitable.

We decided to include also another important part in this chapter related to possible identification of the fractional integral in experimental observations. It is instructive to know that the fractional integral taken in the form of Riemann-Liouville can be realized in nature. The prove of this assessment is not simple, however, we address this part to the readers in order to show "a specific kitchen" where an attentive reader will know about some instruments *how* to realize this statement.

Besides this essential part related to the generalization of the arithmetic operators we decided to include the Secs. 5.5–5.9 related to experimental confirmation of the

DOI: 10.1201/9781003794691-5

Riemann-Liouville fractional integral in the living systems having branching/fractal structure with respect to conducting channels. As an appropriate object for this purpose we chose fresh mushrooms. The complex impedance $Z(j\omega)$ measured for this object during long period of time (93 experiments covering approximately the period of one week) completely confirmed the realization of the R-L integral for description of the complex integral in the frequency domain. This could also be interesting to those readers who wish to enrich them by new and useful knowledge about living systems.

5.1 HOW TO FORMULATE A PROBLEM?

In many cases the mathematics as a branch of natural science is developed in a way of its generalization. In the end of the 17th century French mathematician G.F. L'Hôpital started to think about the generalization of the concept of number for differentiation/integration operators to non-integers or fractions. Fractal geometry (FD) proposed by B. Mandelbrot was appeared after generalization of the concept of the integer Euclidean dimension. Now these generalizations became wide-acknowledged and accepted in many applications. However, as one can see after reading of this chapter the generalization related to fractional integral and derivative contains some new features that can be useful in many applications.

In nowadays the fractional calculus can be considered as the well-developed branch of mathematics, having the numerous applications. However, starting from the key books [2, 3, 4, 5, 6, 7, 8] the basic form of the fractional integral occupying the intermediate position between integer integrals of different orders and the corresponding derivatives not established yet. In our century, this field of mathematics remains the "hot" spot, and many researchers [9, 10, 11, 12, 13, 14, 15, 16] actively working in this field suggested a wide family of mathematical expressions for fractional integrals that creates an additional uncertainty in its definition. The key papers and their careful analysis are collected in the review [1] that covers the recent history and the progress in this field.

Being expressed in different mathematical forms, these fractional operators do *not* have clear physical/geometrical meaning that create some additional obstacles in their practical applications and in creation of some devices that perform the desired and clear operation in treatment of real signals. From our point of view, only the fractional integrals with clear physical meaning will survive and receive a wide propagation among practitioners and engineers. However, the first classification of the different fractional operators and their separation in four classes/criteria $C_{1,2,3,4}$ merits every kind support [1]. This classification will help researchers understand deeper the nature of these complicated operators and extend their understanding among researchers that deal with the application of the fractional mathematics in devices and discovering the operations that can be realized in nature.

From our point of view, only the obvious simplicity in their realization, clear physical meaning, including the understanding of their fractional complex-conjugated values [17], possible relationships with fractal objects [18, 19] will help to wide propagation of this specific (modified in most cases) fractional operators among engineers and researches. Therefore, there is a necessity to suggest some simple forms of the fractional integrals occupying some intermediate position between the first order integer integral and derivative that will be used with certain success in treatment of different data and embedded into different devices.

In the first part of this chapter based on some new principles of the proposed "fuzzy" calculus, occupying an intermediate position between the conventional (+,–) and (**x**, **:**) arithmetic operations, one can derive the fractional operator of new type that differs from the Grunwald-Letnikov definition [2, 3, 4, 5, 6, 7, 8]. It is defined on the discrete set of variables (y_j,x_j ; j=1,2,...,N) and realize the "fuzzy" addition (+) and multiplication (**x**) operators over the field of some mathematical objects, including numbers, vectors, integral and differential operators and matrices. It becomes possible to show that this operator is tightly related with the *continuous* spin (S=1) operator that can accept some continuous set of eigen values from the interval [-1,1] being located in some uncertain states. If spin is subjected to some physical measurement, then it accepts three discrete and certain values (0,±1) only as it is supposed in the quantum mechanics. Based on this simple generalization, one can derive many new formulas that will be useful in the signal processing area and control, where the authors of this book work actively. In particular, one can propose the "perfect" filtering formula that transform *any* deterministic/random function to another arbitrary function and vice versa. Besides this useful expression, an attentive reader will find an expression for the hybrid Fourier transformation formula and the generalized Dirichlet function that can find also their corresponding place in applications.

The rest part of this chapter is organized as follows: In Sec. 5.2, we give the basic expressions for the "fuzzy" calculus and propose new form for the fractional integral. Verification with the criteria listed in paper [1] shows that the proposed fractional integral belongs to the C_1 class. In this part, we show also how to apply this calculus to the known fractional integrals and give useful expression for the "fractional" shifting. We do suppose that this expression will be widely used in treatment of different data. We want to stress here that this formula follows from the factorization of an arbitrary matrix that is realized with the help of the "fuzzy" calculus.

In Sec. 5.3, we test numerically the basic expressions on mimic data. In Sec. 5.4, we repeat in brief the basic results and give the expression for the hybrid F-transform and Dirichlet formula that unify the both transformations (direct and inverse) in one mathematical structure. This interesting hybrid can be applied alongside with other integral transformations as well.

Remaining Secs. 5.5–5.9 form a separate part and related to the experimental confirmation of the fractional R-L integral in living systems.

5.2 THE BASIC OF THE "FUZZY" CALCULUS

We determine the "fuzzy" addition operator, occupying the intermediate position between the conventional addition (+) and subtraction (–) operators by the following equality:

$$Q(\alpha)(A \cap B) \equiv A(Q(\alpha)B). \tag{5.1}$$

Actually, the action of $Q(\alpha)$ is reduced to the multiplication of monotonically increasing/decreasing function from the interval $[-1,1]$ that is applied *only* to an object B located on the right-hand side. Here and below the operator $Q(\alpha)$ can be defined as the *functional operator* because this functional unit initially serves as an operator and finally its action is reduced to multiplication of the function $Q(\alpha)$. We require that $Q(1)=1$ (that corresponds to the addition operation), $Q(0)=0$ that corresponds to the absence of operation over object B (we stress here the difference $A(Q(0)B)=A$, not to zero as it might expect!) and finally $Q(-1)=-1$ that corresponds to subtraction operation between objects A and B. As an object subjected to a "fuzzy" addition we imply numbers, vectors, matrices, integral/differential operators, and their possible combinations. Similarly, one can determine the "fuzzy" multiplication

$$Q(\alpha)(A \cap B) \equiv A\left(B^{Q(\alpha)}\right). \tag{5.2}$$

This operation corresponds to multiplication at $Q(1)=1$, division at $Q(-1)=-1$ and the absence of operation over object B at $Q(0)=0$. As the functional operation $Q(\alpha)$, one can take a monotonically increasing function located in the first and third quarters of the square $[-1 \le \alpha \le 1\ ,\ -1 \le Q(\alpha) \le 1\ ]$. For decreasing functions, one can assign simply the opposite sign and consider this family of functions in the second and the fourth quarters, correspondingly. Below, for simplicity we consider only the increasing family of $Q(\alpha)$ functions. The decreasing part can be considered similarly by the changing of the sign $(+ \to -)$ before operator $Q(\alpha)$. In particular, one can take the function:

$$Q(\alpha) = \frac{\alpha}{|\alpha|} F(\alpha) \tag{5.3}$$

where $F(\alpha)$ determines *any* even function from the interval [-1,1]. For specific demonstration, we choose $F(\alpha) = |\alpha|^{\theta}$ with different values of θ from interval $[0.05,\ 10]$. The family of these functions is shown in Fig. 5.1.

We stress here that this operator should satisfy the requirement $Q(\pm 1) = \pm 1$, $Q(0) = 0$ and, in other aspects, it can be arbitrary and accept nonzero values from the interval (-1,1). This requirement reminds the spin operator $S=1$ with eigen values ($S_z = 0, \pm 1$). As it is known from the matrix theory [20, 21] any smooth function depending on a discrete operator can be decomposed on a polynomial of the (n-1)-th order if an initial matrix operator has n different eigen values. In particular, for $S=1$ we have the polynomial of the second order

$$F(S_z) = A_0 + A_1 S_z + A_2 S_z^2. \tag{5.4}$$

The constants $A_{0,1,2}$ are determined by the eigen values of the function $F(S_z)$. Therefore, for $S=1$ we have

$$\begin{aligned} A_0 &= F(0),\ A_1 = \frac{1}{2}(F(1) - F(-1)), \\ A_2 &= \frac{1}{2}(F(1) + F(-1) - 2F(0)). \end{aligned} \tag{5.5}$$

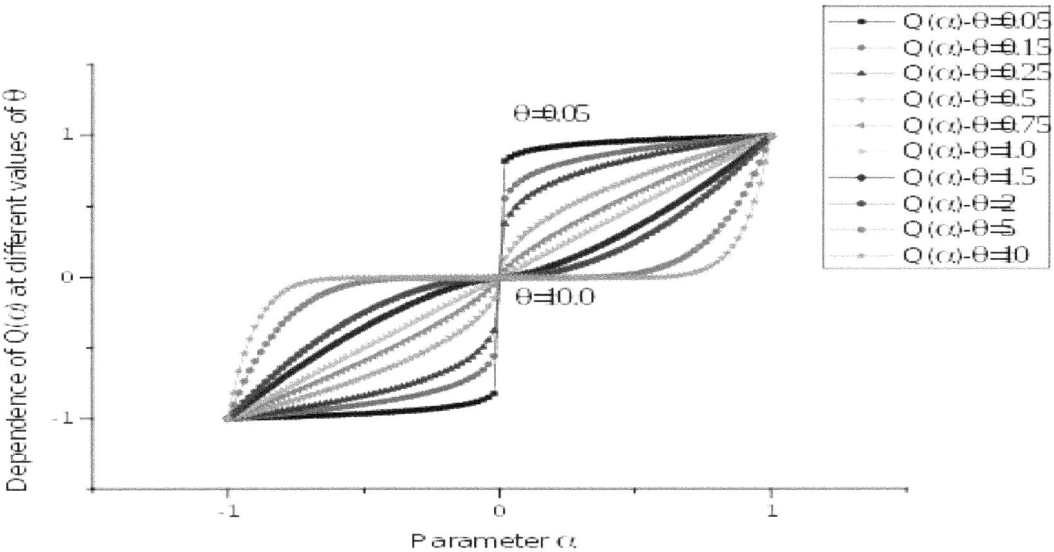

Figure 5.1 The family of the curves calculated for the increasing branch $Q(\alpha) = \frac{\alpha}{|\alpha|}|\alpha|^{\theta}$ and different values of parameters $\alpha \in [-1,1]$ and $\theta \in [0.05,10]$. These parts play different roles. The branch $\alpha \in [-1,0]$ represents itself the smoothing branch, while the second part $\alpha \in [0,1]$ detects the HF (high frequency) fluctuations. The decreasing parts of these curves occupying the second and the fourth quarters are not shown.

These relationships can be generalized if we accept the conception of the *continuous* spin and write the relationship (5.4) as a *postulate* for *any* monotonically increasing/decreasing functional operator $Q(\alpha)$ identifying it with the continuous spin operator. Therefore, we *postulate* that

$$F(Q(\alpha)) = A_0 + A_1 Q(\alpha) + A_2 (Q(\alpha))^2 \tag{5.6}$$

where $Q(\alpha)$ is a monotonically increasing/decreasing function having the properties determined above. Strictly speaking, here and below the equality sign is valid only for the limiting values of $(Q(\alpha) = 0,\pm 1)$. However, we imply that this equality is valid for *all* values of the chosen $Q(\alpha)$ from the interval $[-1,1]$. This postulate reminds the restoration of a polynomial of the second order if three points (in our case the values $A_{0,1,2}$ are known. Based on relationship (5.6) it is easy to show that it commutes with an arbitrary matrix B, i.e.,$\hat{Q}(\alpha)\hat{B} = \hat{B}Q(\alpha)$, where on the right-hand side the functional operator $Q(\alpha)$ loses its operator properties.

In particular, the definition (5.2) in accordance with the basic postulate (5.6) can be rewritten as:

$$AB^{Q(\alpha)} = A + \frac{1}{2}(AB - AB^{-1})Q(\alpha) + \frac{1}{2}(AB + AB^{-1} - 2A)(Q(\alpha))^2 \tag{5.7}$$

The last expression simplifies considerably all algebraic manipulations with the functional operator $Q(\alpha)$. The postulate (5.6) admits the following interpretation if we identify a "fuzzy" operation with continuous spin. In any *uncertain* state the quantum spin operator (S=1) accepts the continuous values located inside the square $[-1 \leq \alpha \leq 1,\ -1 \leq Q(\alpha) \leq 1\ ]$. However, being subjected to any certain measurement it accepts only one eigen value from three possibilities $[0,\pm 1]$ in accordance with the rules of the quantum mechanics. We think that this analogy allows generalizing some class of the Hermitian operators having discrete set of eigen-values eigenvalues for continuous cases. For this generalization, it is necessary to replace the discrete eigenvalue of some Hermitian operator H by its continuous function:

$$\begin{gathered} F(H(\alpha)) = A_0 + A_1\lambda(\alpha) + ... + A_{n-1}[\lambda(\alpha)]^{n-1}, \\ [4pt] \min(\lambda) < \lambda_k(\alpha) < \max(\lambda),\ for\ all\ k = 1,2,...,n. \end{gathered} \tag{5.8}$$

In decomposition (5.8) all eigen values λ_k (k=1,2,...,n) are supposed to be different. We do not analyze this generalization (5.8) because it needs more justified arguments and verification on real data. However, this analogy merits further research if it becomes possible to establish this relationship with another class of discrete operators used in the single processing area and control as it has been done for a couple of "fuzzy" arithmetic operators tightly related with continuous spin S=1.

If $A_0 \neq 0$, then operator (5.6) admits the existence of the *inverse* operator. This operator is found from the condition:

$$\begin{aligned} & Inv(Q(\alpha)) \equiv \left(Q(\alpha)^{-1}\right) = \frac{1}{A_0 + A_1 Q(\alpha) + A_2[Q(\alpha)]^2} = B_0 + B_1 Q(\alpha) + B_2[Q(\alpha)]^2, \\ [4pt] & B_0 = \frac{1}{A_0} = Inv(0),\ B_1 = \frac{1}{2}(Inv(1) - Inv(-1)), \\ & B_2 = \frac{1}{2}(Inv(1) + Inv(-1) - 2 \cdot Inv(0)). \end{aligned} \tag{5.9}$$

After these basic expressions, one can pose the following question: how to find the fractional operator (occupying an intermediate position between the integral and derivative of the first order) in the frame of the "fuzzy" calculus described above?

We consider the discrete expressions for the integral and derivative defined on some discrete set x_j, y_j (j=1,2,...N) of numbers. We subject these conventional operators to the following requirements:

$$\begin{aligned} & Q(-1)J_j = J_j = J_{j-1} + \frac{1}{2}(x_j - x_{j-1}) \cdot (y_j + y_{j-1}), \\ & Q(1)J_j \equiv D_j = \frac{(y_j - y_{j-1})}{(x_j - x_{j-1})},\ Q(0)J_j = 0. \end{aligned} \tag{5.10}$$

Based on the properties defined by requirements (5.1) and (5.2) the action of the operator $Q(\alpha)$ one can present in the form

$$Q(\alpha) J_j = \left(-\frac{1}{2}Q(\alpha) + \frac{1}{2}[Q(\alpha)]^2\right) J_{j-1} + \tag{5.11a}$$

$$\left(\frac{1}{4}Q(\alpha) + \frac{3}{4}[Q(\alpha)]^2\right) \cdot (x_j - x_{j-1})^{-Q(\alpha)} \cdot (y_j - Q(\alpha) y_{j-1}). \tag{5.11b}$$

After taking into account the postulate (5.6) this expression can be simplified and presented as

$$F(Q(\alpha)) J_j \equiv J_j^{\alpha} = \frac{1}{2}(D_j - J_j) Q(\alpha) + \frac{1}{2}(D_j + J_j) Q^2(\alpha) \tag{5.11c}$$

Here the conventional integration and differentiational operators are defined by expression (5.10). These relationships (5.11) can be considered as a natural generalization of the previous limiting cases (5.10) and it can be defined as the *fractional* integral of a new type differed from the Grunwald-Letnikov definition [1, 2, 3, 4, 5, 6, 7, 8]. Based on relationship (5.7) one can simplify the difference $(x_j - x_{j-1})^{-Q(\alpha)}$ and present this operation in the form of the triad (5.6), as well:

$$\begin{aligned} &(\Delta x_j)^{-Q(\alpha)} \equiv 1 + A_1 Q(\alpha) + A_2 \left(Q(\alpha)^2\right), \\ &A_1 = -\frac{1}{2}\left(\Delta x_j - \frac{1}{\Delta x_j}\right), \; A_2 = \frac{1}{2}\left(\Delta x_j + \frac{1}{\Delta x_j}\right) - 1. \end{aligned} \tag{5.12}$$

This relationship helps to obtain the simple relationship (5.11c) from (5.11b). We recommend to a potential reader to obtain (5.11c) from (5.11b) as an exercise.

If we use the criteria given in paper [1], then it is easy to see that expression (5.11) belongs to the class C_1. However, one can extend the applicability of the fractional integral if we consider another general expression:

$$\begin{aligned} &J^{-\nu} D^{Q(\alpha)\nu} (f(x) \cdot g(x)) = \\ &\left[J^{-\nu} + \frac{1}{2}\left(1 - J^{-2\nu}\right) Q(\alpha) + \frac{1}{2}\left(1 - J^{-\nu}\right)^2 (Q(\alpha))^2\right] (f(x) \cdot g(x)). \end{aligned} \tag{5.13}$$

In expression (5.13), we use the following notations: $J^{-\nu}$ is a fractional integral, D^{ν} is a fractional derivative. It is supposed that both operators in (5.13) belong to the $C_{1,2}$ classes, correspondingly. We imply also that combination $J^{-\nu} D^{\nu}=1$ forms the *unit* operator. If we deliberately not to open the product of the functions located on the right-hand side then, having in mind the specific opening of the product in (5.13) in accordance with the general Leibnitz rule [1], one can identify the appliance of expression(5.13) to the classes $C_{1,2}$ as well.

By analogy with this expression, one can write the fractional integral belonging to the $C_{1,2}$ classes in the following form:

$$J^{\nu+Q(\alpha)}f(x) = J^{\nu}f(x) + \frac{1}{2}\left(J^{\nu+1} - J^{\nu-1}\right)f(x)Q(\alpha) + \frac{1}{2}\left(J^{\nu+1} + J^{\nu-1} - 2\cdot J^{\nu}\right)f(x)(Q(\alpha))^{2}. \tag{5.14}$$

As it follows from the previous speculations the functional operator $Q(\alpha)$ loses its operational properties and on the right-hand side it is transformed to usual multiplier.

Another interesting generalization will be related to the fractional shifting formula that can be applied to analysis of discrete data:

$$E_{k+Q(\alpha)} \equiv E_k + \frac{1}{2}(E_{k+1} - E_{k-1})Q(\alpha) + \frac{1}{2}(E_{k+1} + E_{k-1} - 2E_k)(Q(\alpha))^2. \tag{5.15}$$

If we associate the letter E with energy of some quantum spectrum having three discrete levels $E_{k\text{-}1,k,k+1}$, then operator $Q(\alpha)$ receives the following interpretation. It can be interpreted as a contribution of *unobservable* energy states covering the energy interval $E_k = E_{k+1} - E_{k\text{-}1}$. Idea of the fractional shifting has general character and can be realized for any rectangle matrix:

$$M_{i+Q(\alpha),j} \equiv M_{i,j} + \frac{1}{2}(M_{i+1,j} - M_{i-1,j})Q(\alpha) + \frac{1}{2}(M_{i+1,j} + M_{i-1,j} - 2M_{i,j})(Q(\alpha))^2,$$
$$i = 0,1...,I+1,\ j = 1,2,...,J, \tag{5.16}$$

and for the generalized mean value function that is used in the statistics of the fractional moments [22, 23]:

$$G_{n+Q(\alpha)} \equiv \left[\frac{1}{N}\sum_{j=1}^{N}|\Delta x_j|^{n+Q(\alpha)}\right]^{\frac{1}{n+Q(\alpha)}} = G_n + \frac{1}{2}(G_{n+1} - G_{n-1})Q(\alpha) + \frac{1}{2}(G_{n+1} + G_{n-1} - 2G_n)(Q(\alpha))^2,$$
$$\Delta x_j = x_j - \frac{1}{N}\sum_{j=1}^{N}x_j . G_0 = \exp\left(\frac{1}{N}\sum_{j=1}^{N}\ln|\Delta x_j|\right) \equiv \prod_{j=1}^{N}(|\Delta x_j|)^{\frac{1}{N}} - \text{geometric mean}. \tag{5.17}$$

In expression (5.17), G_0 should be considered as the *limiting* case, which after the taking of the corresponding limit ($\lim_{\varepsilon\to 0} G_\varepsilon = G_0$) it coincides with geometric mean.

It is easy to find also the functional operator for a fractal object having some scaling properties. If we subject equation (5.3) to the requirement

$$Q(\alpha\xi) = bQ(\alpha), \tag{5.18}$$

then one can receive the following relationship:

$$Q(\alpha) = \left(\frac{\alpha}{|\alpha|}\right) \cdot |\alpha|^{\theta} = \alpha \cdot |\alpha|^{\theta-1}, \ \theta = \frac{\ln(b)}{\ln(\xi)} > 0. \tag{5.19}$$

We omit here the log-periodic additions because they contain the uncertain oscillations of the functions $\cos(\ln(\alpha)/\ln(\xi))$ and $\sin(\ln(\alpha)/\ln(\xi))$ in the point α =0.

Concluding this section, it is useful to derive the "perfect" filtering formula that can transform *any* arbitrary function (deterministic or random) to another function and vice versa. This expression can be obtained easily from the expression factorizing (dividing) the given matrix. Let us have a rectangle matrix $M(j,m)$ with j rows and m columns: (j=1,2,...,N), m=1,2...,M ($M \leq N$) that is needed to be factorized relatively the indexes (j,m). In accordance with postulate (5.6), one can write:

$$M(j,m) = M_0(j) + M_1(j)Q(\alpha_m) + M_2(j)[Q(\alpha_m)]^2. \tag{5.20}$$

Let us *suppose* that the functions related to discrete values of M are known i.e., $Yup(x_j)$, $Ydn(x_j)$ and $f_c(x_j)$ correspond, accordingly, to the values of $Q(\alpha) = (\pm 1,0)$. From expression (5.5), we have the formula factorizing an array $Y_{j,m}$:

$$\begin{aligned} Y(x_j, v_m) &= f_c(x_j) + \frac{1}{2}(Yup(x_j) - Ydn(x_j))Q(v_m) \\ &+ \frac{1}{2}(Yup(x_j) + Ydn(x_j) - 2f_c(x_j))(Q(v_m))^2. \end{aligned} \tag{5.21}$$

This expression can transform *any* function $Ydn(x_j) \leftrightarrow Yup(x_j)$ through the intermediate/control function $f_c(x_j)$. We should stress again that these functions can have *any* nature i.e., coincide with deterministic function or represent themselves a cloud of random points. As a partial case, one can identify $Yup(x_j)$ with max(Y_j), $Ydn(x_j)$ with min(Y_j) and $f_c(x_j)$ with mean(Y_j), correspondingly. Having these functions, we complete "a gap" between three basic functions that reflect the procedure of the reduction to three incident points considered in the previous chapters. If the functions $Yup(x_j)$, $Ydn(x_j)$, and $f_c(x_j)$ have different number of discrete points then we should choose the most representative/long sampling and fill short samplings by a set of zeroes. In expression (5.21), we chose *another* variable v_m that is related with the variable α_m as:

$$\begin{aligned} &\alpha_m = -1 + 2F(v_m), \\ &0 \leq F(v_m) \leq 1, \ v_m = \frac{m}{M}, \ m = 0, 1, ...M-1, M. \end{aligned} \tag{5.22}$$

In expression (5.22) $F(v_m)$ is an arbitrary function located in the interval [0,1]. The zero value for α is chosen from the condition $F(v_{\bar{m}}) = \frac{1}{2}$. In paper [24] and in Chapter 3 different expressions for $F(v_m)$ were used for enrichment of "poor" samplings. In particular, one can choose the function $F(v) = 3v^2 - 2v^3$ [24] that satisfies to the given conditions - $F(0)$=0, $F(1)$=1, $F(1/2)$=1/2.

Attentive analysis shows that expression (5.21) can be generalized for more complex objects, in particular for tensors having n indexes. Let us determine the n-th index tensor $T(i_1,i_2,\ldots,i_n)$ and suppose that the limits of this tensor together with its mean value for the chosen index i_1 are known, i.e.,

$$T\left(\lambda_{\max}\left(i_i\right),\left(i_2,i_3,\ldots,i_n\right)\right),\ T\left(\lambda_{mean}\left(i_i\right),\left(i_2,i_3,\ldots,i_n\right)\right),\ T\left(\lambda_{\min}\left(i_i\right),\left(i_2,i_3,\ldots,i_n\right)\right). \tag{5.23}$$

The factorization procedure applied to the tensor similar to expression (5.21) allows receiving the following formula:

$$\begin{aligned} T\left(\left[i_1,v_m\right],\left(i_2,\ldots,i_n\right)\right) &= T\left(\lambda_{mean}\left(i_i\right),\left(i_2,\ldots,i_n\right)\right) \\ &+\frac{1}{2}\left[T\left(\lambda_{\max}\left(i_i\right),\left(i_2,\ldots,i_n\right)\right)-T\left(\lambda_{\min}\left(i_i\right),\left(i_2,\ldots,i_n\right)\right)\right]Q\left(v_m\right) \\ &+\frac{1}{2}\left[T\left(\lambda_{\max}\left(i_i\right),\left(i_2,\ldots,i_n\right)\right)+T\left(\ldots\right)-2\cdot T\left(\lambda_{mean}\left(i_i\right),\left(i_2,\ldots,i_n\right)\right)\right]\left(Q\left(v_m\right)\right)^2. \end{aligned} \tag{5.24}$$

The function $Q(v_m)$ in (5.24) is defined by the previous expression (5.23). This expression allows factorizing the multidimensional matrices and can be useful and applicable in analysis of different "BIG DATA" sets. Expression (5.24) can be considered as a recurrence formula if for the index i_f the limits $\left(\lambda_{\min}\left(i_f\right),\lambda_{mean}\left(i_f\right),\lambda_{\max}\left(i_f\right)\right)$ are known *a priory*.

As it has been mentioned above, the presentation of the function $F[Q(\alpha)]$ in the form of triad (5.6) is very convenient and simplify many calculations. In particular, let us consider a polynomial of the n-th order. For it, we have

$$\begin{aligned} &P\left(Q\left(\alpha\right)\right)=\left(Q\left(\alpha\right)\right)^{2n}+a_{2n-1}\left(Q\left(\alpha\right)\right)^{2n-1}+\ldots+a_0 \\ &=a_0+\frac{1}{2}\left(P(1)-P(-1)\right)Q\left(\alpha\right)+\frac{1}{2}\left(P(1)+P(-1)-2a_0\right)\cdot\left(Q\left(\alpha\right)\right)^2. \end{aligned} \tag{5.25}$$

Expression (5.25) is valid also for any smooth function that depends on the functional operator $Q(\alpha)$. In similar manner, one can consider expressions that are a bit more complicated:

$$S(\alpha)=\sum_{n=0}^{N}\left(a_n+Q\left(\alpha\right)\left(b_n\right)^{Q(\alpha)}\right),\ \Pr\left(\alpha\right)=\prod_{n=0}^{N}a_n\left(b_n\right)^{Q(\alpha)}. \tag{5.26}$$

They can be presented in the form of triads (5.6), as well. These presentations we leave as an exercise for an attentive reader.

5.3 VERIFICATION OF THE BASIC EXPRESSIONS IN MIMIC DATA

5.3.1 Verification of expression 5.11c for the fractional integral

In this section, we want to test expression (5.11c) for the chosen mode that is subjected to the influence of a uniform noise. We chose for testing the following expression:

$$Y_j=A\cos(\Omega\cdot x_j-\varphi)+Rnd(x_j,-1,1) \tag{5.27}$$

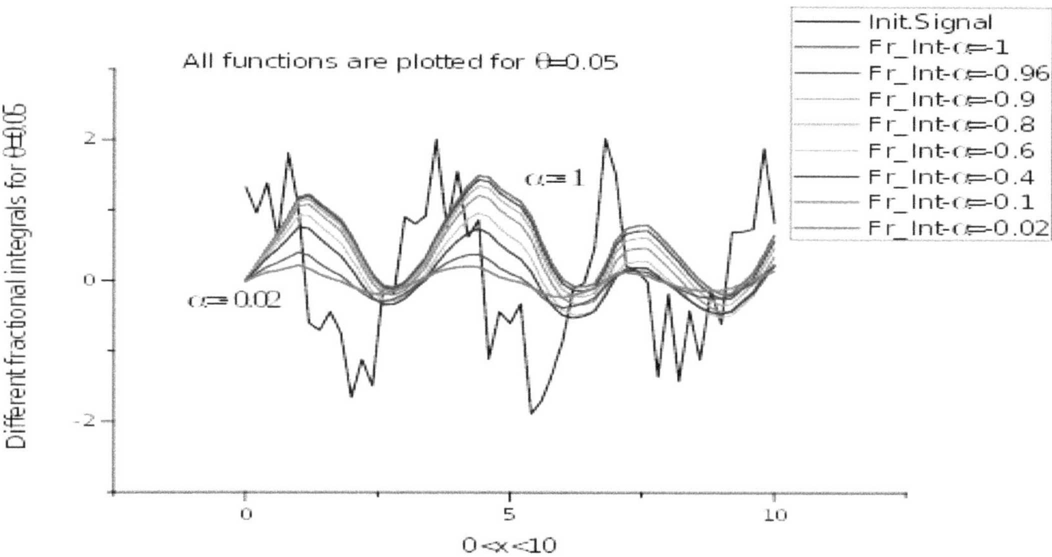

Figure 5.2 Here we demonstrate a family of curves belonging to the smoothing part from the region $[-1,-0.02]$ for small value of the parameter θ =0.05. The black bold line shows the initial curve.

where A=1.3, Ω =3, $x\in[0, 10]$, j=1,2,..., N=100, $Rnd(x,-1,1)$ is uniformly distributed random generator that generates a set of random numbers in the interval $[-1,1]$. As the functional operator $Q(\alpha)$ we chose expression (5.3) with $F(\alpha) = |\alpha|^{\theta}$ for three values of parameter θ =0.05, 1, 5. Interval for α is divided on two parts $\alpha \in [-1,0]$ and $[0,1]$. In the first interval, the fractional integral smoothies the given data, while in the second interval, it becomes close to derivative and exhibits the selection of the hidden HF fluctuations. The plots and necessary explanations are given in Figs 5.2 (a,b,c) ($\alpha \in[-1,0]$, θ =0.05, 1, 5); and Figs. 5.5 (a,b,c) for ($\alpha \in[0,1]$ θ =0.05, 1, 5), correspondingly.

5.3.2 Testing expression (5.21) for the perfect filtering expression

Here we want to demonstrate the effectiveness of expression (5.21) that can transform one function to another one. As $Ydn(x)$ we choose the double rectangle function oriented in opposite directions, $f_c(x)$ function coincides with the random function $Rnd(x$,-1,1) determined in subsection 5.3.1. The final function $Yup(x)$ coincides with the oscillating mode defined by expression (5.27). These three functions $Ydn(x)$, $Yup(x)$, $f_c(x)$ are shown in Fig. 5.8.

The successive stages of transformation $Ydn(x) \to f_c(x) \to Yup(x)$ realized in the frame of expression (5.21) are shown in Figs. 5.8(b,c).

This simple example imitates the situation in data transmission when the initial signal is transformed to a "noise" , which becomes "hidden" for the third side; and on the end of the transmission line, it is transformed again to the known signal, for simplicity expressed in the form of a single harmonic mode. Formula (5.21) will be

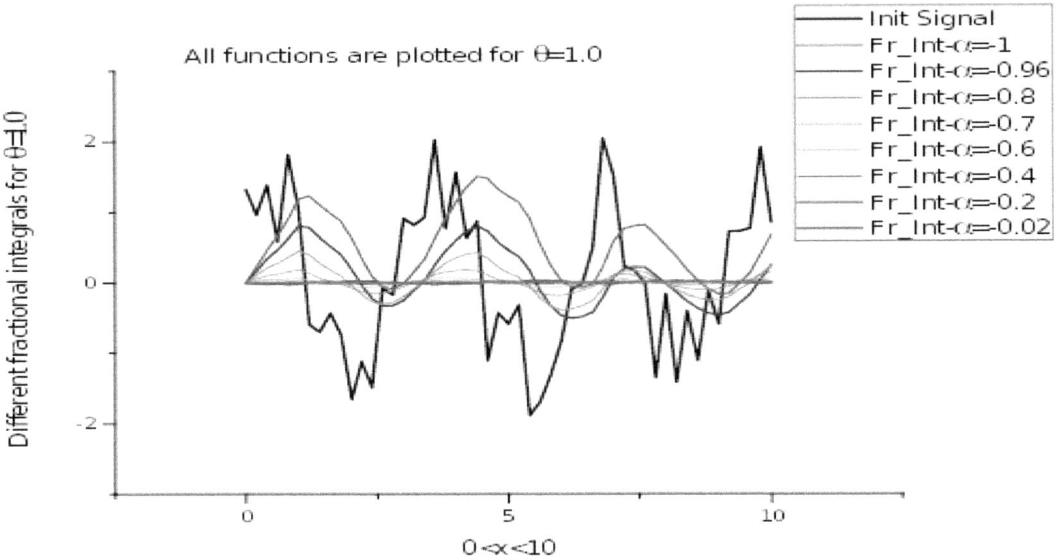

Figure 5.3 Here we show a family of curves belonging to the smoothing part from the same region $[-1,-0.02]$ for central value of the parameter $\theta=1.0$. The black bold line shows the initial curve. In comparison with the previous figure, all smoothed curves tend to zero value much faster. Their ranges become smaller.

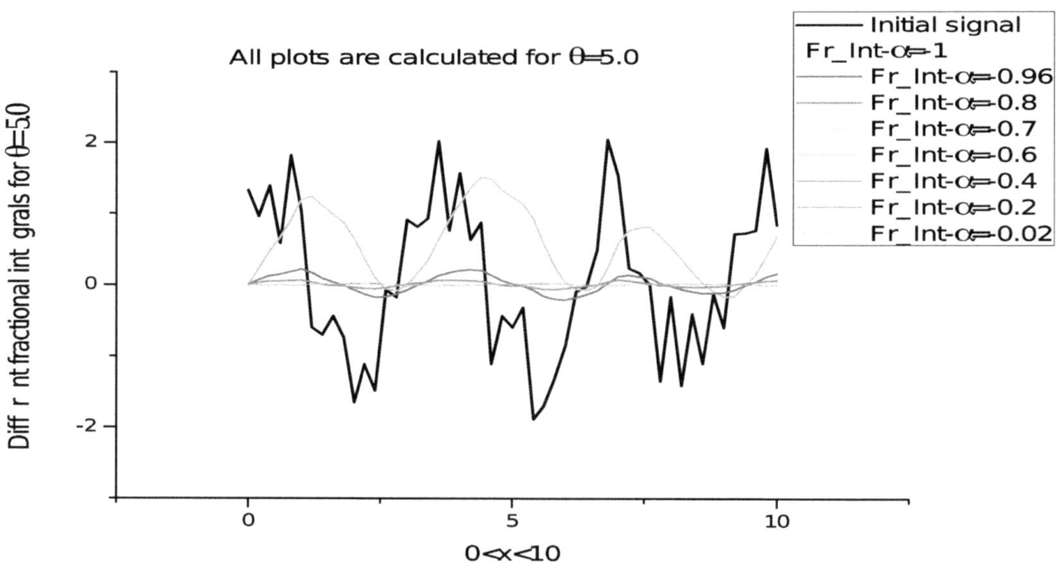

Figure 5.4 Here we show a family of the curves belonging to the smoothing part from the same region $[-1,-0.02]$ for the parameter $\theta=5.0$. Again, the black bold line shows the initial curve. The integration region for this case narrows and the smoothing integration effect decreases. All smoothed curves decrease their amplitudes and tend to zero much faster.

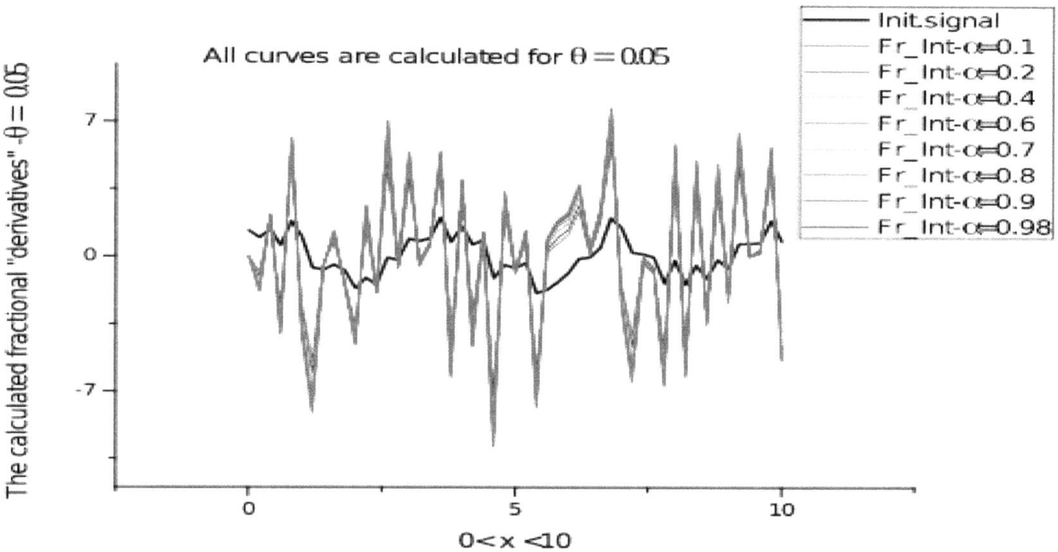

Figure 5.5 Quite opposite phenomenon is observed for the region $\alpha \in [0,1]$, which can be defined as the region of selection of the high frequency (HF) fluctuations. Again, the initial mode distorted by random noise is differentiated by bold curve. The curve that close to derivative curve (α =0.98) is in thick outmost line.

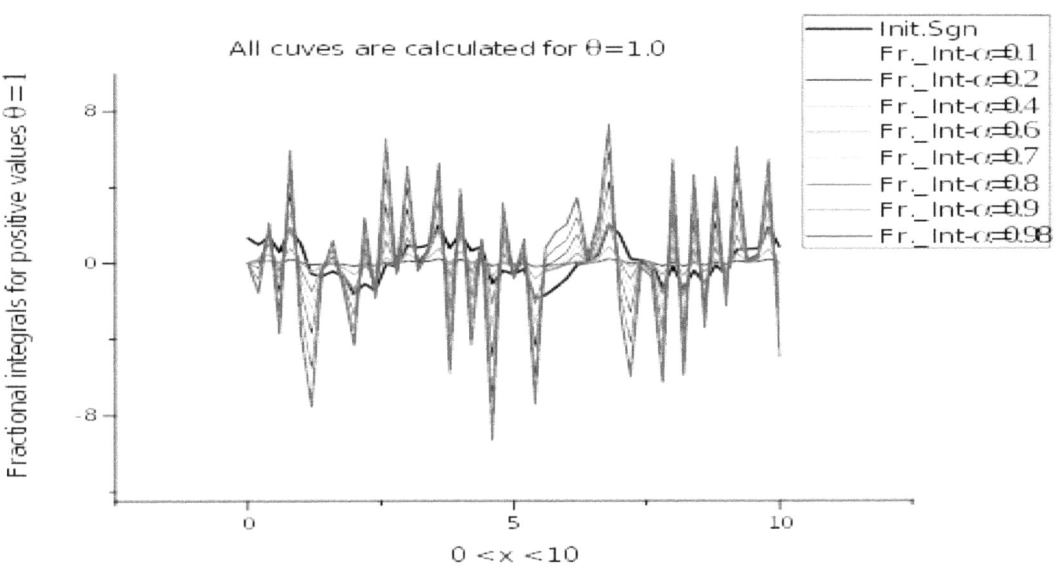

Figure 5.6 In comparison with Fig. 5.5, the effect of selection of the HF fluctuation is expressed more clearly. The amplitudes of these fluctuations become larger. Again, the initial mode distorted by random noise is differentiated by bold curve. The curve that close to derivative curve (α=0.98) is in thick outmost line.

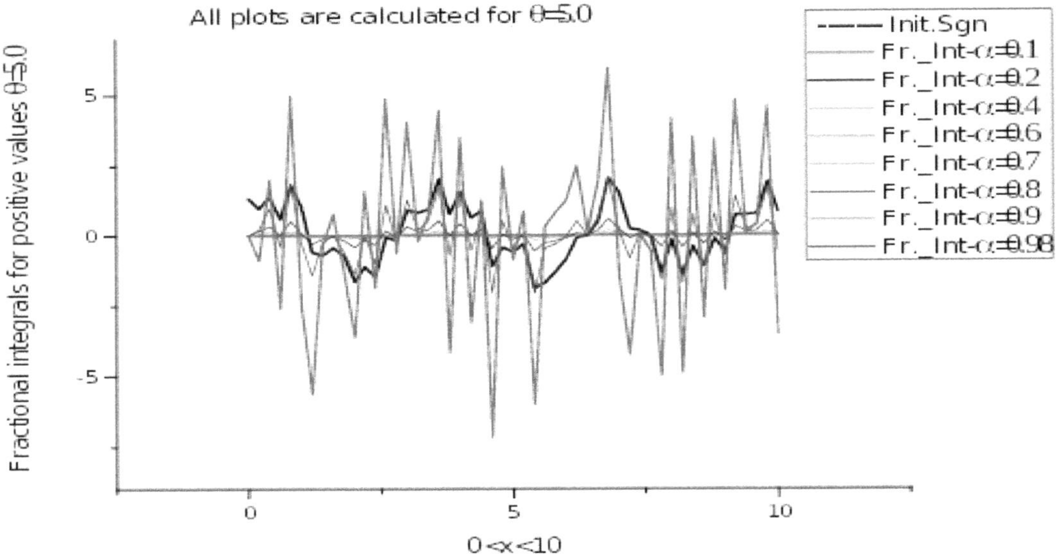

Figure 5.7 In comparison with Fig. 5.6, the effect of selection of the HF fluctuation is remained the same. The amplitudes of these fluctuations become a bit smaller. Again, the initial mode distorted by random noise is differentiated by bold curve. The curve close to derivative curve (α =0.98) is in thick outmost line. This picture can be considered as the final selection of the HF fluctuations that were found on the previous picture 5.6.

helpful also for detection of random uncontrollable fluctuations/factors in the case when we have a "pattern" curve, corresponding to an "ideal" state of the measured equipment. Really, identifying the function $Yup(x)$ with the given pattern/reference curve and forming the difference $DY(x) = Yup(x) - Ydn(x)$, one can receive the desired answer and detect all uncontrollable fluctuations that appear during the cycle of measurements. These fluctuations can be useful for calibration/metrological purposes when the comparison of some tested equipment with the pattern one becomes important procedure.

5.4 SOME REMARKS RELATED TO THE PROPOSED CALCULUS

In this part of the whole chapter (covering Secs. (5.1–5.3), we show some interesting consequences realized in the frame of the proposed "fuzzy" calculus. Actually, this calculus helps to fill a gap between integer operators and discrete indexes, creating a continuous set of unobservable/intermediate states related to conception of the continuous spin. May be for the quantum mechanics these states can be considered as unobservable, while for similar phenomena related to treatment of "big data" they become real and can be used as a useful addition in different applications. In particular, this analogy helps to derive new form for the fractional integral (5.11), to "erode" the discrete indexes in expressions (5.15)–(5.17) and finally find new possibilities for many conventional expressions. From our point of view, the most interesting

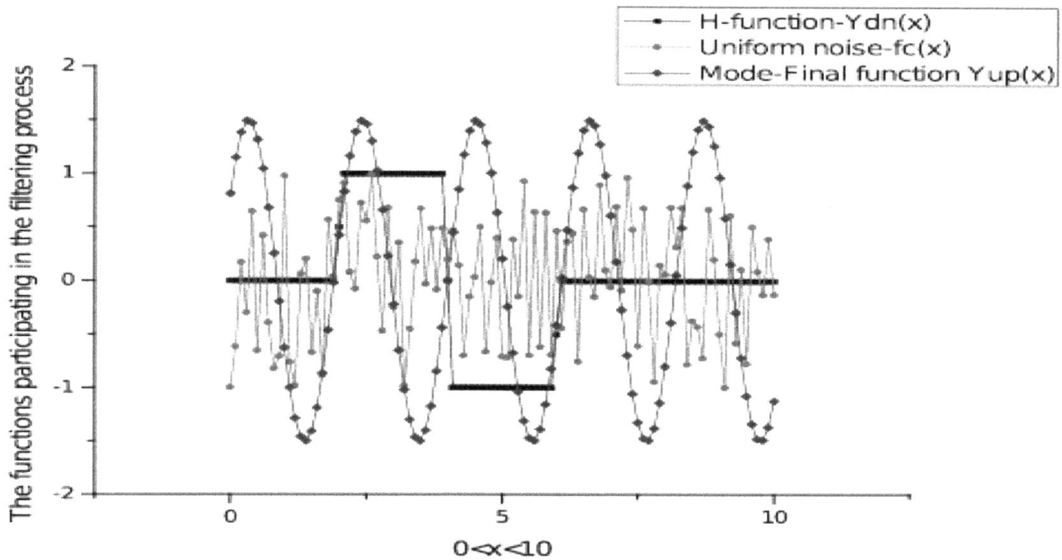

Figure 5.8 Here we demonstrate the functions participating in the transformation (5.21). Initial function $Ydn(x)$ imitates a "signal" expressed in the form of two-step function. The final function $Yup(x)$ imitates the same mode used in the first example and, finally, $f_c(x)$ represents itself the uniform distributed noise $Rnd(x,-1,1)$ used in the previous example. See expression (5.27) for further details.

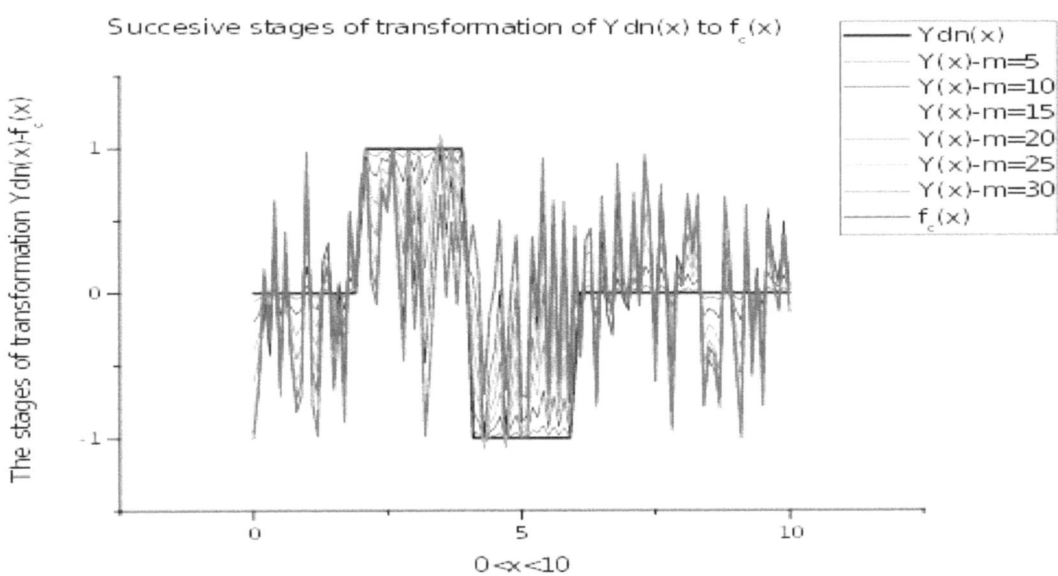

Figure 5.9 Successive stages of transformation of the initial function $Ydn(x)$ expressed in the form of double rectangle signal (shown by the bold black line) to the intermediate function $f_c(x)$ expressed in the form of the uniformly distributed random noise (shown by the thick outmost line).

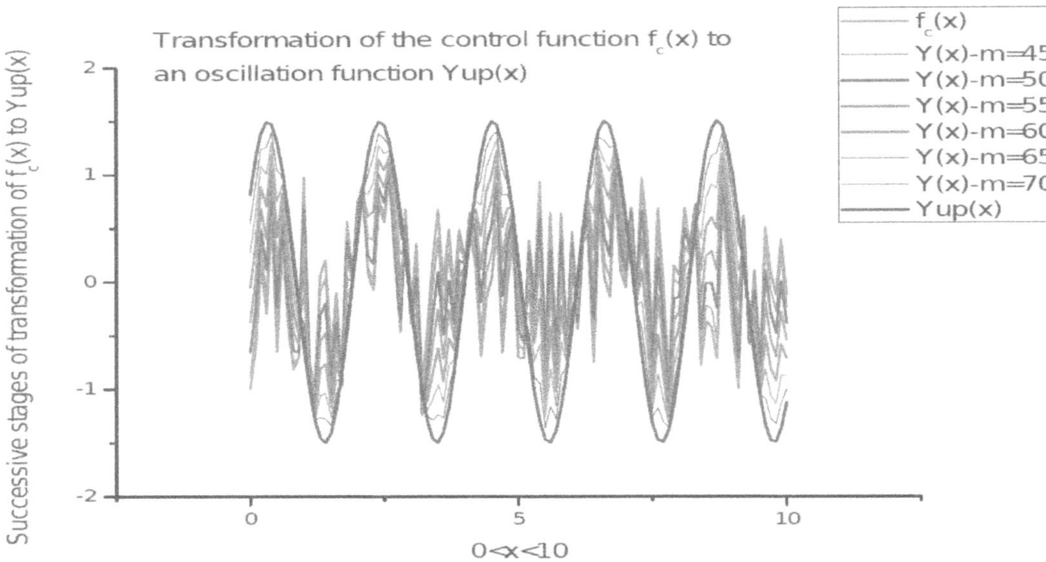

Figure 5.10 Here we show the second stage of transformation of the function $f_c(x)$ to $Yup(x)$ expressed as the single oscillating mode.

conclusion related with expressions (5.21)–(5.23), which allows transforming one arbitrary function to another one and vice versa. It can be applied in many important applications related to calibration of different equipment, in cryptography and in the modern telecommunication area. In conclusion of this section, let us derive the hybrid ("Centaur") formula for the Fourier transformation that unifies direct/inverse expressions in one mathematical unit.

We choose the simplest formula for the direct F-transformation and its inverse expression in the forms that are given in the manual [25]:

$$\begin{aligned} X_m &= \sum_{n=0}^{N-1} x_n \exp\left(-j\frac{2\pi mn}{N}\right), \\ x_n &= \frac{1}{N}\sum_{m=0}^{N-1} X_m \exp\left(j\frac{2\pi mn}{N}\right). \end{aligned} \tag{5.28}$$

Here expression in the first row represents the discrete Fourier transform (DFT), the second formula is associated with the inverse transformation (IDFT). In the frame of "fuzzy" calculus, these expressions can be unified as:

$$\begin{aligned} F(m;Q(\alpha)) &= N(Q(\alpha))\sum_{n=0}^{N-1} x_n \exp\left(-j2\pi\frac{mn}{N}Q(\alpha)\right), \\ N(Q(\alpha)) &= \frac{1}{N}+\frac{1}{2}\left(1-\frac{1}{N}\right)\cdot\left(Q(\alpha)+(Q(\alpha))^2\right). \end{aligned} \tag{5.29}$$

The expression in the second row takes into account the differences in normalizations of the DFT and IDFT, correspondingly. From formula (5.29), it is desirable to receive the generalized Dirichlet formula for the case, when we have "a signal" written as:

$$x_n = Amd_k \exp\left(j\left(\omega_k t_n - \varphi_k\right)\right), \; t_n = t_0 + \left(\frac{n}{N-1}\right)\Delta t, \\ \Delta t = t_{N-1} - t_0, \; n = 0,1,...,N-1, \; Amd_k = \sqrt{Ac_k^2 + As_k^2}. \tag{5.30}$$

Formula (5.30) represents an oscillating mode with the fixed frequency ω_k and phase φ_k. Substitution of (5.30) into (5.29) after some simple algebraic manipulations leads to the following expression

$$F(m;Q(\alpha)) = Amd_k \exp\left[-j\left(\Delta\omega_{m,\alpha,k} - \tilde{\varphi}_k\right)\right] N(Q(\alpha)) \frac{\sin\left(\frac{N}{2}\Delta\omega_{m,\alpha,k}\right)}{\sin\left(\frac{1}{2}\Delta\omega_{m,\alpha,k}\right)}, \\ \Delta\omega_{m,\alpha,k} = \frac{2\pi m}{N} Q(\alpha) - \frac{\omega_k \Delta t}{N-1}, \; \tilde{\varphi}_k = \varphi_k - \omega_k t_0. \tag{5.31}$$

If we take the modulus of 5.31, one can take the generalized Dirichlet function

$$D(m;Q(\alpha)) = Amd_k N(Q(\alpha)) \frac{\sin\left(\frac{N}{2}\Delta\omega_{m,\alpha,k}\right)}{\sin\left(\frac{1}{2}\Delta\omega_{m,\alpha,k}\right)} = \\ D(m;0) + \frac{1}{2}\left(D(m;1) - D(m;-1)\right) Q(\alpha) \\ + \frac{1}{2}\left(D(m;1) + D(m;-1) - 2D(m;0)\right)\left(Q(\alpha)\right)^2. \tag{5.32}$$

The last expression increasing the limits of the conventional Dirichlet function [25] and, because of its importance, needs separate research. The most interesting part of expression (5.32) is related to resonance condition, when the combined frequency equals zero $\Delta\omega_{m.\alpha,k} = 0$, i.e.,

$$\omega_{k,\alpha,m} = \frac{2\pi m}{\Delta t} Q(\alpha)\left(1 - \frac{1}{N}\right), \; 0 < \alpha \leq 1, \; m = 0,1,...,N-1, \tag{5.33}$$

Expression (5.33) determines a surface of the resonance frequencies that is valid only for the *positive* values of from the interval $0 < \alpha \leq 1$.

Concluding this part, one can say that the triad concept increases considerably the possibilities of the "fuzzy" calculus. If some value has three discrete values, then the "fuzzy" calculus allows to "wash away" the given discrete values $(\lambda_{min}, \lambda_{mean}, \lambda_{max})$ and consider all "hidden" values covering the whole range $(\lambda_{max} - \lambda_{min})$. Namely, this idea was used in derivation of expression (5.21) and generalization (5.23) for the multi-component matrices.

Concluding these Secs. (5.1–5.4) one can add some important points to the procedure that is associated with functions matching. In many cases some simple theoretical formulas are obtained at different values of the basic argument, i.e.,

$$F(x)=\begin{cases} f_m(x) & x << x_c \\ f_{\text{int}}(x) & x \simeq x_c \\ f_M(x) & x >> x_c. \end{cases} \tag{5.34}$$

How to match these functions in one "universal" combination in order to fill a possible "gap" between them? In this case, one can apply the basic expression (5.6) with functions that enter to the last expression:

$$\begin{aligned} F(x,\alpha) &= f_M(x)+\frac{1}{2}\left(f_M(x)-f_m(x)\right)Q(\alpha) \\ &+\frac{1}{2}\left(f_M(x)+f_m(x)-2f_{\text{int}}(x)\right)Q^2(\alpha). \end{aligned} \tag{5.35}$$

This combined function can match the limiting functions defined by (5.34). The function $Q(\alpha)$ as before should be located in the interval [−1,1] and has a monotone behavior similar to function defined by expression (5.3). Parameter θ entering into it can serve as the fitting parameter. As an example of application of (5.35) let us consider some instructive case. Let us suppose that we have a mixture of "heavy" particles/ions that follow to the Boltzmann distribution

$$n_k^{(0)}=\exp\left[\frac{\mu_{(0)}-\varepsilon_k^{(A)}}{kT}\right]. \tag{5.36}$$

Here μ is the chemical potential, k is the Boltzmann constant, T temperature, $\varepsilon_k^{(A)}$ - is the energy of the fixed atom/ion that is located in the k-th state. These excited atoms are surrounded by electrons obeying Fermi statistics and emits/absorbs the photons that follow Bose-Einstein statistics. We introduce the following definitions for them

$$n_k^{(\pm1)}=\begin{cases} \left[\exp\left(\dfrac{\varepsilon_k^{(e)}-\mu_{(+1)}}{kT}\right)+1\right]^{-1} \\ \left[\exp\left(\dfrac{\varepsilon_k^{(ph)}-\mu_{(-1)}}{kT}\right)-1\right]^{-1}. \end{cases} \tag{5.37}$$

In the state of the thermodynamic equilibrium the temperature T is remained the same, however, the corresponding energies of electrons $\varepsilon_k^{(e)}$ and photons $\varepsilon_k^{(ph)}$ located in the k-th quantum state together with their chemical potentials $\mu(\pm1)$ can be different. If in this "plasma" combination the number of quantum particles are remained uncertain (the state of nonequilibrium), then one can propose the combined distribution that matches all three types of the known statistics

$$n_k^{(\alpha)} = n_k^{(0)} + \frac{1}{2}\left(n_k^{(1)} - n_k^{(-1)}\right)Q(\alpha) + \frac{1}{2}\left(n_k^{(1)} + n_k^{(-1)} - 2n_k^{(o)}\right)Q^2(\alpha). \quad (5.38)$$

where the number of quantum particles $n_k^{(\alpha)}$ can be written also as

$$n_k^{(\alpha)} = \frac{1}{\left[\exp\left(\frac{\varepsilon_k^{(\alpha)} - \mu_\alpha}{kT}\right) + Q(\alpha)\right]}. \quad (5.39)$$

This combined formula (5.38) needs further research and its analysis for possible applications is beyond the limits of this manual.

5.5 REALIZATION OF THE RIEMANN-LIOUVILLE INTEGRAL IN LIVING SYSTEMS.

In order to finish this chapter related to the receiving of the fractional integral in some new discrete form we want to show that the integral with complex conjugated power law exponents can be realized in reality associated with some living systems. The situation with the fractional calculus in nowadays can be described in brief as follows. The respected mathematicians suggested a lot of expressions related to the fractional integral and derivatives [1, 14]. However, if we have a look at the realizations of the different types of integrals from experimental point of view then one can notice that not any power-law behavior can lead to the desired fractional integral. Possible practical realizations of these simple operators on different materials and with different combinations of RCL circuits are collected in the recent book [26]. This book includes all basic papers covering the 50 years period of intensive research related to applications of the fractional operators in practice. However, we should remind also the papers of the French scientist Jocelyn Sabatier [27, 28, 29], where he proved that not any power-law dependence that is observed in experimental measurements corresponds to some fractional integral or derivative. In this and subsequent sections the authors want to show that if it will follow from the branching/fractal structure of the object one can prove that the fractional integral with complex-conjugated power law exponents can be observed in these living systems. This self-similar structure is observed in many biological objects as leaves, mushrooms, bud flowers, etc. It would be interesting to follow for the temporal evolution of the complex impedance $Z(j\omega)$ of mush-room with respect to time (one week of observations or more) in the frame of some model that takes into account the predominant influence of the branching structure. One of us addressed to Dr. Karabi Biswas (Department of Electrical Engineering, Indian Institute of Technology, Kharagpur, Kharagpur-721302, West Bengal, India) that kindly agreed to measure the complex impedance of some type of mushrooms. This research was presented in a conference paper, however, we decided to include this research to demonstrate the possibility of the reduced fractal model in the fitting of data associated with biological systems as mushrooms. Why this biological object presents a sincere interest of engineers and biologists? The explanation is given below.

Owing to their superior medicinal values and nutritional constituents, mushroom's research on analyzing the primary cause of variations in their water holding capacity are gaining their importance in recent years [30]. Over the years, numerous researches have been carried out in analyzing the dependency of water retention capacity on the structures of mushrooms while investigating the bio-physiological and chemical changes in it. While viewing the cytoplasm and cell wall together as a single phase (gel phase), the cell wall in mushrooms can be thought of as a layered structure of three biopolymers (chitin, betaglucans, and proteins), present with their compositional and structural variants forming an intertwined network and these polymers are cross-linked with each other [31, 32]. While analyzing the water retention capacity, the moisture content in mushrooms has been quantitatively measured under mechanical stress via centrifugation [33]. Aiming to enhance the shelf life of post harvested mushrooms, dependency of cell wall composition, texture, and ultra structure of mushrooms on pH and temperature are observed by acidification or enzymatic treatment of varying intensities [34, 35]. Effect of thermal treatment on cell wall composition and texture of the mushrooms have been observed through blanching and addition of calcium chloride [36]. Under conventional and Ohmic heating while searching for variations in waveform and frequency, the weight and volume loss of mushrooms during blanching was measured [37]. The primary cause of water retention capacity i.e., hydration properties in mushrooms under varying mechanical loads, is explored using a thermodynamic theory (Flory-Rehner) [38]. While viewing mushroom as a gel (a homogeneous mixture of crosslink biopolymers in an aqueous phase) and considering protein denaturation as the primary cause of water holding capacity, the contribution of the gel phase to the water retention capacity after heat treatment of the mushroom is described with Flory-Rehner theory [39]. Change in moisture content of the mushroom tissues under mechanical pressure through centrifugation has been observed [40]. The effects of the cellular phase (cell wall and cellular structures) have been observed through heating, freezing, and milling treatment while predicting the variations in water retention capacity in common white button mushroom (Agaricus bisporus) through the cell membranes integrity loss by measuring the change in electrical conductivity and change of state in structural polymers caused by mainly chitin & proteins of the cell wall [41]. Quantitative evaluation of redistribution of water and water loss in tissue microstructure of mushroom stalks has been performed under the application of pulsed electric field and ultrasound while varying the temperature and monitoring the comparative variations in the distribution of intracellular water to extracellular space [42].

Dependencies of variations in water retention capacity on fibrous material of cell wall, the integrity of cell membrane, and denaturation of proteins, have been studied for rehydrated shiitake mushroom (Lentinulae Dodes), affected by varying hot air drying temperature [30]. This brief review underlines the keen interest of many researchers to this object. In Secs. 5.5–5.9, we propose an accurate model for the different stages of the temporal evolution of mushrooms (belonging to the "Agaricus" family) by introducing new fractal elements with a complex conjugated power law exponents while analyzing the flow of ions through the possible conducting channels, present at the different sections of the mushroom, from its impedance data.

The novelty lies in the modeling of the different evolution stages of mushrooms in terms of loss of moisture content i.e., variations in water holding capacity at different sections/membranes of the mushroom, while modelling the flow of ions through the branching conducting channels. This novelty can be summarized in the following points:

- This approach based on the reduced fractal model (described below) focuses on the highly accurate modeling (0.1%) of the transfer of charge carriers (ions) through the possibly existing conducting channels with 3-D branching structures to different sections of the mushroom under study. Variations in the flow of charges can be attributed to the variations in complex impedance with the changes in pore states in different sections of the mushroom, under the application of alternating electrical excitation of widely varying range of frequency.
- Observation of variations and comparative study of the different parameters of the proposed model is tabulated to identify the different evolutionary stages and to accurately track the abrupt changes in complex impedance from normal to quasi-chaotic behavior of mushroom during the whole series of measurements.
- The abrupt asymptotic temporal variations in evolutionary behavior of mushrooms or any living systems with 3-D branching structures have been treated as purely random. These seemingly non-deterministic variations cannot be tracked owing to their counterintuitive changes from a normal state to quasi chaotic behavior. This eventually results in the requirement of a mathematical model to comprehensively describe the behavioral/adaptive variations.

5.6 COMPLETE THEORY OF THE BRANCHING SYSTEMS IN THE FREQUENCY DOMAIN

We want to generalize the results obtained previously for the propagation of the blow-like signals in fractal media [43] analyzed previously in time-domain for the complex impedance measured in the frequency domain.

We start from the mathematical expression that can describe the complex impedance in fractal media and take into account the branching structure of electric signal/current propagation under the applied voltage.

$$\begin{aligned} P_N(z) &= \prod_{n=-N}^{N} \left(f(z\xi^n)^{n_0(b(z))^n} \right) = \\ &= \exp\left[n_0 \sum_{n=-N}^{N} (b(z))^n \ln\left[f\left(z\xi^n\right)\right] \right] \cong \exp S(z), \\ S(z) &= \lim_{N>>1} S_N(z). \end{aligned} \tag{5.40}$$

Here independent variable $z = i\omega$ associated with complex frequency, relaxation/transfer microscopic function $f(z)$ describes microscopic impedance of some

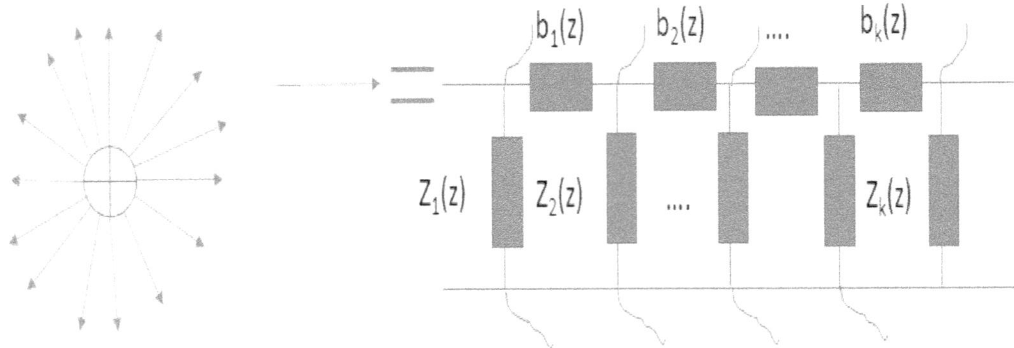

Figure 5.11 Schematic presentation of a model of a 3D-branching system. On the left-cross-section of the bundle of active electric circuits. The same cross-section of electric circuits forming a connected bundle presents this system on the opposite line. Each arrow represents itself a self-similar fractal circuit or long line of the Cauer/Forster's type with complex RC elements and nominals $Z_k(z) = Z_0\xi^k$, $b_k(z) = b_0\xi^k$, (k=0,1,2, ...). Here ξ is the scaling factor, z= $i\omega$. This relaxation/transfer line has a random length and can be connected randomly with other similar lines, filling all 3D-space.

independent block of a complex circuit, the log-periodic function $b(z\xi)$ $=b(z)$ describes a possible branching structure that exists in the complex circuit considered. We take into account the appearance of the branching factor $n_0 b(z)^n$ in the product (5.40). A schematic model of a branching system is shown in Fig. 5.11.

The bio-models considered in details in the review [44] are based on presumably in consideration of 2D-plain RC-models. Highly likely, the electric circuits corresponding to large number of bio-models should have 3D-conducting channels distribution. Expression (5.40) can serve as a specific "reflection" of 3D electric branching structure in frequency domain. We accept it as the working *hypothesis* and pay an attention of our readers that a possible scenario when this *branching* or percolation factor is important can be frequently realized in nature. We will discuss its significance in the last section. When the branching factor $n_0\,(b(z)) \equiv 1$ is absent, we obtain the general result discussed in the paper [45].

$$\lim_{N\to\infty} P_N(z) \equiv P(z) = z^{-\nu}\,\mathrm{Pr}\,(\ln(z))\,,$$
$$v = \ln(\bar{g})\,/\ln(\xi)\,,\ \ f(z)\overset{\mathrm{Re}(z)>>1}{=}\bar{g}. \tag{5.41}$$

In expressions (5.40) and (5.41), ξ is the scaling factor, $\bar{g}$ is the asymptotic limit of the function $f(z)$ at large values of variable z. As it has been done in time domain, we want to show that on the intermediate limits of the frequencies the specific behavior of the microscopic relaxation/transfer function becomes *unimportant*. Based on its asymptotic behavior, it becomes possible to evaluate expression 5.40 and derive rather general expression that can be suitable for the description of the complex impedance in the intermediate range of frequencies for a wide class of biological systems having a branching structure. This possibility will be discussed also in the last section. We *suppose* that the microscopic/transfer function has the following asymptotic behavior

$$\begin{aligned} &\mathrm{Re}|z| << 1 \\ &f(z) \cong c_1 z^{\alpha} + c_2 z^{1+\alpha} + \ldots \\ &f(z) = c_0 + c_1 z + \ldots \end{aligned} \tag{5.42a}$$

$$\begin{aligned} &\mathrm{Re}|z| >> 1 \\ &f(z) \cong A_1 z^{-\gamma} \exp(-rz) + A_2 z^{-\gamma-1} \exp(-2rz) + \ldots \\ &f(z) \cong A_0 + A_1 z^{-1} + \ldots \end{aligned} \tag{5.42b}$$

In contrast to the asymptotic requirements considered in paper [43] and in the book [46], we chose more general behavior at small and large values of variable z that are given by expressions (5.42). In the asymptotic behavior of the function $f(z)$ (the first line of (5.42b) we consider two possibilities (exponential and power-law) scenario, simultaneously. Neglecting the second terms in decompositions (5.42), one can obtain the evaluations of the limiting frequencies that define the mesoscale frequency region where (5.40) is valid

$$\omega_{\min} \cong \left(\min\left(\left|\frac{A_1}{A_0}\right|, \left|\frac{A_2}{A_1}\right|\right)\right) < \omega < \left(\max\left(\left|\frac{c_0}{c_1}\right|, \left|\frac{c_1}{c_2}\right|\right)\right) \cong \omega_{\max}. \tag{5.43}$$

In expression (5.43), we took in account the both decompositions in expressions (5.42a), (5.42b), correspondingly.

At N» 1 the sum $S_N(z)$ can be rewritten as

$$S_N(z) = n_0 \sum_{n=0}^{N} [b(z)]^n \ln|f(z\xi^n)| + n_0 \sum_{n=1}^{N} [b(z)]^{-n} \ln\left|f\left(z\xi^{-n}\right)\right|. \tag{5.44}$$

The second sum by replacement $1/\xi \to \xi$, $1/b(z) \to b(z)$ is reduced to consideration of the first sum; therefore, the second sum can be temporally omitted. If we put $z\xi \to z$ in the first sum, then one can receive the following equality

$$S(z\xi) = \frac{1}{b(z)} S(z) - \frac{n_0 \ln|f(z)|}{b(z)} + n_0 [b(z)]^N \ln\left|f\left(z\xi^{N+1}\right)\right|. \tag{5.45}$$

We *suppose* that $b(z\xi) = b(z)$ is a log-periodic function and it can be presented by the log-periodic decomposition. Therefore, the log-periodic function $b(z)$ is limited and lies in the interval

$$\begin{aligned} b(z) &= bc_0 + \sum_{k=1}^{K} [bc_k \cos(\Omega_k \ln z) + bs_k \sin(\Omega_k \ln z)], \\ \Omega_k &= \frac{2\pi \cdot k}{\ln \xi}. \end{aligned} \tag{5.46a}$$

$$|b(z)| \le \max b(z) - \min b(z) \equiv Range(b(z)) = b_R. \tag{5.46b}$$

If $b_R < 1$, then the contribution of the last term in (5.45) will be very small for any values of z. Taking into account the asymptotic values of the function $f(z)$ at small and large values of z (one takes into account the first lines in the decompositions (5.42a,5.42b), correspondingly), we obtain the following functional equation that is valid for the intermediate range of scales (5.43)

$$S(z\xi) = \frac{1}{b(z)} S(z) + \mu \ln(z) + gz + X_0. \tag{5.47}$$

Here the parameters,μ,g and X_0 are defined by the flowing expressions

$$\mu = -n_0\left(\frac{\alpha}{b_{\min}} - \frac{\gamma}{b_{\max}}\right),\ g = -\frac{n_0 r}{b_{\max}},\ X_0 = -\ln\left(\frac{c_1 A_1}{b_{\max} b_{\min}}\right). \tag{5.48}$$

The solution of equation (5.47) obtained with the help of the uncertain coefficients method can be written as

$$\begin{aligned} &S(z) = z^{\nu(z)} \Pr(\ln z) + C_1(z) \ln z + C_2(z) \cdot z + C_3(z), \\ &C_1(z) = \frac{\mu}{1 - 1/b(z)},\ C_2(z) = \frac{g}{\xi - 1/b(z)},\ C_3(z) = \frac{X_0 - C_1(z) \ln \xi}{1 - 1/b(z)}, \\ &\nu(z) = -\frac{\ln(b(z))}{\ln \xi},\ b(z\xi) = b(z). \end{aligned} \tag{5.49}$$

The function $b(z)$ is limited and, therefore, the log-periodic functions in (5.49) can be replaced by its range b_R from (5.46b). The case $b_R = 1$ requires special research. The details are given in the Mathematical Appendix in the end of this chapter. We want to stress also the fact that the general solution (5.49) depends strongly on the character of asymptotic behavior on the ends of the frequency interval $(\omega_{min}, \omega max)$ correspondingly. If we take the second lines of the decompositions in equations (5.42a) and (5.42b), then one obtains the following functional equation

$$S(z\xi) = \frac{1}{b(z)} S(z) + S_0,\ S_0 = -n_0\left(\frac{\ln|c_0|}{b_{\min}} + \frac{\ln|A_0|}{b_{\max}}\right). \tag{5.50}$$

with the following solution:

$$\begin{aligned} &S(z) = z^{\nu(z)} \Pr(\ln(z)) + C_0(z),\ P(z) = \exp(S(z)), \\ &\nu(z) = -\frac{\ln(b(z))}{\ln(\xi)},\ C_0(z) = \frac{S_0}{1 - (b_R)^{-1}},\ b_R \ne 1, \\ &S(z) = \Pr(\ln(z)) + \mu \ln(z),\ \mu = \frac{S_0}{\ln \xi},\ b_R = 1. \end{aligned} \tag{5.51}$$

It is interesting to note that a couple of power-law exponents $v = \ln(b_R)/\ln(\xi)$ and μ is appeared in *all* solutions; before (μ) or inside (v) exponent, correspondingly. Other combinations (the first line in (5.42a and the second line in (5.42b)) or vice versa) give equation (5.47) but with other values of constants μ, g, X_0 entering in (5.48). We do not show them and give a chance for our readers to receive them, if necessary. From the physical point of view, the power-law exponents ν and μ have a clear meaning. The exponent ν related with the branching/percolation phenomenon and can have log-periodic corrections, while the real power-law exponent μ tightly related with peculiarities of the microscopic transfer function $f(z)$ on the limiting values of the frequencies defined by expression (5.43).

Concluding this section, one can write at least four functions that can be used for the fitting purposes being applied to real data.

The hypothesis $\mathrm{H}_1(b_R \neq 1)$

The first line of (5.42a) is combined with the first line of asymptotic decomposition (5.42b). The same structure of solution is conserved if we consider the first line of (5.42a) with the second line of (5.42b) but with other values of the constants.

$$\begin{aligned} &P(z) = \exp(S(z)), \\ &S(z) = z^{\nu}\Pr(\ln z) + C_1 \ln z + C_2 z + C_3, \\ &\nu = -\frac{\ln(b_R)}{\ln \xi},\ C_1 \cong \frac{\mu}{1-(b_R)^{-1}},\ C_2 \cong \frac{g}{\xi-(b_R)^{-1}}, \\ &C_3 = \frac{X_0 - C_1 \ln \xi}{1-(b_R)^{-1}}. \end{aligned} \tag{5.52}$$

Here, for simplification of the fitting procedure we replace approximately $b(z)$ in (5.49) by its limiting value b_R from (5.46b).

The hypothesis $\mathrm{H}_2(b_R = 1)$

The first line of (5.42a) is combined again with the first line of asymptotic decomposition (5.42b). The same structure of solution is conserved if we combine the first line of (5.42a) and the second line of (5.42b), however, with other values of the constants

$$\begin{aligned} &S(z) = \Pr(\ln z) + a_1 \ln z + a_2 \ln^2 z + G \cdot z, \\ &a_1 \cong \frac{X_0}{\ln \xi} - \frac{\mu}{2},\ a_2 \cong \frac{\mu}{2 \cdot \ln \xi},\ G = \frac{g}{\xi - 1}. \end{aligned} \tag{5.53}$$

These values associated with asymptotic values that enter in expression (5.62) in the Mathematical Appendix after the Sec. 5.9

Here we want to notice that in addition to the power-law exponent a_1 new unusual term $\ln^2(\mathrm{z})$ is appeared. Another peculiarity that is appeared with hypothesis H_2 is that the power-law exponent a_1 can accept *arbitrary* values exceeding the conventional interval [-1,1]. Really, the key role in a_1 belongs to the ratio $X_0/\ln(\xi)$ that can be located out the limits of the available interval [-1,1]. See for additional details the Mathematical Appendix below.

The hypothesis $\mathrm{H}_3(b_R \neq 1)$

In this case the second line of (5.42a) is combined with the second line of asymptotic decomposition (5.42b).

$$S(z) = z^{v} Pr(\ln(z)) + C_0,$$
$$v = -\frac{\ln(b_R)}{\ln(\xi)}, C_0 \cong \frac{S_0}{1-(b_R)^{-1}}, S_0 = -n_0\left[\frac{ln|c_0|}{b_{min}} + \frac{ln|A_0|}{b_{max}}\right]. \tag{5.54}$$

The hypothesis $H_4(b_R = 1)$

$$S(z) = Pr(ln(z)) + \mu\ln(z),$$
$$\mu \cong \frac{S_0}{\ln(\xi)}. \tag{5.55}$$

As we can notice from the analysis of these expressions, the log-periodic corrections are conserved for *all* 4 working hypotheses. The log-periodic function $\Pr(\ln z)$ is defined similar to decomposition (5.46a)

$$\Pr(\ln z) = Ac_0 + \sum_{k=1}^{K}\left[Ac_k \cos\left(\Omega_k \ln z\right) + As_k \sin\left(\Omega_k \ln z\right)\right],$$
$$\Omega_k = \frac{2\pi \cdot k}{\ln \xi}, \ \Pr\left(\ln z \pm \ln \xi\right) = \Pr\left(\ln z\right). \tag{5.56}$$

Which criterion can be used in order to select the best fitting hypothesis among the proposed ones?

We will choose the hypothesis containing minimal values of the fitting parameters entering in (5.56). In other words, among the proposed hypotheses (5.52–5.55) we will choose the hypothesis providing the most accurate fit with minimal number of K and, besides this criterion, the chosen hypothesis allows the fitting parameter a_1 to be located out from admissible interval $[-1,1]$. Therefore, the most probable hypothesis (5.52) will correspond to these two preliminary chosen requirements.

5.7 EXPERIMENTAL DETAILS

Fresh mushroom (belonging to the "Agaricus" family) is plucked from the field at 5 am in the morning and impedance measurement stated at 10.45 am using an impedance analyzer (Make Novocontrol, Model Alpha A). The electrode (needle type gold electrode) connection is as shown in the Fig. 5.12. Frequency sweep is applied for 10^2 Hz to 4 MHz, where for each sweep 93 impedance data are collected. For one measurement it takes about 30 min. Before providing trigger for the next measurement, the mushroom is in relaxation state for at least 30 min. The applied signal is sinusoidal with amplitude of 3V RMS (root mean squares).

On the first day 10 sets of impedance data are noted. Using the similar procedure at total of 93 measurements are carried out for 12 days. During measurement the room temperature is kept constant at 25±1°C. However, the humidity of the air varied from 46 % to 79 % as no facility to control the humidity is present in the laboratory where the measurement was carried out. Figure 5.12a shows the photograph of the

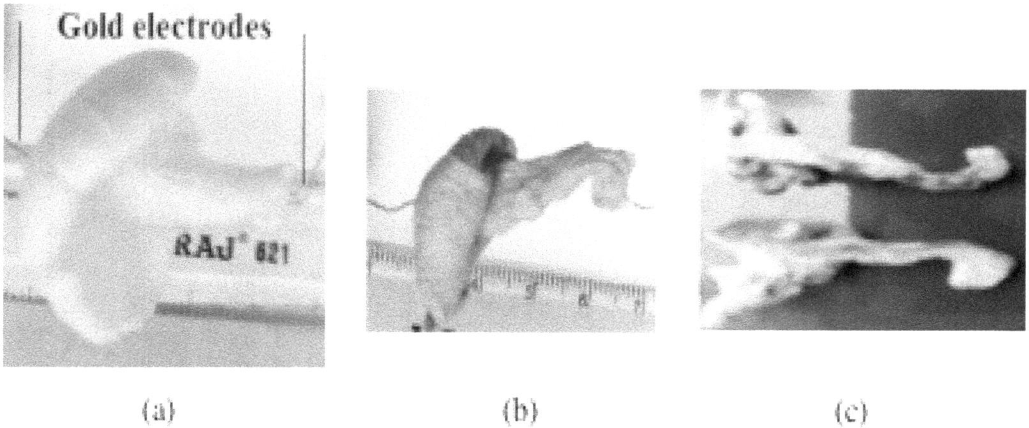

Figure 5.12 Possible scheme of connection of a mushroom stem to AC-generator for measuring the complex impedance. The value of the voltage equals 3 V. On the left figure 5.12a a fresh mushroom is shown, on the middle figure the same mushroom after 12 days of impedance measurements (in total 93 measurements were performed). The golden electrodes (with penetration about 5 mm from both sides into the mushroom body) are seen. The right-most figure is the photograph of the mushroom bisected lengthwise after 19 days of measurement. This bisection is shown on the right-hand figure.

mushroom taken on first day and Fig. 5.12b is for the 12th Day. After completing the measurement, the mushroom was bisected lengthwise and Fig. 5.12c shows the photograph of the same.

5.8 THE FITTING FUNCTION AND THE PROPOSED ALGORITHM

In order to modify expression (5.53) for the fitting purposes we replace the complex frequency $z{=}i\omega$ in (5.53) and extract the real and imaginary parts, accordingly. After some simple algebraic manipulations, we obtain finally

$$\begin{aligned}
\mathrm{Re}(S(i\omega)) &= \left(A_0^{(r)} - \frac{\pi^2}{4}a_2\right) + \sum_{k=1}^{K}\left(Ac_k^{(r)}\cos(\Omega_k x) + As_k^{(r)}\sin(\Omega_k x)\right) + a_1 x + a_2 x^2,\\
\mathrm{Im}(S(i\omega)) &= \left(\frac{\pi}{2}a_1\right) + \sum_{k=1}^{K}\left(Ac_k^{(i)}\cos(\Omega_k x) + As_k^{(i)}\sin(\Omega_k x)\right) + \pi\cdot a_2 x + G\exp(x),\\
\Omega_k &= \frac{2\pi\cdot k}{\ln\xi},\ x = \ln\omega.
\end{aligned} \tag{5.57}$$

After realization of the fitting of expressions in (5.57), the final fit is expressed as

$$\begin{aligned} &\mathrm{Re}\left(Z\left(\omega\right)\right)-i\,\mathrm{Im}\left(Z\left(\omega\right)\right)= \\ &=\exp\left(\mathrm{Re}\,S\left(i\omega\right)\right)+i\exp\left(\mathrm{Im}\,S\left(i\omega\right)\right). \end{aligned} \tag{5.58}$$

We should note also that large values proportional to $sinh(\pi\Omega_k/\ln(\xi))$, $cosh(\pi\Omega_k/\ln(\xi))$ in sums (5.57) are included into the constants $Ac_k^{(r,i)}, As_k^{(r,i)}$ and, therefore, these constants can accept the large values. Analysis of expressions (5.57) shows that the fitting functions are *not* independent from each other. Therefore, the proposed algorithm is reduced to initial fit of the real part $\mathrm{Re}(S(i\omega))$ and after evaluation of the fitting constants $a_{1,2}$ one can fit the function $\mathrm{Im}(S(i\omega))$.

Based on the values of these parameters as $\ln(\xi)$, $a_{1,2}$ and G, one can find the key parameters as μ, g and X_0 from (5.64) and, finally, one can evaluate the following combinations if we suppose that the branching parameter n_0 equals approximately two.

$$\begin{aligned} &D_{\alpha\gamma}\equiv\alpha-\gamma=-\mu=-2a_2\ln(\xi), \\ &X_0=-\ln\left(c_1A_1\right)=\left[a_1\ln(\xi)+a_2\ln^2\left(\xi\right)\right], \\ &r=-g=G\left(\xi-1\right),\ at\ n_0\cong 2. \end{aligned} \tag{5.59}$$

Expression (5.59) allows to evaluate the difference between the power-law exponents $\mathrm{D}_{\alpha\gamma}$, the values X_0 and parameter r that enter to the asymptotic decompositions (5.42). Besides the fitting constants $Ac_k^{(r,i)}, As_k^{(r,i)}$ that we do not have a possibility to show them all 93 (number of experiments)$\times$(K=38-43)2 =7630. However, one can select the most interesting temporal stages of the mushroom impedance evolution.

In Fig. 5.12(a,b) we show the pictorial evolution of a mushroom during 12 days of continuous measurements. On the left-hand side the fresh mushroom is shown (the first day), while on the right-hand side we show the same mushroom in the 13-th day of its evolution. Figures 5.13 and 5.14 show the fit of the selected hypothesis (5.57) and (5.58) to the first measurement. It is interesting to notice that in this case the parameter $\mathrm{D}_{\alpha\gamma}$ close to zero, the parameters X_0 and r are also small. This stage in the total impedance evolution we determine as a "normal" one. Quite opposite picture is observed starting from the 11-th measurement. For this case, the calculated value of the difference $\mathrm{D}_{\alpha\gamma}$ considerably exceeding the unit value, the same behavior is observed for the parameter X_0 and value of r that exceed the values in the normal state on some orders of its magnitude. This state we determine as a quasi-chaotic state. On all periods of the impedance evolution, we observe the alternation of quasi-chaotic states with "normal" ones.

Figs. 5.13–5.19, confirm these hidden peculiarities. The first measurement Fig. 5.13(a,b) can be combined with Fig. 5.19(a,b) and defined as the normal state, while the curves depicted in Figs. 5.15 and 5.16 (11-th measurement) as well as

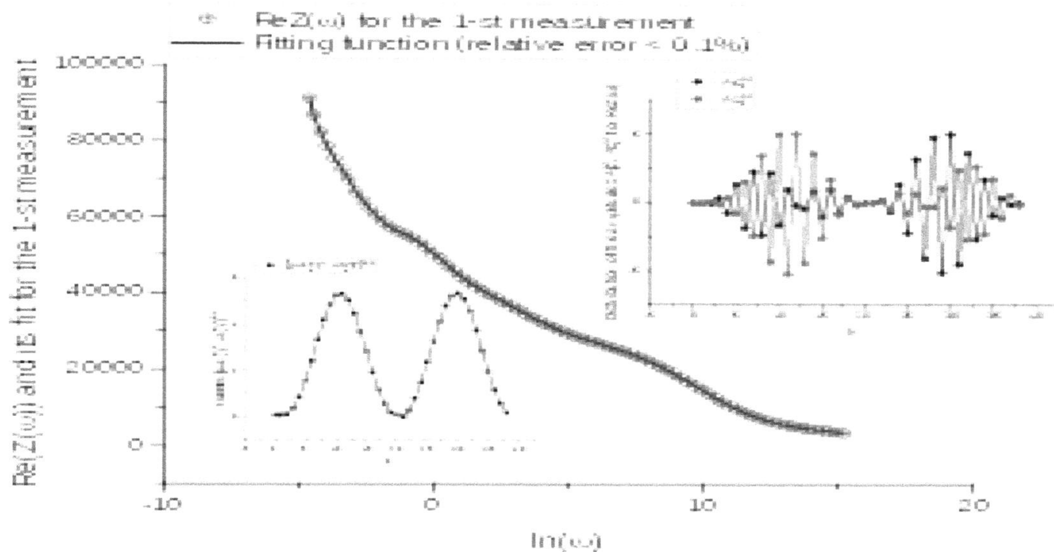

Figure 5.13 The fit of the ReZ(ω) for first measurement to hypothesis (5.57) and (5.58). The small picture above demonstrates the distribution of the amplitudes $Ac_k^{(r)}, As_k^{(r)}$ that figure in (5.57). The small picture below shows the distribution of the modulus $Md_k = \left(\left(Ac_k^{(r)}\right)^2 + \left(As_k^{(r)}\right)^2\right)^{1/2}$. One can notice from this figure the distribution of amplitudes $Ac_k^{(r)}, As_k^{(r)}$ is not random.

Figs. 5.17 and 5.18 (86-th experiment) can be referred as quasi-chaotic state. All set of these calculated parameters $D_{\alpha\gamma}$, X_0 and r for all 93 measurements are shown in the Table 5.1. In Fig.5.21 we show the behavior of the dimensionless "frequency" Ω, the parameter $x = \ln(\omega)$ is also dimensionless because the current frequency ω is normalized to 1 Hz. In Fig. 5.22 we show also chaotic behavior of the ranges of the measured impedance. As it follows from this figure the ranges of the real part are predominant. The opposite picture is observed in analysis of the maximal values $\max(Ac_k^{(r)}, As_k^{(r)})$, $\max(Ac_k^{(i)}, As_k^{(i)})$ that is shown in Fig. 5.23. The distribution $\max\left(Ac_k^{(i)}, As_k^{(i)}\right)$ is predominant and noncompatible with the central plot, therefore, it is shown above in Fig. 5.23. All these figures illustrating the realized fit are shown below.

5.9 ANALYSIS OF THE MEASURED DATA AND DISCUSSION OF THE OBTAINED RESULTS

How to understand the impedance behavior of the measured mushroom (at least qualitatively) based on analysis of the total set of 93 experiments?

We *suppose* that the structure of a living mushroom is similar to a sponge structure and contains a random system of the open/closed pores. Each micropore surrounded by the corresponding membranes, in turns, contains a specific nutrient

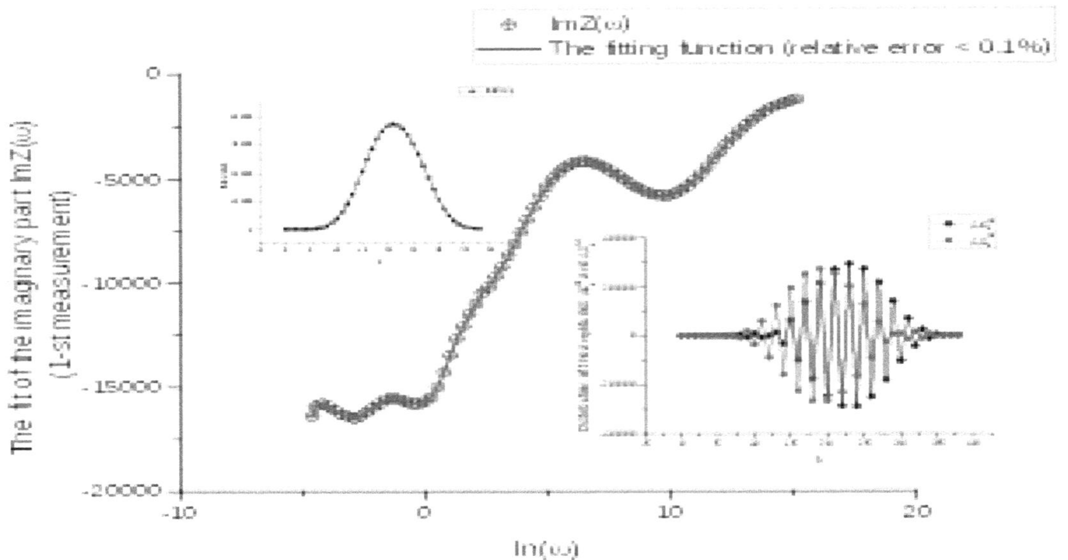

Figure 5.14 The fit of the ImZ(ω) for the first measurement to hypothesis (5.57) and (5.58). The small picture below demonstrates the distribution of the amplitudes $Ac_k^{(i)}, As_k^{(i)}$ that figure in (5.57). The small picture above shows the distribution of the modulus $Md_k = \left(\left(Ac_k^{(i)}\right)^2 + \left(As_k^{(i)}\right)^2\right)^{1/2}$. One can notice from the both figures that the distributions of amplitudes $Ac_k^{(r,i)}, As_k^{(r,i)}$ are not random.

solution/free water with the dissolved ions (H^+, K^+, Na^+, Ca^{++} and etc.) inside useful minerals together with necessary organic substances that are necessary for supporting the vital activity of a mushroom. In the beginning of measurements, we have excess of the solution and, therefore, the value of the resistance between applied electrodes is low (kilo-Ohms). In the beginning of the second day the conductivity is decreasing and to the 11-th measurement we observe the first anomaly (quasi-chaotic behavior), when the character of the asymptotic is suddenly/abruptly changed. We associate these changes with the fact that part of the closed pores is opened, new portion of free carriers is added and, therefore, the total conductivity is increased. As one can see from the Fig. 5.22 the resistance variations of the ranges ReZ(ω) and ImZ(ω) have chaotic character and the periods of the resistance increasing/decreasing are random. Even in the end of impedance evolution some closed pores are conserved and the mushroom at whole still keeps its relative humidity before its final stage, when it becomes completely dry. As data collected in Table 5.1 shows that these abrupt asymptotic changes take place over the whole period of measurements. Three key parameters as $D_{\alpha\gamma}$, X_0 and r having different sensitivity are correlated with each other and simultaneously indicate to this asymptotic anomaly. It is interesting to notice also that in the initial impedance curves this quasi-chaotic behavior is unnoticeable. We deliberately demonstrate the pair combination of the Figs. 5.13 and 5.14, Figs. 5.19 and 5.20 taken in the beginning and in the end of experiment for

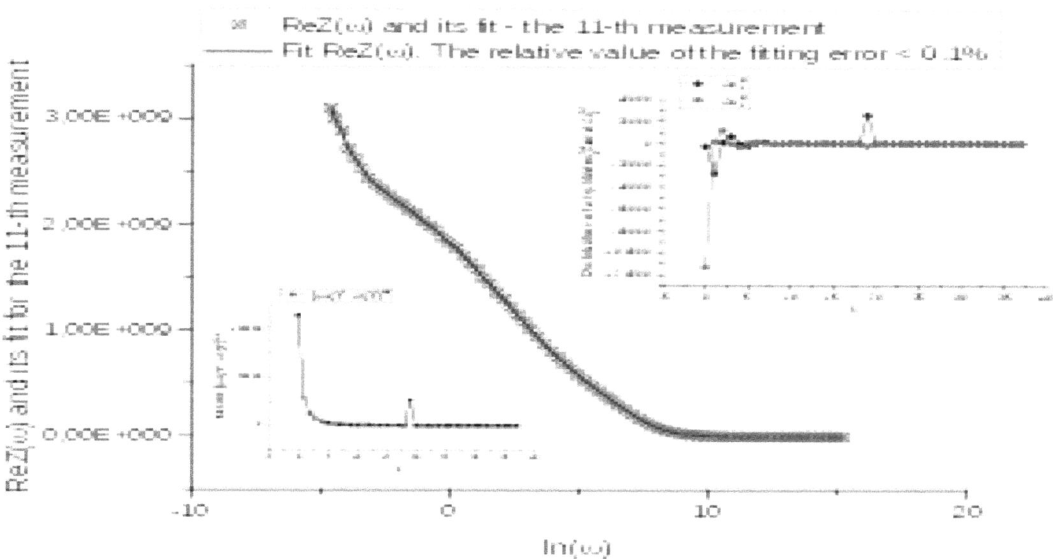

Figure 5.15 The fit of the ReZ(ω) for the 11-th measurement to hypothesis (5.57) and (5.58). The small picture above demonstrates the distribution of the amplitudes $Ac_k^{(r)}, As_k^{(r)}$ that figure in (5.57). The small picture below shows the distribution of the modulus $Md_k = \left(\left(Ac_k^{(r)}\right)^2 + \left(As_k^{(r)}\right)^2\right)^{1/2}$. One can notice from this figure the distribution of amplitudes $Ac_k^{(r)}, As_k^{(r)}$ looks as quasi-random.

comparison. The first pair is defined as a normal, while the second pair has abnormal asymptotic behavior. Analysis of data placed in Table 5.1 shows that the total number of abnormal state equals approximately 16%. Therefore, the process of the closed pores opening is continued throughout the whole stage of its temporal evolution.

Highly likely, each fresh mushroom from the "conductive" point of view represents itself a unique system with certain ratio of the closed pores. These pores form a specific "stock" that keeps nutrient solution that is added (in some moments of time) to the system of open channels. The external voltage 3V serves as a "brutal" electric force that opens these pores under compulsion. The open pores represent themselves a system of open channels that help to deliver the nutrient solution to the parts of mushroom that are needed to growth and for the keeping its vital activity. This qualitative (and we admit is not complete) picture outlined above is consistent for explanation of key points that are appeared in this short analysis. Concluding this section, we want to highlight the following results that were founded in this research.

1. Discovery of a new fractal element with complex-conjugated power-law exponents that can exist in living systems.

2. The proposed theory of the blow-like signals was applied successfully in the frequency domain for description of evolution of 3D-branching system as a fresh mushroom.

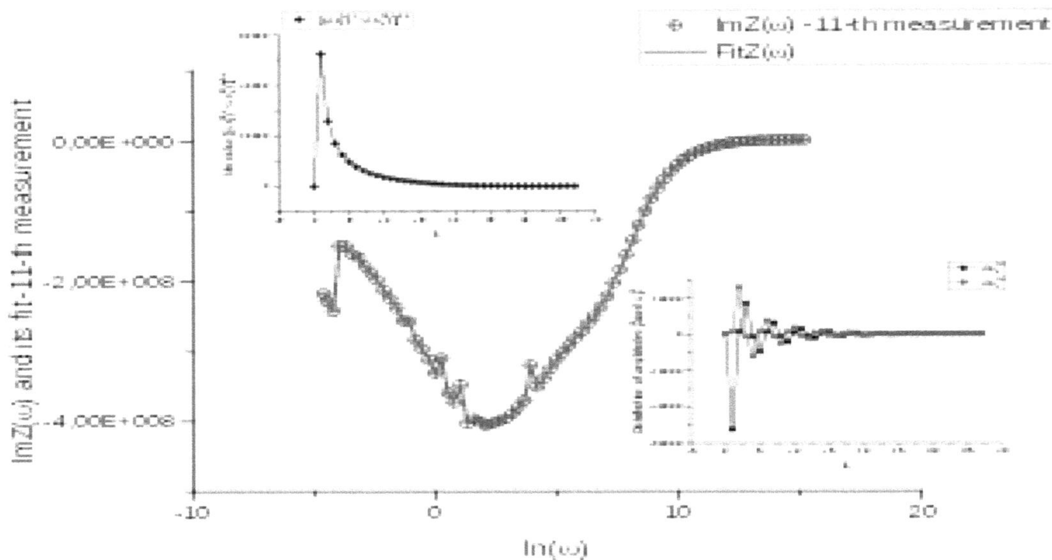

Figure 5.16 The fit of the ImZ(ω) for the 11-th measurement to hypothesis (5.57) and (5.58). The small picture below demonstrates the distribution of the amplitudes $Ac_k^{(i)}, As_k^{(i)}$ corresponding to quasi-chaotic behavior that figure in (5.57). The small picture above shows the distribution of the modulus $Md_k = \left(\left(Ac_k^{(r)}\right)^2 + \left(As_k^{(r)}\right)^2\right)^{1/2}$. One can notice from this figure that the distribution of amplitudes $Ac_k^{(r)}, As_k^{(r)}$ looks as monotone function.

3. All measured data (even including outliers) are fitted completely to the proposed fitting functions (expressions 5.57 and 5.58) with the value of the relative error less than 0.1%.

4. We give a qualitative explanation of the conductive processes that do not contradict with the measured impedance data.

5. In the first time we discovered the additional term $\ln^2(\omega)$ in behavior of the fitting function (5.57). This term can be important in construction of new types of the fractional integrals that enrich the exiting theory of the fractional calculus.

- All quasi-chaotic states having asymptotic anomaly, when the parameters $D_{\alpha\gamma}$. X_0 exceeding the unit values are highlighted. We also highlighted the parameter $(r \times 10^7)$ that exceeds ten values. The total number of quasi-chaotic states equals approximately (a1, a2..... a16) that foot up to 17%.

- The tabulation helps to identify the different evolutionary stages and to accurately track the abrupt changes in complex impedance from normal to quasi-chaotic behavior of mushroom during measurement.

Finishing this chapter our potential reader can stimulate his interests to this interesting branch of science that contains many interesting tasks for solution.

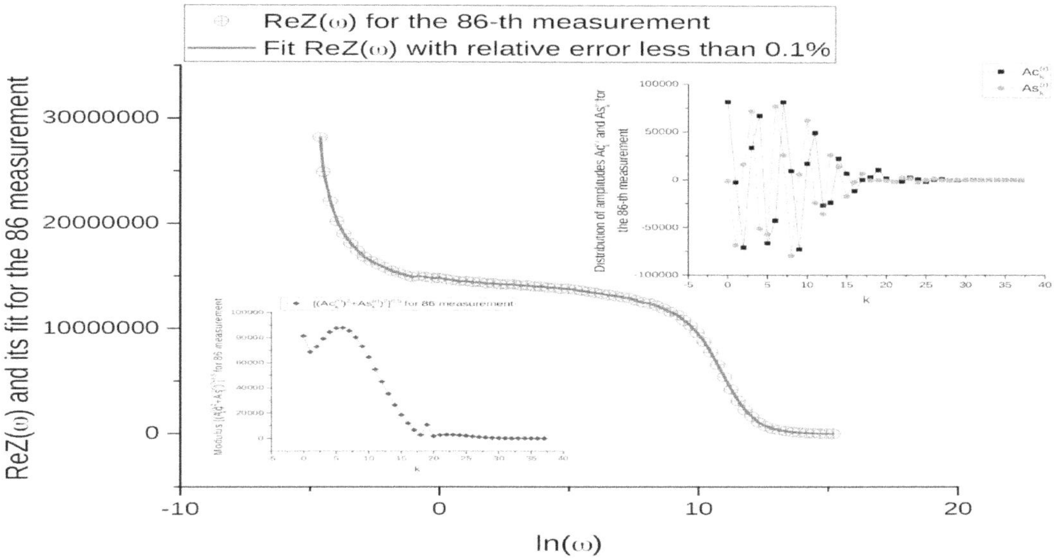

Figure 5.17 The fit of the ReZ(ω) for the 86-th measurement to hypothesis (5.57) and (5.58). The small picture above demonstrates the distribution of the amplitudes $Ac_k^{(r)}, As_k^{(r)}$ that enter in (5.57). The small picture below shows the distribution of the modulus $Md_k = \left(\left(Ac_k^{(r)}\right)^2 + \left(As_k^{(r)}\right)^2\right)^{1/2}$. One can notice from this figure the distribution of amplitudes $Ac_k^{(r)}, As_k^{(r)}$ looks as quasi-random

1. Based on the "fuzzy" calculus presented in the first part it becomes possible to "stretch" many ordinary formulae out from their ordinary limits.

2. One can try to find the physical/geometrical interpretation of many mathematical expressions related to the fractional operators. We recommend our readers to read the books [4, 5] and have a look on the review [14].

3. It would be important to find for the mathematical expressions of the fractional integrals and their realization in nature. Only these forms of the fractional integral confirmed by experimental observations will survive and find a wide propagation among engineers and practitioners.

4. An intriguing question for our readers can be formulated as: Does any physical conservation law (as found for energy, momentum, etc.) expressed in the form of the fractional integral exist or not? The answer for this key question will be very important for the further development of the fractional calculus that exists in the present form.

MATHEMATICAL APPENDIX

1.The case b_R=1.

In this case we should into account the second sum, in expression (5.44). We put again zz and present expression (5.44) in the form

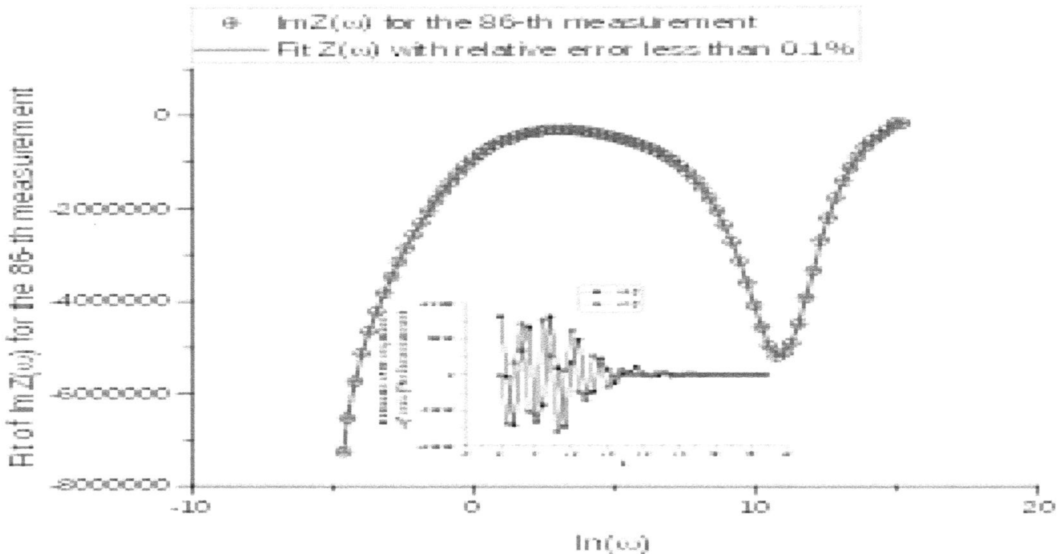

Figure 5.18 The fit of the ImZ(ω) for the 86-th measurement to hypothesis (5.57) and (5.58). The small picture below demonstrates the distribution of the amplitudes $Ac_k^{(i)}, As_k^{(i)}$ that enter in (5.58). The distribution of the modulus $Md_k = \left(\left(Ac_k^{(i)}\right)^2 + \left(As_k^{(i)}\right)^2\right)^{1/2}$ is not shown because of the lack of place. The curves shown in Figs. 5.15 and 5.16 as well as Figs. 5.17 and 5.18 are different, however, only the fitting parameters can fix their anomalous. The both figures can be classified as an "abnormal" state.

$$S(z\xi) = S(z\xi) - \frac{n_0}{2}\left[\ln\left[f(z\xi)\, f\left(z\xi^{-1}\right)\right]\right] + \frac{n_0}{2}\left[\ln\left[f\left(z\xi^{N+1}\right) f\left(z\xi^{-(N+1)}\right)\right]\right] - \frac{n_0}{2}\ln\left[f(z)\right]. \tag{5.60}$$

Substitution of the asymptotic values (5.42a) and (5.42a) presented by their first lines into (5.60) leads to the following equation

$$S(z\xi) = S(z) + \mu \ln z + gz + X_0 \tag{5.61}$$

Here the parameters in (5.61) are defined by the following expressions

$$\mu = -\frac{n_0}{2}(\alpha - \gamma),\; g = -\frac{n_0 r}{2},\; X_0 = -\frac{n_0}{2}\ln(c_1 A_1) \tag{5.62}$$

The solution of the functional equation (5.61) can be written as

$$S(z) = \Pr(\ln z) + a_1 \ln z + a_2 \ln^2 z + Gz \tag{5.63}$$

Using the method of uncertain coefficients one can find the unknown parameters $a_{1,2}$ and G substituting (5.63) into (5.61).

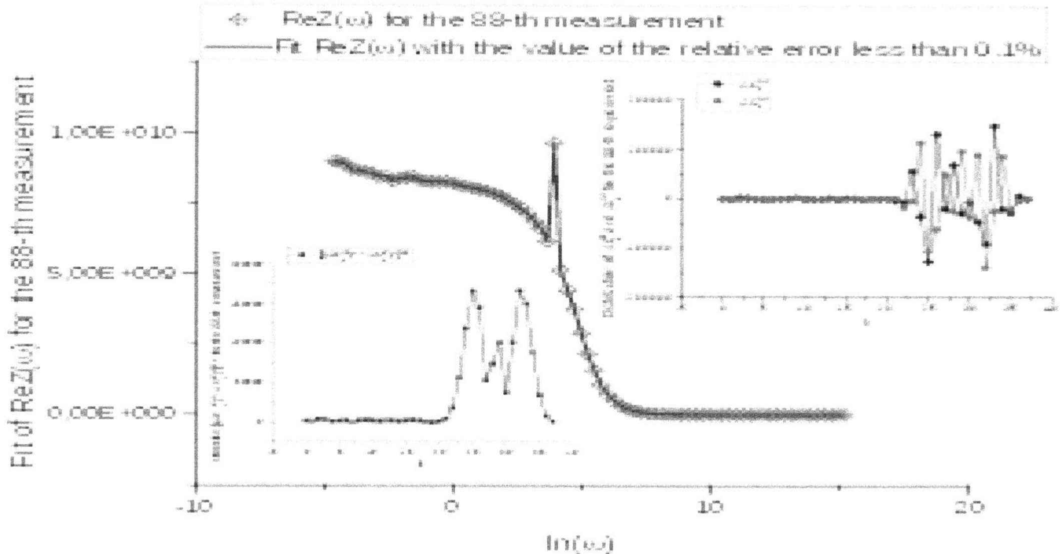

Figure 5.19 The fit of the ReZ(ω) for the 88-th measurement to hypothesis (5.57) and (5.58). The small picture above demonstrates the distribution of the amplitudes $Ac_k^{(r)}, As_k^{(r)}$ that enter in (5.57). The small picture below shows the distribution of the modulus $Md_k = \left(\left(Ac_k^{(r)}\right)^2 + \left(As_k^{(r)}\right)^2\right)^{1/2}$. One can notice from this figure, the distribution of amplitudes $Ac_k^{(r)}, As_k^{(r)}$ looks as an almost "normal".

$$G = \frac{g}{\xi - 1},\ a_1 = \frac{X_0}{\ln\xi} - \frac{\mu}{2},\ a_2 = \frac{\mu}{2\ln\xi} \tag{5.64}$$

The parameters in (5.64) tightly associated with asymptotic decompositions (5.61) again if we take into account the link between (5.64) and (5.62). From the physical point of view, this transfer/relaxation process corresponds to the slowed-down case. From (5.64) it follows

$$\begin{aligned} &P(z) = P_0 \cdot z^{a_1} \exp\left[\Pr(\ln z)) + a_2 \ln^2 z + Gz\right], \\ &\Pr(\ln z) = \sum_{k=1}^{K} \left[Ac_k \cos(\Omega_k \ln z) + As_k \sin(\Omega_k \ln z)\right], \\ &\Omega_k = \frac{2\pi}{\ln\xi} k, P_0 = \exp(Ac_0). \end{aligned} \tag{5.65}$$

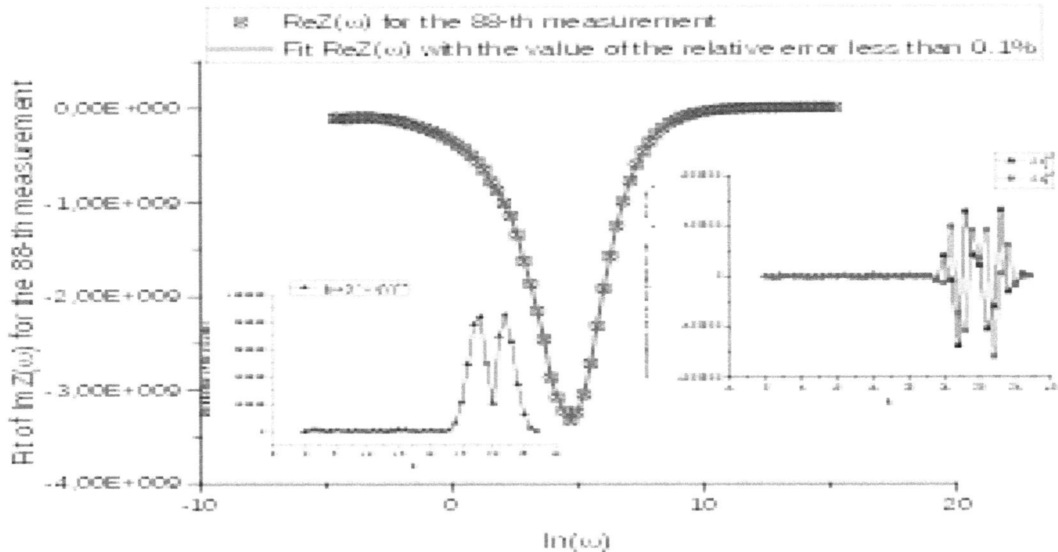

Figure 5.20 The fit of the ImZ(ω) for the 88-th measurement to hypothesis (5.57) and (5.58). The small picture in the right demonstrates the distribution of the amplitudes $Ac_k^{(i)}, As_k^{(i)}$ that enter in (5.57). The small picture below shows the distribution of the modulus $Md_k = \left(\left(Ac_k^{(i)}\right)^2 + \left(As_k^{(i)}\right)^2\right)^{1/2}$. One can notice from this figure the distribution of amplitudes $Ac_k^{(i)}, As_k^{(i)}$ looks as a random, however, these curves depicted in Figs. 5.12(a,b) can be classified as the "normal" ones.

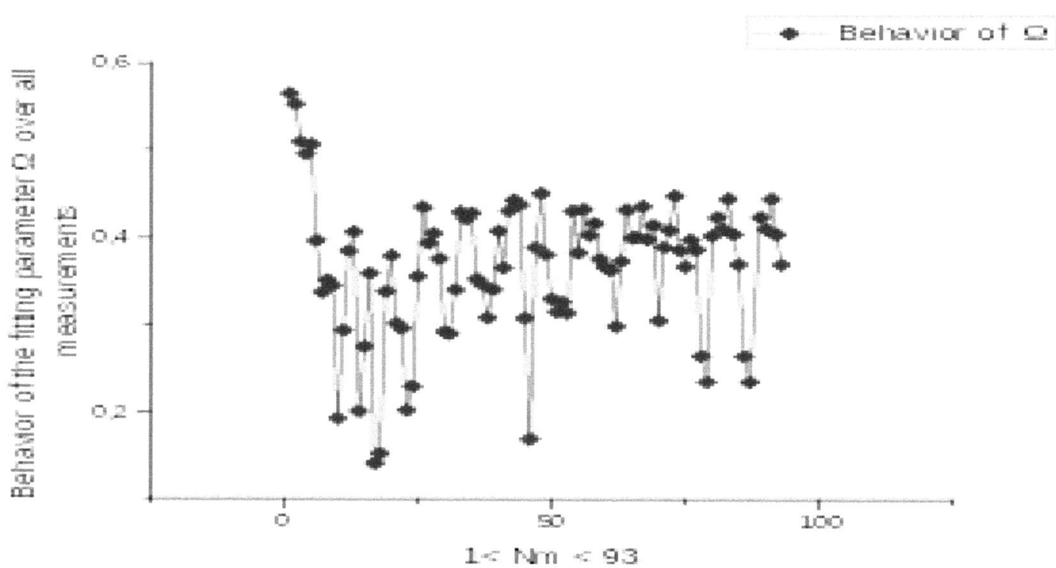

Figure 5.21 Behavior of the parameter Ω over 93 measurements that enters into the fitting function (5.57).

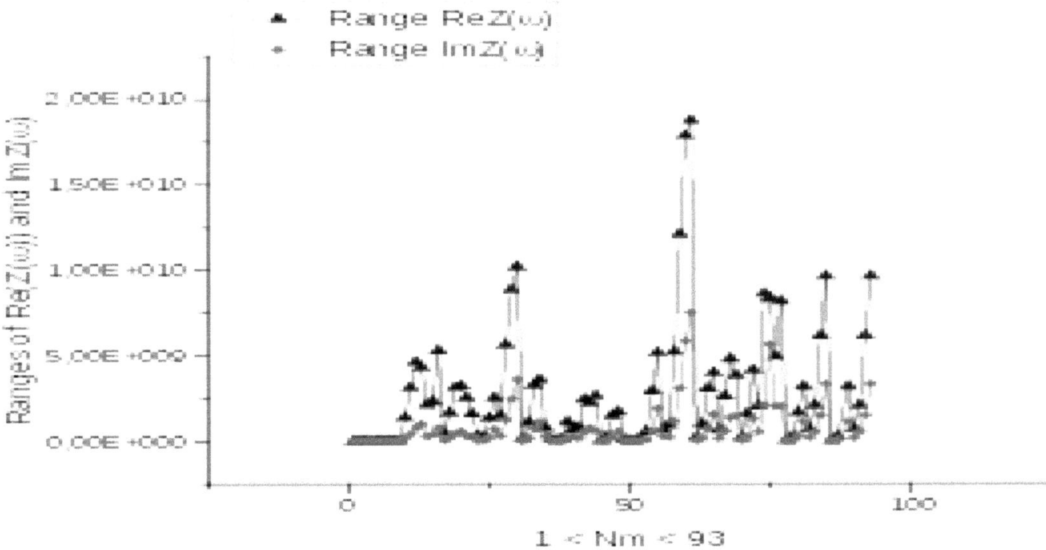

Figure 5.22 This figure shows the distribution of the ranges of two measured impedances. As one can notice from this figure, the "oscillations" of the resistance measured in Ohms is *not* monotone. In the beginning of the measurements the complex impedance starts from the values of kilo-Ohms and achieves the values of tens Giga-Ohms. In the beginning fresh mushroom has a sufficient amount of free water, then it gradually looses it.

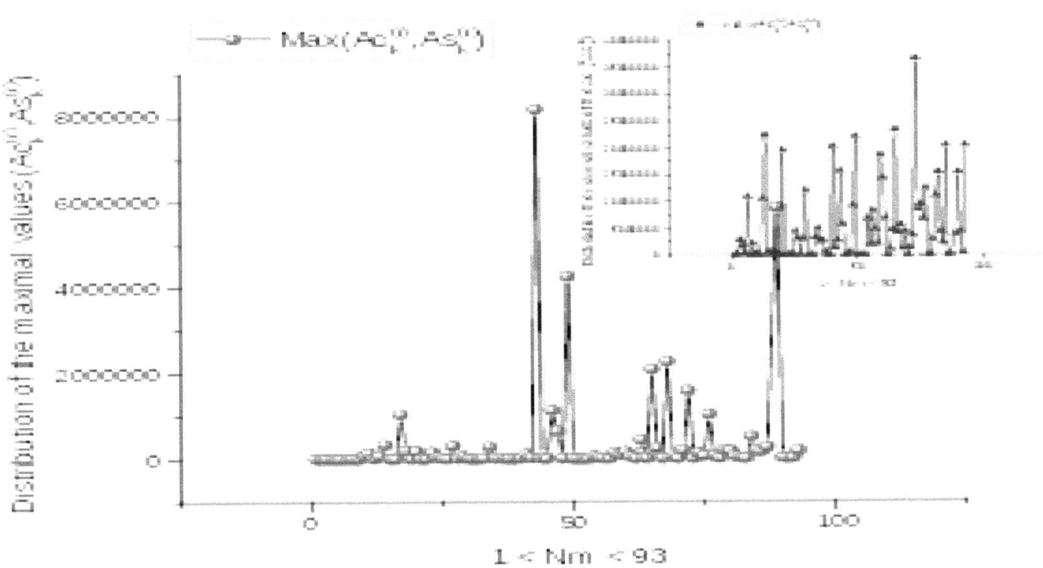

Figure 5.23 Distribution of the maximal values for amplitudes $\max\left(Ac_k^{(r)}, As_k^{(r)}\right)$, $\max\left(Ac_k^{(i)}, As_k^{(i)}\right)$ that enter into decomposition (5.57). Because of their incomparable values these distributions are placed in different plots. The distribution $\max\left(Ac_k^{(i)}, As_k^{(i)}\right)$ is shown above.

TABLE 5.1 The values of the key parameters $D_{\alpha\gamma}, X_0, r$ for all 93 measurements. Because of their incomparable values they are given in the form of a table.

Nm	$D_{\alpha\gamma}$	X_0	$r\times10^7$	Nm	$D_{\alpha\gamma}$	X_0	$r\times10^7$
1	0.00136	0.00535	7.59075	46-a11	**-14.4759**	**72.6377**	**2093.24**
2	0.00156	0.00486	7.68921	47	0.00461	0.01369	0.75358
3	0.00163	0.00545	5.98135	48	0.00714	0.00951	-0.68073
4	0.00163	0.00627	5.78207	49	-0.01295	0.09148	**5.98633**
5	0.00183	0.00562	5.89107	50	0.0138	-0.05097	0.18997
6	0.00118	0.00603	1.90004	51	0.02723	-0.13112	**-30.5648**
7	0.00111	0.00586	0.62521	52	-0.01291	0.09231	2.33275
8	0.00124	0.00605	0.83091	53	-0.00413	0.05772	**-39.1192**
9	0.00129	0.00647	0.81781	54	0.0062	0.01291	0.51946
10-a1	-0.25627	**-95.7165**	**-8459.15**	55	0.00457	0.02248	1.64575
11-a2	**229.898**	**-1202.65**	**26.2545**	56	0.00668	0.00422	-1.95488
12	0.00462	0.01742	2.46087	57	0.0052	0.00919	0.02229
13	0.00567	0.01541	3.50101	58	0.00564	0.01856	1.641
14-a3	**-7.04744**	**2.6204**	**40220**	59	0.00355	0.03008	1.98967
15	-0.05813	0.99746	**-1698.44**	60	0.00254	0.03743	2.45821
16	0.00333	0.02219	1.89878	61	0.0025	0.03698	2.27366
17-a4	**-10.5496**	**-586.908**	**876.232**	62-a12	0.8756	**-4.47842**	**-28.8134**
18-a5	**5.84809**	**67.8227**	**-1296.59**	63	0.00488	0.00888	0.56848
19	0.00566	0.00596	0.76391	64	0.00631	0.01362	0.76277
20	0.00461	0.01671	2.07886	65	0.00586	0.01645	1.13276
21	0.19125	**-1.39944**	-8.01823	66	0.00567	0.00497	-0.27815
22-a6	**-1.38604**	**9.40757**	**-124.235**	67	0.00648	0.01187	0.23362
23-a7	**-1.66128**	**-10.6591**	**18814.1**	68	0.00568	0.01696	1.20528
24-a8	**1.02934**	**-36.8452**	**32298.6**	69	0.00626	0.01567	1.36716
25	0.00374	0.01454	0.9329	70	0.10189	-0.49526	**-30.6999**
26	0.0065	0.01291	1.19148	71	0.00455	0.01383	0.70199
27	0.00462	0.01387	0.75707	72	0.00585	0.01688	1.34923
28	0.00521	0.02024	2.39703	73	0.00682	0.00967	-2.20055
29	0.00381	0.02754	2.23495	74	0.00454	0.02408	1.9821
30	0.01018	-0.02999	**17.6169**	75	0.00301	0.03113	1.78053
31-a9	**4.46966**	**-23.5596**	**-38.5828**	76	0.00552	0.01918	1.44149
32	0.00572	0.00277	0.39777	77	0.00468	0.02278	1.66786
33	0.00617	0.01474	1.88604	78-a13	**-1.71125**	-0.36092	**-1197.51**
34	0.00587	0.01795	2.25394	79-a14	**-165.793**	**1077.35**	**-1243.04**
35	0.00692	0.00286	-1.49903	80	0.00511	0.01239	0.65334
36	0.00621	-0.0079	-0.09173	81	0.00623	0.01568	1.53001
37	0.00865	-0.02174	-0.56424	82	0.00569	0.00618	-0.62172
38	0.02923	-0.10702	**-32.9457**	83	0.00678	0.01021	-0.65132
39	0.00659	-0.0015	0.18293	84	0.00551	0.01902	1.37817
40	0.00587	0.00652	-0.14099	85	0.00319	0.03107	1.81752
41	0.00397	0.01205	0.60282	86-a15	**18.2169**	**-448.576**	**8974.87**
42	0.00624	0.01232	1.4637	87-a16	**-42.4566**	**165.728**	**-11594.9**
43	0.00401	0.02705	3.10601	89	0.00595	0.01696	1.41064
44	0.00642	0.01243	0.50933	90	0.00569	0.00618	-0.62172
45-a10	0.05062	-0.29225	**-25.593**	91	0.00678	0.01021	-0.65132
92	0.00543	0.01945	1.60103	93	0.00319	0.03107	1.81752

BIBLIOGRAPHY

[1] G. S. Teodoro, J. T. Machado, and E. C. De Oliveira, "A review of definitions of fractional derivatives and other operators," *Journal of Computational Physics*, vol. 388, pp. 195–208, 2019.

[2] S. G. Samko, A. A. Kilbas, and O. I. Marichev, *Fractional Integrals and Derivatives: Theory and Applications.* Gordon and Breach Science Publishers, 1993.

[3] R. L. Magin, *Fractional calculus in bioengineering.* Begell House Redding, 2006.

[4] V. E. Tarasov, *Fractional Dynamics: Applications of Fractional Calculus to Dynamics of Particles, Fields and Media.* Nonlinear Physical Science, Springer, 2010.

[5] V. V. Uchaikin, *Fractional Derivatives for Physicists and Engineers: Background and Theory.* Nonlinear Physical Science, Springer, 2013.

[6] A. A. Kilbas, H. M. Srivastava, and J. J. Trujillo, *Theory and Applications of Fractional Differential Equations*, vol. 204 of *North-Holland Mathematics Studies.* Elsevier, 2006.

[7] R. Hilfer, ed., *Applications of Fractional Calculus in Physics.* World Scientific, 2000.

[8] V. S. Kiryakova, *Generalized fractional calculus and applications.* CRC Press, 1993.

[9] G. Jumarie, "Table of some basic fractional calculus formulae derived from a modified Riemann-Liouville derivative for non-differentiable functions," *Appl. Math. Lett*, vol. 22, pp. 378–385, 2009.

[10] M. Caputo and M. Fabrizio, "A new definition of fractional derivative without singular kernel," *Progr. Fract. Differ. Appl*, vol. 1, no. 2, pp. 1–13, 2015.

[11] K. S. Miller and B. Ross, *An introduction to the fractional calculus and fractional differential equations.* Wiley, 1993.

[12] M. D. Ortigueira, "Fractional central differences and derivatives," *IFAC Proceedings Volumes*, vol. 39, no. 11, pp. 58–63, 2006.

[13] M. D. Ortigueira and J. T. Machado, "A critical analysis of the caputo–fabrizio operator," *Communications in Nonlinear Science and Numerical Simulation*, vol. 59, pp. 608–611, 2018.

[14] M. Ortigueira, D. Valerio, and J. Machado, "Variable order fractional systems," *Commun. Nonlinear Sci. Numer. Simul*, vol. 71, pp. 231–243, 2019.

[15] H. J, "Essai sur l'étude des fonctions données par leur développement de taylor," *J. Pure Appl. Math*, vol. 4, pp. 101–186, 1892.

[16] W. Chen, "Time-space fabric underlying anomalous diffusion," *Chaos Solitons Fractals*, vol. 28, pp. 923–929, 2006.

[17] J. Machado and D. Baleanu, "On local fractional operators view of computational complexity: Diffusion and relaxation defined on cantor sets," *Thermal Science*, vol. 20, pp. S755–S767, 2016.

[18] N. R and L. M. A, "Is there a geometrical/physical meaning of the fractional integral with complex exponent," *Journal of Non-Crystalline Solids*, vol. 351, pp. 2888–2899, 2005.

[19] R. Nigmatullin and D. Baleanu, "New relationships connecting a class of fractal objects and fractional integrals in space," *Fractional Calculus and Applied Analysis*, vol. 16, pp. 1–26, 2013.

[20] F. R. Gantmacher, "The theory of matrices. volume one," *Translated by KA Hirsch, Chelsea Publishing Company, Printed in USA, Card Nr. 59-11779, ISBN: 8284-0131-4*, 1960.

[21] R. Bellman, *Introduction to matrix analysis.* SIAM, 1997.

[22] R. R. Nigmatullin, "The statistics of the fractional moments: Is there any chance to "read quantitatively" any randomness?," *Signal Processing*, vol. 86, no. 10, pp. 2529–2547, 2006.

[23] R. Nigmatullin, S. Osokin, and S. Nelson, "Application of fractionalmoments statistics to data for two-phase dielectric mixtures," *IEEE Transactions on Dielectrics and Electrical Insulation*, vol. 15, pp. 1385–1392, 2008.

[24] R. R. Nigmatullin and P. Agarwal, "Direct evaluation of the desired correlations: Verification on real data," *Physica A: Statistical Mechanics and its Applications*, vol. 534, p. 121558, 2019.

[25] R. Lyons, *Understanding Digital Signal Processing.* Prentice Hall PTR Publication, 2001.

[26] A. K. Gil'mutdinov, P. A. Ushakov, and R. El-Khazali, *Fractal Elements and their Applications.* Analog Circuits and Signal Processing, Springer, 2017.

[27] J. Sabatier, C. Farges, and V. Tartaglione, "Some alternative solutions to fractional models for modelling power law type long memory behaviours," *Mathematics*, vol. 8, no. 2, p. 196, 2020.

[28] J. Sabatier, "Beyond the particular case of circuits with geometrically distributed components for approximation of fractional order models: Application to a new class of model for power law type long memory behavior modelling," *Journal of Advanced Research*, vol. 25, pp. 243–255, 2020.

[29] J. Sabatier, "Modelling fractional behaviours without fractional models," *Frontiers in Control Engineering*, p. 7, 2021.

[30] A. Y. Ramteke, A. Nayak, A. Sagar, S. K. Kesari, and P. W. Ramteke, "Recent advances in mushrooms preservation," *International Journal of Chemical Sciences*, vol. 8, no. 2, pp. 2376–2381, 2020.

[31] G. Michalenko, H. Hohl, and D. Rast, "Chemistry and architecture of the mycelial wall of agaricus bisporus," *Microbiology*, vol. 92, pp. 251–262, 1976.

[32] S. Ifuku, R. Nomura, M. Morimoto, and H. Saimoto, "Preparation of chitin nanofibers from mushrooms," *Materials*, vol. 4, pp. 1417–1425, 2011.

[33] H. Donker and H. V. As, "Cell water balance of white button mushrooms (agaricus bisporus) during its post-harvest lifetime studied by quantitative magnetic resonance imaging," *BiochimicaetBiophysicaActa (BBA)-General Subjects*, vol. 1427, pp. 287–297, 1999.

[34] M. Brennan, G. Le Port, and R. Gormley, "Post-harvest treatment with citric acid or hydrogen peroxide to extend the shelf life of fresh sliced mushrooms," *LWT-Food Science and Technology*, vol. 33, pp. 285–289, 2000.

[35] S. Zivanovic, R. Buescher, and S. Kim, "Mushroom texture, cell wall composition, color, and ultrastructure as affected by ph and temperature," *Journal of Food Science*, vol. 68, pp. 1860–1865, 2003.

[36] S. Zivanovic and R. Buescher, "Changes in mushroom texture and cell wall composition affected by thermal processing," *Journal of Food Science*, vol. 69, pp. Q44–Q49, 2004.

[37] A. Sensoy and S. Sastry, "Ohmic blanching of mushrooms," *Journal of food process engineering*, vol. 27, pp. 1–15, 2004.

[38] R. Van Der Sman, E. Paudel, A. Voda, and S. Khalloufi, "Hydration properties of vegetable foods explained by flory-rehner theory," *Food research international*, vol. 54, pp. 804–811, 2013.

[39] P. E, B. R, and V. D. S. R, "Change in waterholding capacity in mushroom with temperature analyzed by Flory-Rehner theory," *Food and Bioprocess Technology*, vol. 8, pp. 960–970, 2015.

[40] E. Paudel, R. Boom, and R. Van Der Sman, "Effects of porosity and thermal treatment on hydration of mushrooms," *Food and Bioprocess Technology*, vol. 9, pp. 511–519, 2016.

[41] E. Paudel, R. M. Boom, E. van Haaren, J. Siccama, and R. G. van der Sman, "Effects of cellular structure and cell wall components on water holding capacity of mushrooms," *Journal of Food Engineering*, vol. 187, pp. 106–113, 2016.

[42] N. Dellarosa, D. Frontuto, L. Laghi, M. Dalla Rosa, and J. G. Lyng, "The impact of pulsed electric fields and ultrasound on water distribution and loss in mushrooms stalks," *Food chemistry*, vol. 236, pp. 94–100, 2017.

[43] R. Nigmatullin, V. Toboev, P. Lino, and G. Maione, "Reduced fractal model for quantitative analysis of averaged micro motions in mesoscale: Characterization of blow-like signals," *Chaos, Solitons & Fractals*, vol. 76, pp. 166–181, 2015.

[44] M. M, S. L, M. A, R. A, and E. A, "Fractional-order bio-impedance modeling for interdisciplinary applications: A review," *IEEE Access*, 2021.

[45] R. R. Nigmatullin and A. Le Mehaute, "Is there geometrical/physical meaning of the fractional integral with complex exponent?," *Journal of Non-Crystalline Solids*, vol. 351, no. 33-36, pp. 2888–2899, 2005.

[46] R. R. Nigmatullin, P. Lino, and G. Maione, *New Digital Signal Processing Methods: Applications to Measurement and Diagnostics*. Springer Cham, 1 ed., 2020.

CHAPTER 6

Hereditary Principled Methods

In this chapter, we want to show how to derive the equation with memory if the initial equation do not have it. As an example, we chose the well-known Newtonian equation for a particle that moves under the applied external force. Then we generalize this idea and consider more complex cases related to the system of equations and balance equation. Again, for initial systems of balance equation we show how and in what cases one might expect the equation with memory. The final part of this chapter will be devoted to the case when the conventional Hurst exponent that appears in many experimental observations can be generalized and finally, we obtain again the memory with complex-conjugated power-law exponents. As it was done in the previous chapter, we will deal with the Riemann-Liouville (RL) fractional integral. Other problems associated with more sophisticated forms of the fractional integral expect more attentive and capable readers. This chapter serves as an invitation for our young researchers where we demonstrate simple mathematics that helps to modify the well-known equation and derive from them the desired memory. We do not doubt that our readers can apply this approach to other equation and understand deeper in what cases and how a possible memory will appear in equation that initially do not have it.

6.1 IS IT POSSIBLE TO DERIVE NEWTONIAN EQUATIONS WITH MEMORY?

The fractional calculus is as old as the classical ones and it deals with derivative and integral of any order [1, 2, 3, 4, 5, 6, 7, 8, 9, 10, 11, 12]. The properties of the fractional derivatives and integrals are not as general as the classical ones; therefore, we expect at the first sight that this kind of calculus will have more potential applications than we know at the present time. However, the applications of the fractional calculus reveal some rich type of problems, which can be solved by using the fractional calculus. The fractional calculus describes better the behavior of complex systems involving memory effect. During the last decades the researchers tried to build fractional models in several areas of science and engineering. The validation of

DOI: 10.1201/9781003794691-6

the fractional calculus will come when a series of phenomena will be described only within this type of calculus. This question is as deep as the one related to the physical meaning of the fractional integrals and derivatives. In our opinion the answer to this question is one of the key points of the fractional calculus. One way is to look to some behaviors of the complex phenomena, to apply the fractional techniques and to conclude that within this technique the results are better and look more general to the classical ones. Another way is more solid and more accurate to deal with physical non-local phenomena and it can be summarized as follows: we have to start from the basic laws of physics and general concepts of physics, to fractionalize them properly and to prove that there is a physical meaning of the obtained results. In other words, we have to find a class of physical phenomena which requires naturally the fractional calculus derivatives and integrals.

Quite recently, [13] V.V. Uchaikin published his book entitled as "The Methods of the Fractional Derivatives", where he collected the numerous examples of application of the mathematical apparatus of the fractional calculus starting from nanophysics, from one side and moving to the fractional structure of the Universe, from another one. In consideration of many physical systems two basic points have to be taken into account: the importance and influence of the memory (nonlocal) effects in time and space and the finding of the general physical meaning of non-integer operators of differentiation and integration. These two basic ideas will help to make a decisive step for penetration of the mathematics of the fractional calculus into 'a body' of natural sciences. In spite of the fact that many researches are trying to apply the mathematical methods of the fractional calculus in their own investigations (bibliography of the papers cited in [13] exceeds more than 1000 papers, reviews, and books) the absence of the clear physical/geometrical meaning serves a specific obstacle on the way of applications. One of us (RRN) [14, 15] suggested a rather simple explanation of the fractional integral with real and complex power-law exponents that will help on the first stage to more comprehensive interpretations of these non-integer operators. The meaning of the fractional differentiation operation can be expressed by the following sentences:

The R-L fractional integral represents itself the *continuous* analog of the convolution operation of a smooth function with a fractal *discrete* set (we add here similar to the known Cantor sets) that modeling some self-similar process. If the discrete structure of the fractal set is not essential, as the averaging procedure leads to the conventional definition of the R-L integral with real power-law exponent. If the localization of neighboring stages of the process depends essentially on discrete structure, then the imaginary part of the exponent of the RL integral reflects the influence of the "fuzzy" discrete structure of the self-similar process that is conserved in the continual limit.

In this chapter we are going to show *how* to derive the Newtonian equation with memory taking as an example the relative movement of the layers of viscous liquid relatively each other. This simple model should give a key idea for derivation of equation with memory if initially the memory effect seems to be absent. Then our reader will see how to generalize this idea for more complex cases.

6.2 FORMULATION OF A PROBLEM AND PHYSICAL MODEL

It is well known that initially the Newtonian equation were proposed as equation without memory

$$m\frac{d\mathbf{v}}{dt} = \mathbf{F}(\mathbf{r}, \mathbf{v}, t). \tag{6.1}$$

For one particle we can formulate the following question: Is it possible to derive equation of the following type

$$m\frac{d\mathbf{v}}{dt} = \int_0^t K(t-\tau)\mathbf{F}(\mathbf{r}, \mathbf{v}, t)d\tau \tag{6.2}$$

when memory effects are taken into account? If this type of equation can be derived, then it will open new possibilities for understanding the classical mechanics. Here we want to describe a simple mechanical model that allows in realization of the desired transition from (6.1) to (6.2).

Description of the mechanical model. Let us consider N layers of viscous and heterogeneous liquid (layer polymer) and the movement of each the k-th laminar layer with density ρ_k (k=1,2,...,N) is described by the following Newton's equation

$$\begin{aligned}
&\rho_1\frac{d\mathbf{v}_1}{dt} = \gamma_{1,2}\mathbf{v}_2,\\
&\ldots\\
&\rho_k\frac{d\mathbf{v}_k}{dt} = \gamma_{k,k+1}\mathbf{v}_{k+1},\\
&\ldots.\\
&\rho_{N-1}\frac{d\mathbf{v}_{N-1}}{dt} = \gamma_{N-1,N}\mathbf{v}_N,\\
&\rho_N\frac{d\mathbf{v}_N}{dt} = \mathbf{F}\left(|\mathbf{r}_{1,N}|, |\mathbf{v}_{1,N}|, t\right).
\end{aligned} \tag{6.3}$$

The physical meaning of this system of equations is the following. The movement of each layer is described presumably by movement of the neighboring layers located on the right-hand side. The first and the Nth layers occupy the final left and right-hand positions, correspondingly. So, the numeration of the layers is started from the left to the right. Therefore, we take into account the influence of the layers, which are located closer to the acting external force (more precise to the density of the force). This force depends of the relative distance $|\mathbf{r}_{1,N}| = |\mathbf{r}_1 - \mathbf{r}_N|$ and relative velocity $|\mathbf{v}_{1,N}| = |\mathbf{v}_1 - \mathbf{v}_N|$ with respect to the relative movement and location of the first layer. Let us try to exclude the intermediate variables in order to reduce the influence of the external force to the effective movement of the first layer. Using the mathematical induction method, it is easy to derive the equation of motion for the velocity $\mathbf{v}_1(t)$. It accepts the form

$$\rho_1 \frac{d\mathbf{v}_1}{dt} = \frac{\gamma_{12}\gamma_{23} \cdot \ldots \gamma_{N-1,N}}{\rho_2\rho_3 \ldots \rho_N} \frac{1}{(N-2)!} \int_0^t (t-\tau)^{N-2} F(|\mathbf{r}_{1,N}|, |\mathbf{v}_{1,N}|)\, d\tau + \sum_{k=2}^{N} \mathbf{v}_k(0) \frac{t^{k-2}}{(k-2)!}. \tag{6.4}$$

Therefore, we obtain Newton's equation with memory and the influence of all intermediate layers is reduced to the influence as some external factors. In [16] the classical derivatives were replaced by the fractional ones and some exact solutions were found. Based on the results obtained in the book [14] and later in paper [15] of one of the authors (RRN) is it easy to transform the integer integral into the *fractional* one if one supposes that the temporal interaction between layers represents itself a self-similar process and distributed over the Cantor set and its possible generalizations. Because of importance and generality of this process it merits the separate consideration.

6.3 THE CONSIDERATION OF SOME SELF-SIMILAR PROCESS THAT LEADS TO THE FRACTIONAL INTEGRAL

If we to take the Laplace-image of equation (6.4) then the further consideration is reduced to consideration of the convolution of the smothered external force with memory function that it will be presented in more general form

$$\rho \frac{d\mathbf{v}}{dt} = \int_0^t K(t-\tau)\mathbf{F}(\tau) d\tau \Leftrightarrow \rho[p\mathbf{v}(p) - \mathbf{v}(0)] = K(p)\mathbf{F}(p). \tag{6.5}$$

We consider the memory function $K(p)$ as the result of some self-similar process that takes place in the physical system considered. In this case the memory function $K(p)$ is written in the form of the product [15]

$$K(p) \equiv K^{(N)}(z) = \prod_{n=-(N-1)}^{N-1} \hat{g}(z\xi^n) = \prod_{n=0}^{N-1} \hat{g}(z\xi^n) \prod_{n=1}^{N-1} \hat{g}\left(z\xi^{-n}\right), \tag{6.6}$$

where $z{=}p\tau_0$ – is the dimensionless parameter of the Laplace transform (τ_0 determines the characteristic time of the self-similar process at $n = 0$), ξ - defines the value of the scaling factor. The product (6.6) defines the Cantor set with M bars or its possible generalization. In particular, if the self-similar process is considered as the distributed over the Cantor set with M bars that the function $g(z)$ accepts the form

$$g_M(z) = \frac{1}{M} \frac{1 - \exp\left(-\frac{zM}{M-1}\right)}{1 - \exp\left(-\frac{z}{M-1}\right)}. \tag{6.7}$$

In relationship (6.7) $z=pT(1\text{-}\xi)$, where T determines the temporal segment where the Cantor set is located, the value M determines the number of the Cantor's bars. The product (6.6) satisfies the following exact functional equation:

$$K^{(N)}(\xi z)=\frac{g\left(z\xi^{N}\right)}{g\left(z\xi^{-N+1}\right)}K^{(N)}(z). \tag{6.8}$$

For further investigation of the product (6.8) we suppose that the function $g(z)$ has the following decompositions

for $\mathrm{Re}(z) \ll 1$

$$\hat{g}(z)=1+c_1z+c_2z^2+\ldots \tag{6.9a}$$

for $\mathrm{Re}(z) \gg 1$

$$\hat{g}(z)=\bar{g}+\frac{A_1}{z}+\frac{A_2}{z^2}+\ldots \tag{6.9b}$$

Keeping the limiting values for both sides, from (6.8) we obtain the simplified functional equation

$$K(\xi z)=\frac{1}{\bar{g}}K(z) \tag{6.10}$$

The general solution of the functional equation (6.10) has the form [14, 15]

$$\lim_{N\to\infty}K^{(N)}(z)\equiv K_v(z)=\frac{\mathrm{Pr}_v(\ln(z))}{z^v} \tag{6.11}$$

where

$$v=\frac{\ln(\bar{g})}{\ln(\xi)}=\frac{\ln(1/\bar{g})}{\ln(1/\xi)}=\frac{\ln(1/M)}{\ln(1/\xi)}. \tag{6.12}$$

Here we took into account the limiting value for the chosen Cantor set with M bars from (6.7)

The function $\mathrm{Pr_v}[\ln(z)]$ is a log-periodic function with the period $(\ln\xi)$

$$\mathrm{Pr}_v(\ln(z)\pm\ln\xi)=\mathrm{Pr}_v(\ln(z)). \tag{6.13}$$

So, we proved that the general solution for a self-similar process has the structure expressed by relationship (6.11). The further investigations realized in paper [15] showed that the log-periodical function is essential only for the cases when we have strong correlations (memory effects) between the neighboring location of the discrete stages of the self-similar process considered. When the self-similar process is presumably random (the neighboring stages of the process do not correlate with each other) then in the infinite series

$$\begin{aligned} K_v(z)&=\sum_{n=-\infty}^{\infty}C_n\exp\left[\left(\ln\left(\frac{1}{\bar{g}}\right)+2n\pi i\right)\frac{\ln(z)}{\ln(\xi)}\right]\\ &=\sum_{n=-\infty}^{\infty}C_n\exp\left[(-v+i\Omega_n)\ln(z)\right], \end{aligned} \tag{6.14}$$

only the zero-Fourier component in (6.14) is conserved

$$K_v(z) \cong C_0 z^{-v}, C_0 = \langle \Pr_v(\ln(z)) \rangle = \int_{-a}^{a} \Pr_v(\ln z + x\ln\xi)dx. \tag{6.15}$$

Using the obvious relationship

$$\frac{t^{\alpha-1}}{\Gamma(\alpha)} =: p^{-\alpha} \tag{6.16}$$

we obtain from (6.4)

$$\begin{aligned} \rho_1 \frac{d\mathbf{v}_1}{dt} = \frac{\gamma_{12}\gamma_{23}\cdots\gamma_{N-1,N}}{\rho_2\rho_3\ldots\rho_N} \frac{1}{\Gamma(N-2+v)} \int_0^t (t-\tau)^{N-2+v} \ldots \\ \ldots F\left(|\mathbf{r}_{1,N}|, |\mathbf{v}_{1,N}|\right) d\tau + \sum_{k=2}^{N} \mathbf{v}_k(0) \frac{t^{k-2}}{(k-2)!}. \end{aligned} \tag{6.17}$$

Here the power-law exponent is supposed to be located in the interval [0,1]. $\Gamma(x)$ – is the ordinary gamma-function. So, based on the results of this section (see more details in [14, 15]) it is possible to derive the Newtonian equation with memory. The memory function for some self-similar process is expressed in the form of the desired Riemann-Liouville integral.

6.4 POSSIBLE GENERALIZATIONS

It is possible to generalize the system of equation (6.3) taking into account the interaction between velocities of *all* layers that are located on the right-hand side with respect to the initial layer

$$\begin{aligned} &\rho_1 \frac{d\mathbf{v}_1}{dt} = \gamma_{1,2}\mathbf{v}_2 + \gamma_{1,3}\mathbf{v}_3 + \ldots + \gamma_{1,N}\mathbf{v}_N, \\ &\ldots \\ &\rho_k \frac{d\mathbf{v}_k}{dt} = \gamma_{k,k+1}\mathbf{v}_{k+1} + \ldots + \gamma_{k,N}\mathbf{v}_N, \\ &\ldots\ldots \\ &\rho_N \frac{d\mathbf{v}_N}{dt} = \mathbf{F}\left(|\mathbf{r}_{1,N}|, |\mathbf{v}_{1,N}|, t\right). \end{aligned} \tag{6.18}$$

It is easy to prove that this system also leads to the structure (6.5), where the memory function has the following form

$$K(t) = \sum_{k=1}^{N-2} r_k \frac{t^k}{k!}, \tag{6.19}$$

where the coefficients r_k form a certain ratio from the coefficients of interaction $\gamma\{k\}$ and densities $\rho\{k\}$. The concrete form of these coefficients is not so important for this analysis. For us it is important only the fact that this model leads again to the structure of the Newtonian's equation with memory. As before, the intermediate layers form a specific 'field' of interaction between the applied force and the velocity of the first layer. For attentive readers as an exercise, we suggest to verify this statement and derive (6.19) from (6.18) by the mathematical induction method.

6.5 CONCLUSION RELATED TO THESE PRELIMINARY RESULTS

The fractional calculus has several deep open problems. Despite of the fact that several attempts were made to clarify the geometrical and the physical meaning of the fractional integral up to now there is no general answer to this problem. Therefore, all models based on the replacement of the classical derivatives with the fractional ones have in some cases a fundamental question from the beginning: what is the physical meaning of the model?

Fractional calculus is an emerging field and the corresponding models are non-local ones. Therefore, it is natural to ask if we can find easily a model, which is non-local and exists in our daily life. The first thing is to ask: what is meaning of the fractional Newtonian mechanics and what is the non-local counterpart of the second law of mechanics. Going back to the roots of the fractional mechanics we can ask ourselves to find a process which is governed by a non-local Newton's second law. Actually, the R-L integral (the fractional derivative is obtained easily by using the fractional derivative) is obtained after a repetitive process which leads naturally to a convolution of an external force with memory function generated by external field that relates the acceleration with the force. We want to stress here again that memory function is formed as a result of elimination of intermediate layers. These layers form a specific field of interaction that relates the external force with movement of the first layer.

Having these results in mind we investigated in the first part of the chapter the Newton's equation applied to consideration of movement of N layers of viscous and heterogeneous liquid and the corresponding movement of each k-th layer (k=1,2…,N) with a specified density. By using the mathematical induction method, we have proved that the second law is given by (6.4), which is transformed to the Newton's equation with memory. However, the kernel contains only integer power. The next step was to find the extension to the fractional ones. The main problem is associated with selection of a way that we have to use, namely, a natural mathematical transformation of the integer number N into a real one or to prove that there exists a physical phenomenon, which is behind the corresponding fractional generalization.

Then one can prove that the result presented by (6.4) can be generalized to the fractional one under the assumption that the temporal interaction between the layers represents itself some general self-similar process and it is distributed over the Cantor set and its possible generalizations.

The results obtained in this chapter open the door for discussion the origin of the fractional Newton's equation as a natural and real extension of the classical counterpart. One can underline again that the results obtained here give a strong impact that the fractional calculus is a powerful natural mathematical tool to describe complex phenomena with memory effect.

After this preliminary analysis one can make the following step in generalization of the relatively simplified system of equation (6.3) and (6.18). What we can add more in order to make the features of this problem more important. From the beginning it was clear that dealing with the physical meaning of fractional operators is not an easy task, and nowadays it has become clear that there is not just one unique interpretation of them. Thus, finding different physical interpretations of various operators becomes a worthy adventure within the fractional calculus universe [17]. The variety of operators proposed in fractional calculus, with their associated advantages and disadvantages, raised a serious classification problem which is going to be resolved during the near future [8, 9]. In order to put some order and structure on the constantly evolving shape of fractional calculus with many different kernels and operators emerging, the notion of general classes of operators was proposed in [18] as a way to unify pure and applied viewpoints in this field. Some specific classes have been considered in recent works such as [19, 20, 21], and their properties investigated in general to provide specific results which are useful in applications.

The problem that is considered in this part of this chapter can be formulated as follows: is it possible to derive the generalized memory connecting the first particle with the last particle if initially they are not connected directly with each other? The relationship between them forms a specific temporal chain connecting only the nearest neighbors. We want to show that in this simple case, when the particles are related with each other by means of linear balance equation, these can be combined for a new form of the Newtonian equation, where all intermediate connections between neighboring particles form the generalized memory. This generalized memory is not simple as one might expect; the interval between local time events *cannot* form the convolution equation. Based on this simple balance equation and its possible generalizations for nonlinear cases, one can derive nontrivial and very general expressions for memory kernels that connect the ends of this specific chain. The temporal influence between the first particle and the final Nth particle is *not* simple and form different types of memories that can lead to fractional integrals with different kernels. The content of this chapter was stimulated by the recent review [18], listing various types of fractional integral and derivative operators. In this part of this chapter, we consider more general balance equation in order to understand more deeply the mathematical structure of the generalized memory that can connect the final ends of the temporal chain connecting the initial and the final Nth particle subjected under the influence of an arbitrary force $\mathbf{F}(t; \mathbf{r}_N; \mathbf{v}_N)$.

6.6 BALANCE EQUATION AND THEIR RECURRENCE SOLUTIONS

The first type of a balance equation that we are going to analyze in this chapter can be written in the form

$$\begin{aligned}
&\frac{d\mathbf{p}_1}{dt} = -\gamma_1(t)\mathbf{p}_1 + \gamma_2(t)\mathbf{p}_2,\\
&\ldots\\
&\frac{d\mathbf{p}_k}{dt} = -\gamma_k(t)\mathbf{p}_k + \gamma_{k+1}(t)\mathbf{p}_{k+1},\\
&\ldots\\
&\frac{d\mathbf{p}_{N-1}}{dt} = -\gamma_{N-1}(t)\mathbf{p}_{N-1} + \gamma_N(t)\mathbf{p}_N,\\
&\frac{d\mathbf{p}_N}{dt} = \mathbf{F}\left(t, \mathbf{p}_N, \mathbf{r}_N\right).
\end{aligned} \tag{6.20}$$

In this system of equation written for the vector momenta $\mathbf{p}_k$ of some particles, the parameters $\gamma_k \equiv \gamma_k(t)$, $k = 1, 2..., N$ are arbitrary functions depending on the temporal variable t. In further, we omit this variable t for simplicity and will restore it in the final expressions. The "mass" of the particles we associate with the given momentum, therefore they can be arbitrary. A "force" (written as an arbitrary external factor) in the final equation $\mathbf{F}_N(t, \mathbf{r}_N, \mathbf{v}_N)$ is expressed by an arbitrary function, also. In order to obtain the final solution and relate the motion of the final particle (k=N) with the first one (k=1) we present the "solution" for the k-th particle in the form

$$\begin{aligned}
&\mathbf{p}_k(t) = \int_{t_0}^{t} \gamma_{k+1}(\tau) \cdot \mathbf{p}_{k+1}(\tau) M_k(\tau; t) d\tau\\
&M_k(\tau; t) = \exp\left[-\int_{\tau}^{t} \gamma_k(u) du\right].
\end{aligned} \tag{6.21}$$

We put $\mathbf{p}_k(t_0)$=0 and omit the term associated with homogeneous solution in order to stress the influence of the memory effect. We should stress also that temporal variable t is an arbitrary variable with respect to the and satisfy to the inequality $\tau \le t < \infty$. For further manipulations and transformations of the system (6.20), it is convenient to define the following function

$$\begin{aligned}
Q_{k+1,k}\left(\tau_{k+1}; \tau_k\right) &= \gamma_{k+1}\left(\tau_{k+1}\right) M_k\left(\tau_{k+1}, \tau_k\right)\\
&\equiv \gamma_{k+1}\left(\tau_{k+1}\right) \exp\left[-\int_{\tau_{k+1}}^{\tau_k} \gamma_k(u) du\right].
\end{aligned} \tag{6.22}$$

With the help of expression (6.22), the recurrence relationship connecting the neighboring momenta in expression (6.21) accepts the following form

$$\begin{aligned}
&\mathbf{p}_1\left(\tau_1\right) = \int_{t_0}^{\tau_1} \mathbf{p}_3\left(\tau_3\right) \tilde{Q}_{3,1}\left(\tau_3; \tau_1\right) d\tau_3,\\
&\tilde{Q}_{3,1} = \int_{t_0}^{\tau_1} Q_{3,2}\left(\tau_3; \tau_2\right) \cdot Q_{2,1}\left(\tau_2; \tau_1\right) d\tau_2 \equiv Q_{3,2}\left(\tau_3; \tau_2\right) I_2 Q_{2,1}\left(\tau_2; \tau_1\right).
\end{aligned} \tag{6.23}$$

The symbol I_2 means the integration over the variable τ_2 in the limits of the following senior variable $[t_0,\ \tau_1]$. The last expression allows to relate the first particle with the final particle using the mathematical induction method. The final solution can be written in the form

$$\begin{aligned}
&\mathbf{p}_1(\tau_1) = \int_{t_0}^{\tau_1} \mathbf{p}_N(\tau_N)\,\tilde{Q}_{N,1}(\tau_N;\tau_1)\,d\tau_N \\
&\tilde{Q}_{N,1}(\tau_N;\tau_1) \\
&= \prod_{p=0}^{N-3} Q_{N-p,N-p-1}(\tau_{N-p};\tau_{N-p-1})\,I_{N-p-1}Q_{N-p-1,N-p-2}(\tau_{N-p-1};\tau_{N-p-2}) \\
&= Q_{N,N-1}(\tau_N;\tau_{N-1})\,I_{N-1}Q_{N-1,N-2}(\tau_{N-1};\tau_{N-2})\ldots Q_{3,2}(\tau_3;\tau_2)\,I_2Q_{2,1}(\tau_2;\tau_1).
\end{aligned} \tag{6.24}$$

In order to attach to equation (6.24) its final form one can present the solution of the system (6.20) as

$$\begin{aligned}
&\mathbf{p}_1(t) = \int_{t_0}^{t} \tilde{Q}_{N,1}(\tau_N;t)\,\mathbf{p}_N(\tau_N)\,d\tau_N \\
&\frac{d\mathbf{p}_N}{dt} = \mathbf{F}(t,\mathbf{p}_N,\mathbf{r}_N).
\end{aligned} \tag{6.25}$$

The system (6.25) proves the statement made above. All particles are related with each other by means of the generalized memory function that has the form expressed by relationship (6.24). It is obvious that the solution for any particle is expressed by similar expression. For us it is important that the structure of the generalized memory function $\tilde{Q}_{N,1}(\tau_N, t)$ and its structure at certain conditions can lead to the specific classes of the fractional integrals proposed in the review [18].

It is interesting to consider another balance equation, when the first momenta is perturbed by the same force $\mathbf{F}(t,\, \mathbf{p}_1,\, \mathbf{r}_1)$

$$\begin{aligned}
&\frac{d\mathbf{p}_1}{dt} = \mathbf{F}(t,\mathbf{p}_1,\mathbf{r}_1) \\
&\ldots \\
&\frac{d\mathbf{p}_k}{dt} = -\gamma_k(t)\mathbf{p}_k + \gamma_{k-1}(t)\mathbf{p}_{k-1}, \\
&\ldots \\
&\frac{d\mathbf{p}_{N-1}}{dt} = -\gamma_{N-1}(t)\mathbf{p}_{N-1} + \gamma_{N-2}(t)\mathbf{p}_{N-2}, \\
&\frac{d\mathbf{p}_N}{dt} = -\gamma_N(t)\mathbf{p}_N + \gamma_{N-1}(t)\mathbf{p}_{N-1}.
\end{aligned} \tag{6.26}$$

Making the same manipulations with the second type of balance equation we obtain another relationship connecting the first perturbed momenta with the final one. In order to see the difference between the systems (6.20) and (6.26) let us write down the intermediate equation

$$\mathbf{p}_k(\tau_k) = \int_{t_0}^{\tau_k} \mathbf{p}_{k-1}(\tau_{k+1})\,Q_{k-1,k}(\tau_{k-1};\tau_k)\,d\tau_{k-1} \tag{6.27}$$

with

$$
\begin{aligned}
Q_{k-1,k}\left(\tau_{k-1};\tau_k\right) &= \gamma_{k-1}\left(\tau_{k-1}\right) M_k\left(\tau_{k-1},\tau_k\right) \\
&\equiv \gamma_{k-1}\left(\tau_{k-1}\right)\exp\left[-\int_{\tau_{k-1}}^{\tau_k}\gamma_k(u)du\right].
\end{aligned} \tag{6.28}
$$

Using the recurrence relationship, we obtain finally

$$
\begin{aligned}
\tilde{Q}_{1,N}\left(\tau_1;\tau_N\right) =& Q_{1,2}\left(\tau_1;\tau_2\right) I_2 Q_{2,3}\left(\tau_2;\tau_3\right) \\
& \ldots Q_{N-2,N-1}\left(\tau_{N-2};\tau_{N-1}\right) I_{N-1} Q_{N-2,N-1}\left(\tau_{N-1};\tau_N\right).
\end{aligned} \tag{6.29}
$$

Based on this equation, one can write

$$
\begin{aligned}
\frac{d\mathbf{p}_1}{dt} &= F\left(t,\mathbf{p}_1,\mathbf{r}_1\right) \\
\frac{d\mathbf{p}_N}{dt} &= \int_{t_0}^{t}\tilde{Q}_{1,N}\left(t;\tau_N\right)\mathbf{p}_1\left(\tau_N\right)d\tau_N.
\end{aligned} \tag{6.30}
$$

Because of the difference between the intermediate memories (6.24) and (6.29), one can conclude that the final equations (6.25) and (6.30) are *different*. It means that in general case. $\tilde{Q}_{N,1}\left(\tau_N,t\right) \neq \tilde{Q}_{1,N}\left(t,\tau_N\right)$. What is the meaning of this important inequality? If the forces act independently, say, the first force acts from the end of the temporal chain, while the second force perturbs the chain starting from the beginning, then, in this case, the influence of the final momenta $p_N(t)$ to the first one $p_1(t)$ expressed by the function $Q_{N,1}(\tau_N;\tau_1)$ does *not* coincide with the function $Q_{1,N}(\tau_1;\tau_N)$ that connects the first perturbed momenta $p_1(t)$ with the final one $p_N(t)$. This inequality forces to reconsider the third Newton's law that is valid for particles interacting with each other *instantaneously*. The presence of the memory having different structure can modify this law.

6.7 PARTIAL CASES AND POSSIBLE GENERALIZATION

In this section, we consider one important generalization that allows to propagate the results obtained in the previous sections for *nonlinear* case. We consider the balance equation for some physical value $y(t)$ (it can be the impulse projection, velocity, the probability density etc.,) that is associated with other similar values by nonlinear balance equation of the type:

$$
\begin{aligned}
\frac{dy_1}{dt} &= -\gamma_1(t)y_1 + \gamma_2(t)y_1^{\alpha_1}y_2^{1-\alpha_2}, \\
&\ldots \\
\frac{dy_k}{dt} &= -\gamma_k(t)y_k + \gamma_{k+1}(t)y_k^{\alpha_k}y_{k+1}^{1-\alpha_{k+1}}, \\
&\ldots \\
\frac{dy_{N-1}}{dt} &= -\gamma_{N-1}(t)y_{N-1} + \gamma_N(t)y_{N-1}^{\alpha_{N-1}}y_N^{1-\alpha_N}, \\
\frac{dy_N}{dt} &= \Phi_N\left(t,y_N,\mathbf{r}_N\right).
\end{aligned} \tag{6.31}
$$

Here the functions $\gamma_k(t)$ (k=1,2,...,N) has a sense as before (they regulate the nonstationary balance between different parts of the given system), the fractional power-law parameters α_k are responsible for nonlinear exchange between the species/"particles" defined by the functions $y_k(t)$. If we change the corresponding variables $u_k(t) = y_k^{1-\alpha_k}(t)$ then the system of balance equation accepts the form

$$
\begin{aligned}
&\frac{du_1}{dt} = -\gamma_1(t)u_1 + \gamma_2(t)u_2(t),\\
&\cdots\\
&\frac{du_k}{dt} = -\gamma_k(t)u_k + \gamma_{k+1}(t)u_k(t),\\
&\cdots\\
&\frac{du_{N-1}}{dt} = -\gamma_{N-1}(t)u_{N-1} + \gamma_N(t)u_N,\\
&\frac{dy_N}{dt} = \Phi_N(t, y_N, \mathbf{r}_N).
\end{aligned}
\tag{6.32}
$$

As one can notice from system (6.32), it has the same structure as the system (6.20) and all procedure applied for derivation of the generalized memory from the system (6.20) can be completely applied to system (6.32). In particular, if the last equation in system (6.32) has the form

$$
\frac{dy_N(t)}{dt} = f_N(t)y_N(t) + g_N(t)y_N^{\alpha_N}(t) \tag{6.33}
$$

then making the same change of variables $u_N(t) = y_N^{1-\alpha_N}(t)$ one can reduce this equation to the linear form of the type

$$
\frac{du_N(t)}{dt} = f_N(t)u_N(t) + g_N(t). \tag{6.34}
$$

Equation (6.34) together with system (6.32) considerably generalize the previous approach applied for the linear case.

Alongside with the nonlinear structure (6.31) describing the exchange balance between "parts" of the whole system one can propose other structures that generate finally the generalized memory functions. Let us start from the simplest form

$$
\begin{aligned}
&\frac{dp_k(t)}{dt} = -\gamma_k(t)p_k(t) + \gamma_{k+1}(t)p_{k+1}(t)p_k(t),\\
&\frac{dp_N(t)}{dt} = F(p_N, t, \{r\}), k = 1, 2, \ldots, N-1.
\end{aligned}
\tag{6.35}
$$

The physical meaning of the parameters in (6.35) is the same as before, however, the character of exchange between the chosen species is *different*. In the last equation, we define the fictitious "force" that interacts only with the final "particle" $p_N(t)$. The system (6.35) allows to separate the fixed variable $p_k(t)$ and present it in the form

$$
\begin{aligned}
p_k(t) = p_k(0)\exp\left[\int_{t_0}^{t}\left[-\gamma_k(\tau)+\gamma_{k+1}(\tau)p_{k+1}(\tau)\right]d\tau\right] \equiv \\
\equiv p_k(0)E_{k,k+1}(t).
\end{aligned}
\tag{6.36}
$$

In order to find the desired recurrence relationship connecting the different parts of the generalized memory we introduce the following notations:

$$
E_{k,k+1}(\tau_k) = \exp\left[\int_{t_0}^{\tau_k}\left[-\gamma_k(\tau_{k+1})+\gamma_{k+1}(\tau_{k+1})p_{k+1}(0)E_{k+1,k+2}(\tau_{k+1})\right]d\tau_{k+1}\right].
\tag{6.37}
$$

This equation connects by the recurrence relationship the successive parts of the generalized memory. By complete analogy with this case, one can consider the nonlinear "balance" equation of the type

$$
\begin{aligned}
\frac{dp_k(t)}{dt} &= -\gamma_k(t)p_k^{\alpha_k}(t)+\gamma_{k+1}(t)p_{k+1}^{\alpha_{k+1}}(t)p_k^{\alpha_k}(t) \\
\frac{dp_N(t)}{dt} &= F\left(p_N, t, \{r\}\right), k = 1, 2, \ldots, N-1
\end{aligned}
\tag{6.38}
$$

Using new set of variables $u_k(t) = p_k^{\alpha_k}(t)$ the first part of system (6.38) for $k =$ 1,2,...,N-1 one can reduce to the differential equation with separable variables

$$
\frac{1}{\alpha_k}u_k^{\left(\frac{1}{\alpha_k}-2\right)}\frac{du_k(t)}{dt} = -\gamma_k(t)+\gamma_{k+1}(t)u_{k+1}(t),
\tag{6.39}
$$

From this simple differential equation, one can receive the desired recurrence relationship connecting different parts of the generalized memory for the set of variables $u_k(t)$, $k = 1, 2, ..., N-1$.

$$
u_k(\tau_k) = u_k(0)\left[1+\frac{1-\alpha_k}{(u_k(0))^{\frac{1}{\alpha_k}-1}}\left\{\int_{t_0}^{\tau_k}\left[-\gamma_k(\tau_{k+1})+\gamma_{k+1}(\tau_{k+1})u_{k+1}(\tau_{k+1})\right]\right\}d\tau_{k+1}\right]^{\frac{\alpha_k}{1-\alpha_k}}.
\tag{6.40}
$$

These recurrent links (6.37) and (6.40) help to establish the desired relationships that exist between the different fractional integral structures proposed in review [18]. As one can see these fractional integrals can be generated from the generalized memory that gains, in turn, from the different balance equations written above.

6.7.1 Some partial cases

Let us consider some partial cases that follows from general equation (6.20) and (6.21). We put $\gamma_1(t) = \gamma_2(t) = ... = \gamma_N(t)$ and in other cases these functions can be arbitrary. In this case, the kernel $Q_{k+1,k}$ accepts the following form

$$Q_{k+1,k}(\tau_{k+1};\tau_k) = \gamma(\tau_{k+1})\exp\left[-\int_{\tau_{k+1}}^{\tau_k}\gamma(u)du\right]. \tag{6.41}$$

Let us calculate the structure in expression (6.41)

$$\begin{aligned}\tilde{Q}_{3,1}(\tau_3;\tau_1) &= Q_{3,2}(\tau_3;\tau_2)\,I_2 Q_{2,1}(\tau_2;\tau_1)\\ &= \int_{t_0}^{\tau_1}\gamma(\tau_3)\exp\left[-\int_{\tau_3}^{\tau_2}\gamma(u)du\right]\gamma(\tau_2)\exp\left[-\int_{\tau_2}^{\tau_1}\gamma(u)du\right]d\tau_2\\ &= \gamma(\tau_3)\exp\left[-\int_{\tau_3}^{\tau_1}\gamma(u)du\int_{t_0}^{\tau_1}\gamma(\tau_2)\,d\tau_2 \equiv (J\gamma)_{\tau_1}Q_{3,1}(\tau_3;\tau_1).\right.\end{aligned} \tag{6.42}$$

As it follows from (6.42) the combination of integrals inside the exponential functions does *not* depend on the intermediate variable τ_2 and this property facilitates considerably further calculations. From the result of application of relationship (6.42) for evaluation of the kernel $\tilde{Q}_{N,1}(\tau_N;\tau_1)$, we obtain finally the following compact expression

$$\begin{aligned}&\tilde{Q}_{N,1}(\tau_N;\tau_1) = Q_{N,N-1}(\tau_N;\tau_{N-1})\,I_{N-1,N-2}Q_{N-1,N-2}(\tau_{N-1};\tau_{N-2})\\ &\ldots Q_{3,2}(\tau_3;\tau_2)\,I_2Q_{2,1}(\tau_2;\tau_1) = \left[\int_{t_0}^{\tau_1}\gamma(u)du\right]_{\tau_1}^{N-2} Q_{N,1}(\tau_N;\tau_1),\\ &N = 2,3,\ldots.\end{aligned} \tag{6.43}$$

Therefore, the basic equation (6.25) accepts the following form

$$\begin{aligned}&\mathbf{p}_1(t) = \int_{t_0}^{t}(J\gamma)_t^{N-2}Q_{N,1}(\tau_N;t)\,\mathbf{p}_N(\tau_N)\,d\tau_N,\\ &\frac{d\mathbf{p}_N}{dt} = \mathbf{F}(t,\mathbf{p}_N,\mathbf{r}_N).\\ &(J\gamma)_t = \int_{t_0}^{t}\gamma(u)du,\; Q_{N,1}(\tau_N;t) = \gamma(\tau_N)\exp\left[-\int_{\tau_N}^{t}\gamma(u)du\right].\end{aligned} \tag{6.44}$$

Equation (6.44) represents itself the generalized memory for this partial case.

In order to see another possible option for further evaluation of the generalized memory one can rewrite the general equation (6.20) and present it in the form

$$\begin{aligned}&\frac{d\mathbf{p}_k}{dt} = -W(t)\mathbf{p}_k + U_{k+1}(t)\mathbf{p}_{k+1},\\ &k = 1,2\ldots,N-1\\ &\frac{d\mathbf{p}_N}{dt} = \mathbf{F}(t,\mathbf{p}_N,\mathbf{r}_N).\end{aligned} \tag{6.45}$$

In this balance equation we suppose that all momenta have the unified expenditure/relaxation rate W, however, the energy receipts can be different and expressed by the set of the functions $U_k(t)$ (k=1,2,...,N-1). In this case the local memory has the following structure

$$Q_{k+1,k}(\tau_{k+1};\tau_k) = U_{k+1}(\tau_{k+1})\exp\left[-\int_{\tau_{k+1}}^{\tau_k} W(u)du\right] \tag{6.46}$$

$$\begin{aligned}
&\tilde{Q}_{3,1}(\tau_3;\tau_1) = Q_{3,2}(\tau_3;\tau_2)\, I_2 Q_{2,1}(\tau_2;\tau_1) = \\
&= \int_{t_0}^{\tau_1} U_3(\tau_3)\exp\left[-\int_{\tau_3}^{\tau_2} W(u)du\right] U_2(\tau_2)\exp\left[-\int_{\tau_2}^{\tau_1} W(u)du\right] d\tau_2 \\
&= U(\tau_3)\exp\left[-\int_{\tau_3}^{\tau_1} W(u)du\right]\int_{t_0}^{\tau_1} U_2(\tau_2)\, d\tau_2 \equiv (JU_2)_{\tau_1} Q_{3,1}(\tau_3;\tau_1).
\end{aligned} \tag{6.47}$$

This property (that is independent from the exponential functions on the intermediate temporal parameter τ_2) helps to facilitate further calculations. Realizing similar calculations as in (6.44),

$$\begin{aligned}
&\tilde{Q}_{N,1}(\tau_N;\tau_1) = Q_{N,N-1}(\tau_N;\tau_{N-1})\, I_{N-1,N-2} Q_{N-1,N-2}(\tau_{N-1};\tau_{N-2}) \\
&\ldots Q_{3,2}(\tau_3;\tau_2)\, I_2 Q_{2,1}(\tau_2;\tau_1) \\
&= \left[\left(\int_{t_0}^{\tau_1} U_2(u)du\right)\left(\int_{t_0}^{\tau_1} U_3(u)du\right)\ldots\left(\int_{t_0}^{\tau_1} U_{N-1}(u)du\right)\right] U_N(\tau_N) \\
&\ldots \exp\left[-\int_{\tau_N}^{\tau_1} W(u)du\right] \\
&\equiv \left[\prod_{k=2}^{N-1}\left(\int_{t_0}^{\tau_1} U_k(u)du\right)\right] Q_{N,1}(\tau_N;\tau_1) \\
&N = 2,3,\ldots
\end{aligned} \tag{6.48}$$

Obviously, when all parameters in the balance equation (6.45) become equal each other: i.e., $U_2(t){=}U_3(t){=}\ldots U_N(t)$ the memory kernel of (6.46) coincides with solution presented by (6.44). This expression generalizes expression (6.43) and finally the generalized Newton's equation accepts the following structure

$$\begin{aligned}
\mathbf{p}_1(t) &= \int_{t_0}^{t} \tilde{Q}_{N,1}(\tau_N;t)\, \mathbf{p}_N(\tau_N)\, d\tau_N \\
\frac{d\mathbf{p}_N}{dt} &= \mathbf{F}(t,\mathbf{p}_N,\mathbf{r}_N).
\end{aligned} \tag{6.49}$$

This system with the memory defined by expression (6.45) generalizes all previous results and serves as a possible source for derivation of the different fractional integrals.

Let us consider some interesting cases when the kernel in (6.25) can be *separable.*

Case 1. We put $\gamma(t) = \frac{d}{dt}\Phi(t)$.

and calculate other functions forming the memory in (6.44)

$$\begin{aligned}
&(J\gamma)_t = \Phi(t) - \Phi(t_0) \\
&Q(\tau_N;t) = \left(\frac{d}{d}\Phi(t)\right)_{\tau_N} \exp\left[-\Phi(t) + \Phi(\tau_N)\right].
\end{aligned} \tag{6.50}$$

Therefore, in this case the memory is separated and the first row of (6.44) accepts the following form

$$\mathbf{p}_1(t) = [\Phi(t) - \Phi(t_0)]^{N-2} \exp[-\Phi(t)] \int_{t_0}^{t} \Phi'(\tau_N) \exp[\Phi(\tau_N)] \mathbf{p}_N(\tau_N) d\tau_N. \quad (6.51)$$

Case 2. We put $\gamma(t) = \frac{d}{dt}(\ln(f(t))) = \frac{f'(t)}{f(t)}$.
and similarly, calculate other functions in (6.44)

$$(J\gamma)_t = \ln\left[\frac{f(t)}{f(t_0)}\right], \quad Q(\tau_N; t) = \frac{f'(\tau_N)}{f(t)} \quad (6.52)$$

$$\mathbf{p}_1(t) = \frac{1}{f(t)}\left[\ln\left(\frac{f(t)}{f(t_0)}\right)\right]^{N-2} \int_{t_0}^{t} f'(\tau_N) \mathbf{p}_N(\tau_N) d\tau_N. \quad (6.53)$$

In what cases one can receive the desired difference leading to the fractional integral? Coming back to equation (6.44) one can notice that when all functions $\gamma(t) = \gamma_0$. In this case we have

$$J\gamma_t = \int_{t_0}^{t} \gamma(u) du \Rightarrow \gamma_0 (t - t_0)^{-v},$$
$$0 < v = \ln(1/M)/\ln(1/\xi) < 1. \quad (6.54)$$

Here we applied the ideas outlined in the Sec. 6.3 when the averaging procedure realized over Cantor set having M bars leads to the RL-fractional integral. Taking into account the relationship (6.54), we obtain expression for $Q_{N,1}(\tau_N, t)$

$$Q_{N,1}(\tau_N, t) = \gamma_0 \exp\left[-\gamma_0 (t - \tau_N)^{-v}\right]. \quad (6.55)$$

Taking into account the last expressions we obtain finally

$$\mathbf{p}_1(t) = \gamma_0^{N-1} \int_{\tau_N}^{t} (t - \tau_N)^{-v(N-2)} \exp\left[-\gamma_0 (t - \tau_N)^{-v}\right] \mathbf{p}_N(\tau_N) d\tau_N. \quad (6.56)$$

Here we replace t_0 for τ_N supposing that this temporal parameter coincides with the initial time.

As an example, we suggest to derive the fractional integral from relationship (6.48) when $U_2(u) = ... = U_{N-1}(u) = U_0 = const$ We omit here the log-periodic corrections in expressions for the fractional integrals. They considered in paper [22].

6.8 THE GENERALIZED HURST EXPONENT

In this chapter we want to show *how* to extract some deterministic information from long-time series. We concentrated presumably on generalization of the well-known Hurst empirical law and consider a set of reduced parameters for quantitative description of the long-time series. Usually, these long-time series are considered as a specific response of a complex system (economical, geophysical, electromagnetic, and other systems), where the successive fixation of external factors becomes impossible. We considered three hypotheses. The first one contains one power-law exponent, the second one incorporates two power-law exponents and in many cases these power-law exponents are complex-conjugated. The third hypothesis contains three power-law exponents and two of them are complex-conjugated, as well. These hypotheses describe with acceptable accuracy (relative error does not exceed 1%) a wide set of trendless sequences (TLS) associated with radiometric measurements. We should notice also that the R/S ratio covers all set of the reduced measurements (not only the asymptotic region) and these fitting parameters can be used as the reduced parameters for description of the given data. We should stress also that this general approach can be applied also for other TLS, as well. The generalized kernel expressed in the form of the Hurst exponent can serve as a memory for description of the fractal Brownian motion. Not going into more specific details that are described well in the book [23] we want to show *how* derive a new type of memory from the segments of random long time series. As an example we took the radiometric data proving that this generalization is valid. How to formulate the partial problem that we are going to solve in this chapter?

Any complex system represents itself a *diffusive* system, where the direct / deterministic physical laws are absent for their accurate description and, therefore, for further predictions. In other words, the direct fitting function derived from the first principles (or from an appropriate simple model) and corresponding to the deterministic law is *absent*. In these cases, the correlation analysis connecting different physical and other values starts to play a significant role. The drawbacks of the conventional correlation analysis were expanded in the previous chapter 5. The trendless sequences (TLS) defined frequently as long-time series (LTS) or a "noise" and describing fluctuations originated from different sources, represent the most difficult response for analysis and extraction the reliable and stable information. However, during the last 20–30 years many researchers are trying to extract a reliable information from TLS. We remind here the progress in analysis of the TLS achieved in electrochemistry [23, 24, 25, 26, 27, 28] and other branches of sciences thanks to the works of Prof. S. F. Timashev, who proposed and developed the flicker-noise spectroscopy [29, 30] and Prof. R.M. Yulmetyev, who applied the Mori-Zwanzig formalism [31, 32, 33] for description of many phenomena in different complex systems, having diffusive nature. We should notice also a monograph written by one of us (YQC) with his pupils [34] associated with analysis of different types of model noises, having presumably a fractal/scaling origin. From another side, one can remind the β-distribution approach [35], when analysis of real temporal-noises in biological systems is reduced to almost 10-20 robust parameters associated with parameters of beta-distributions and their fitting functions. We should stress also that this bell-like curve approach is based on the generalization of the detrended fluctuation analysis (DFA) developed by Peng et al., [36]. In this chapter, we concentrate presumably on the generalization of the Hurst empirical law playing an important role in analysis of the different time-series. As it is known [23] that this relationship characterizes the growth of the cumulative averaged values and can be expressed approximately as

$$\frac{R(x)}{S(x)} \overset{x>>1}{\cong} C \cdot x^{H}. \tag{6.57}$$

Here C is a positive constant and power-law exponent H is located in the interval [0,1]. This empirical law discovered by Hurst has asymptotic character and is valid for the relatively segments of some temporal series x. We suppose also that the variable x can coincide with temporal value and, in general case, can be arbitrary coinciding with Cartesian coordinates, with wave length and etc.

The basic aim of this chapter is to generalize expression (6.57) and propagate this relationship for segments of temporal series x having an *arbitrary* length. An attentive analysis of the previous results obtained in the books [23, 34] allows to suggest three hypothesis that cover completely the chosen segment x:

$$\begin{aligned}
&GH(1) \equiv y_1(x; A, B, H) = \frac{R(x)}{S(x)} \cong A + B \cdot x^{H} \\
&GH(2) \equiv y_2\left(x; B_{0,1,2} H_{1,2,}\right) = \frac{R(x)}{S(x)} \cong B_0 + B_1 \cdot x^{H_1} + B_2 \cdot x^{H_2} \\
&GH(3) \equiv y_3\left(x; A_{0,1,2,3} H_{1,2,3}\right) = \frac{R(x)}{S(x)} \cong A_0 + A_1 \cdot x^{H_1} + A_2 \cdot x^{H_2} + A_3 x^{H_3}.
\end{aligned} \tag{6.58}$$

We should stress here that these hypotheses are not only mechanical continuation of the power-law exponents. The second and the third hypotheses contain *complex* Hurst power-law exponents that cannot be provided by the first hypothesis. Analysis of real data shown below demonstrates an interesting relationship, namely, in order to provide the accuracy of the fitting procedure in the cases of the second/third hypotheses two Hurst exponents should be always complex-conjugated. In order to fit completely the given segment by these fitting functions we use the eigen-coordinates method, that was described in details in the paper [37] and in the book [38]. However, in the Mathematical Appendix of this chapter we give the desired transformations that allows to transform these non-linear fitting functions to the basic linear relationship

$$Y\left(x_j\right) = \sum_{s=0}^{S-1} C_s X_s\left(x_j\right). \tag{6.59}$$

The relationship (6.59) allows to apply the linear least squares method (LLSM) and find the desired fitting constants in (6.58) and relate the constants C_s in (6.59) with initial fitting constants (6.59) for GH(1), 5 for GH(2), and 7 for GH(3) in (6.58) by some non-linear relationships.

6.8.1 Description of algorithm

In order to prepare the desired ratio $R(x)/S(x)$ reflecting a specific and "universal" response of some complex system and described by the given TLS, we divide the given LTS on the segment having the fixed length a. Then we calculate the ratio

$R(a)/S(a)$. Then we successively increase the length a, $2a$, $\ldots ka$, up to the final length Ka which satisfies to the condition $K = [N/a]$, where N – is the total number of data points, $[\ldots]$ is the symbol of taking the integer part. For each segment we calculate $R(ka)/S(ka)$ $(k=1,2,\ldots,K)$ and finish these calculations up to the final value $k=K$. Then we try to fit the ratios $R(ka)/S(ka)$ for each k with the help of the hypotheses (6.58) with respect to segments $x_1,x_2,\ldots,x_K$. If the first hypothesis provides the fit with relative error less 1% then it is accepted. This function is defined by the conventional expression

$$\mathrm{Err}(y,f) = \left(\frac{\mathrm{Stdev}(y-f)}{\mathrm{mean}(|y|)}\right)\cdot 100\%$$
$$\mathrm{Stdev}(Y) = \left(\frac{1}{N}\sum_{j=1}^{N}(Y_j)^2\right)^{1/2}, \quad \mathrm{mean}(Y) = \frac{1}{N}\sum_{j=1}^{N}(Y_j) \tag{6.60}$$

Here y_j $(j=1,2\ldots,N)$ is the measured function and f_j is the corresponding fitting function from (6.58). For more complex hypotheses we use the same criterion associated with the value of the relative error. In order to accelerate these calculations, we apply the procedure, which was used in the book [38] as reduction to three incident points (RTIP). The essence of this effective procedure is the following. We strongly recommend to use it for any LTS. We choose a segment of the length L that contains $y_1,y_2,\ldots,y_L = \{y\}_L$ data points and then select only three points ($\max(\{y\}_L)$, $\mathrm{mean}(\{y\}_L)$, $\min(\{y\}_L)$) that are invariant relatively $(L)!$ permutations corresponding to the chosen segment. Therefore, one can obtain the *compressed* LTS containing three subsequences of the length $[N/L]$ that demonstrate the distributions of the maximal, mean and minimal values of the initial TLS, correspondingly. The evaluation of correlations made with the help of the statistics of the fractional moments [38] shows that the compression in $L=10$ times is sufficient, because it leaves the correlation coefficient in the interval [0.9–0.975]. In this section we put also the initial length $a \cong L$ equaled to the value of the compression coefficient. The reason of this equality is related to the fact for keeping the values of correlations of the compressed TLS, from one side, and, from another side, the length $a \cong 10$ gives a sufficient number of points for calculating the desired ratio $R(a)/S(a)$. Finishing this section, one can describe the algorithm that will be used for calculation of the fitting parameters, in (6.58). These three (A,B,H), $5(B_{0,1,2},H_{1,2})$ and $7(A_{0,1,2,3},H_{1,2,3})$ parameters in (6.58) can be considered as the *reduced* parameters reflecting the behavior of some complex system, if responses of the chosen system are presented in the form of the TLS. Therefore, the algorithm is simple, incorporates three steps only and cannot create uncontrollable errors related to its application.

- *Step 1.* Selection of L and application of the RTIP-procedure, if necessary.
- *Step 2.* Calculation of the ratios $R(ka)/S(ka)$ up to $K=[N/a]$ with $a=L$.
- *Step 3.* The fit of the obtained ratios to the hypothesis in (6.58) with the help of the EC-method.

TABLE 6.1 First hypothesis statistics.

	Fit error	H	A
Mean	1.0%	-1.0	3.19
Std	0.4%	0.1	0.02

The basic linear relationship (6.59) for these non-linear hypotheses are given in the Mathematical Appendix in the end of this chapter. The value of the relative/dimensionless error is defined by expression (6.60).

6.8.2 The description of the data

We tested the aforementioned hypothesis on some examples of TLS which are common in practical applications. First, we analyzed noise from radio receivers of Adalm Pluto SDR and National Instruments USRP-2930. Pluto Software Defined Radio (SDR) is a programmable radio device, which can be used both in transmitter (TX) and receiver (RX) modes. The receiver path features 12-bit Analog-to-Digital Converter (ADC) and a cascade of analog filters. NI USRP-2930 is also a programmable radio device capable of receiving and sending RF signals. Compared to Pluto SDR it has a more precise ADC (14 bits) and a less noisy receiver path.

We recorded the receiver noise both from Pluto SDR and from USRP, i.e., the output of ADC when there are no signals being transmitted. The recorded signal contains airwave noise, the noise of amplifiers and the noise of ADC.

In addition to this, we recorded audio noise from Audio-Technica AT2020USB+ microphone in a soundproof music recording studio. Since it is a stereo microphone, there are two noise records: for the left and the right channel. In this study we subtracted left channel signal from the right channel one and analyzed the difference.

To summarize the data description, we used a total of three datasets of different origin: a) phase of the Adalm Pluto SDR RX noise b) phase of the NI-USRP 2930 RX noise c) differential noise of AT2020USB+ microphone.

6.8.3 The treatment of the selected data

The limited size of this section does not allow to present all fitting parameters. However, based on the criterion associated with minimal value of the fitting error less than 2%, one can demonstrate the obtained results. Tables 6.1–6.3 present the key parameters corresponding to hypotheses (6.58) tested on data described above. First hypothesis showed quite a good fitting on Pluto SDR noise, second hypothesis was tested on NI USRP noise, and the audio noise of AT2020USB+ was analyzed using the third hypothesis. Some parameters have relatively standard deviation and therefore they are not included in the tables. The fit to the proposed hypothesis is shown in Figs. 6.1–6.3.

To summarize the data description, we used three datasets of different origin:

1. phase of the Adalm Pluto SDR receiver noise. (Hyp.1)

TABLE 6.2 Second hypothesis statistics.

	Fit error	B0	H1, real	H2, real	H1, complex	H2, complex
mean	1.0%	4.17	0.59	-0.52	-0.05	0.55
std	0.3%	0.4	0.42	0.38	0.07	0.15

TABLE 6.3 Third hypothesis statistics.

	Fit error	B2	B3	H1	H2	H3
mean	2.0%	0.048	-0.05	-0.16	-0.04	-2.4
std	0.5%	0.5	0.59	0.25	0.4	0.5

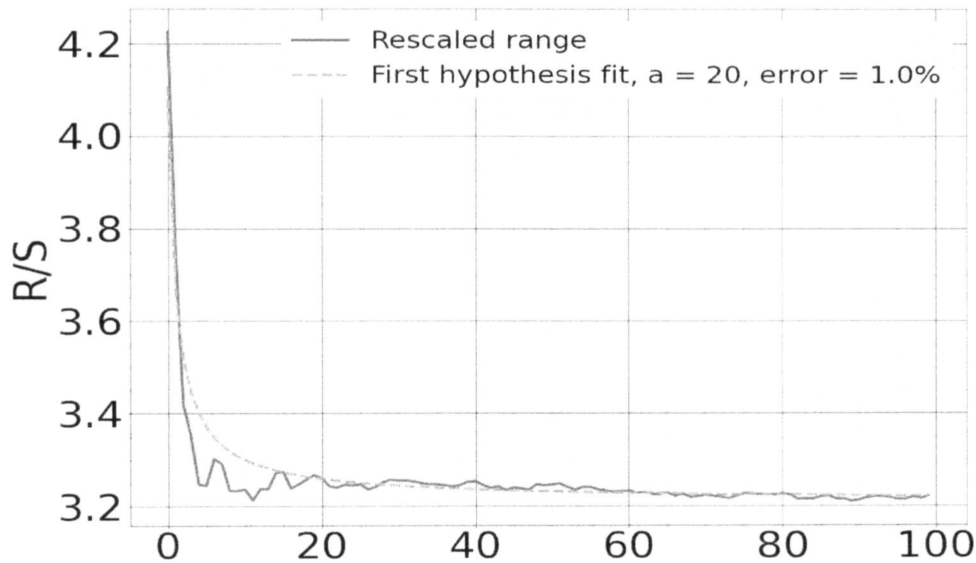

Figure 6.1 Rescaled range fitting of Adalm Pluto SDR noise phase (Hyp 1).

2. phase of the NI-USRP 2930 receiver noise. (Hyp.2)
3. differential noise of AT2020USB+ microphone. (Hyp.3)

6.9 RESULTS AND DISCUSSIONS

In this chapter we consider two origins of appearance of the memory. The first origin is explained in Secs. 6.1– 6.7. We show here *how* to derive the memory from the system of equations that initially does not have a memory. The problem is not solved completely. Probably the proposed equation contains some interesting combinations that can lead to other forms of the fractional integrals that are proposed by mathematicians without discussion their origin. This research is waiting its continuation by ambitious young researchers.

In the second part 6.8, we show how to transform TLS to the desired R/S ratio. This ratio reflects the scaling properties of the complex system and can be

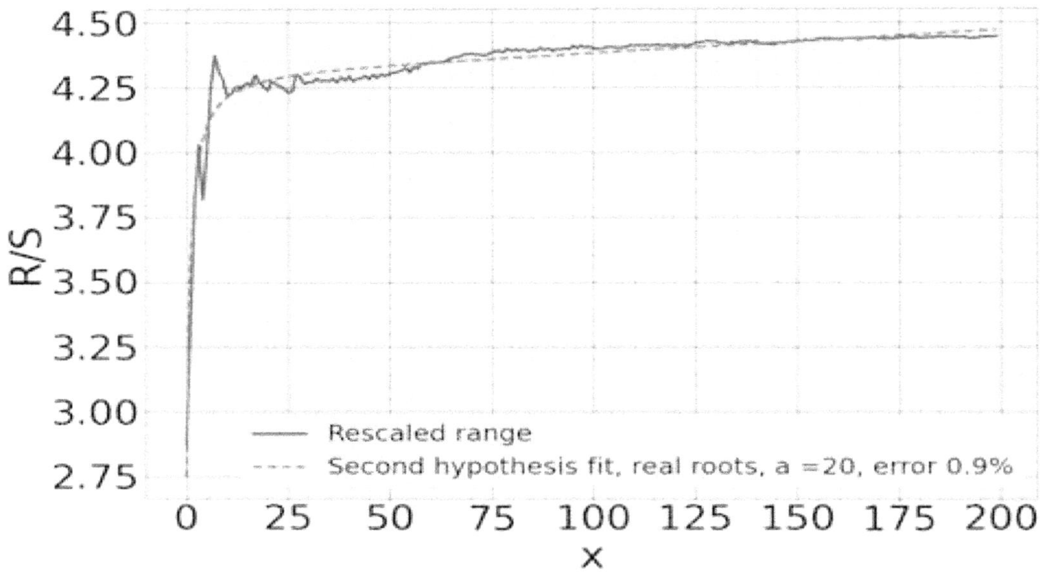

Figure 6.2 Rescaled range fitting of NI USRP noise phase (Hyp2) Real part.

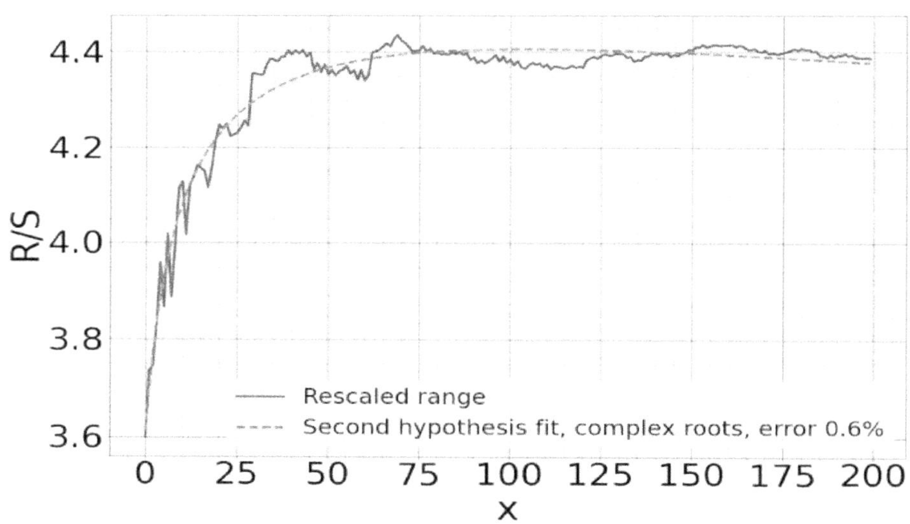

Figure 6.3 Rescaled range fitting of NI USRP noise phase (Hyp2.) Imaginary part.

applied as a reduced procedure to analyze various TLSs. The fitting parameters presented by the fitting functions in (6.58) can be used as quantitative parameters for further classification of the compared TLS. Such approach may be useful for ex-

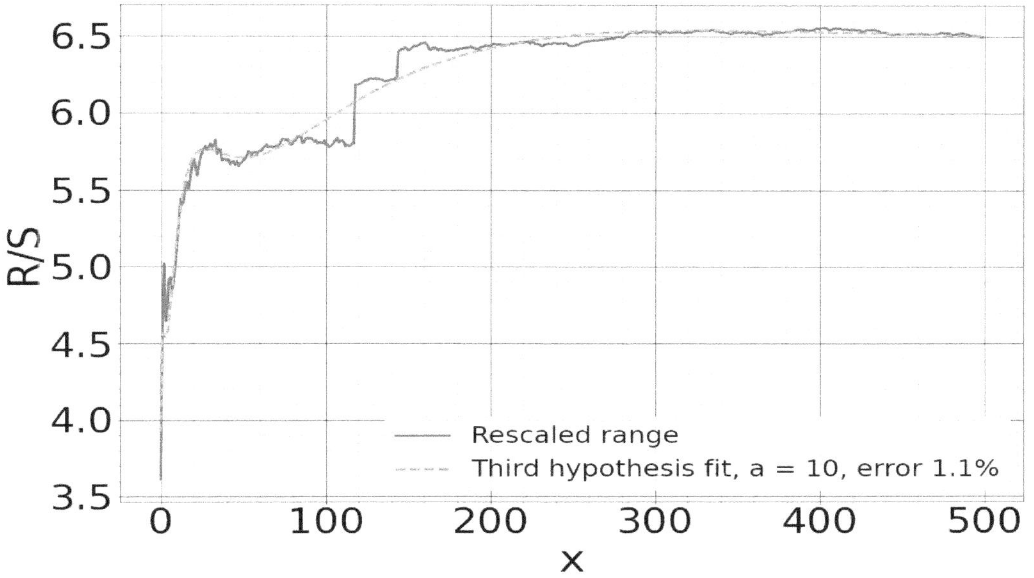

Figure 6.4 Rescaled range fitting of AT2020USB+ differential noise (Hyp 3).

ample in the Radiometric Identification task, when the transmitter devices are to be classified based on the difference of the received and the reference signal. The new set of parameters may also show good results in a task of weak signal detection. The most interesting feature that follows from the real data considered is related to the appearance of complex-conjugated power-law exponents. These values can change essentially the analysis of the long-ranged temporal series and understanding of the conventional Brownian motion discussed in [23]. These new relationships associated with the influence of the complex power-law exponents on the different types of random/complex motions merits further research. The examples shown in this section definitely demonstrate their appearance in real data analysis and their possible corrections should be taken into account. This chapter deliberately is not fully finished. Two parts related to analysis of possible memory in complex systems are needed in further research. We invite our potential readers to continue this research related to derivation/appearance memory in complex systems.

Our ambitious reader probably noticed that the authors of this book invite you to consider the following problem: is it possible to find the systems of equations without memory to the set of the present fractional integrals that mathematicians proposed in their theoretical papers? Does this desired system (of a balance or with other structure) equations exist? If our reader receive any answer (positive or negative) he would be able to select the desired fractional integrals tightly related with a system of equations for practitioners and practical engineers.

MATHEMATICAL APPENDIX

1. The eigen-coordinates for the hypothesis GH1 $y(x)$

It is easy to notice that this function satisfies to the differential equation of the first order

$$xy' = H(y - A) \tag{6.61}$$

As one can notice from (6.61) two parameters H and A enter in (6.61) by linear way. The integration by parts allows to present the basic relationship (6.59) in the form

$$\begin{aligned}
&Y(x) = C_0X_0 + C_1X_1,\\
&Y(x) = x \cdot y - \langle x \cdot y \rangle\\
&X_0(x) = \int_0^x y(u)du - \langle \ldots \rangle, C_0 = H + 1\\
&X_1(x) = x - \langle \ldots \rangle, C_1 = -A \cdot H\\
&\langle M \rangle = \frac{1}{N}\sum_{j=1}^{N} M_j.
\end{aligned} \tag{6.62}$$

After evaluation of the parameters A and H, one calculates easily the parameter B as the slope from (6.58) between functions $y_1(x)$ and x^H.

For the hypothesis GH(2), one can notice that the function $y_2(x)$ satisfies to the Euler's equation.

$$D^2y_2(x) + r_1Dy_2(x) + r_0y_2(x) = K, \quad D = x\frac{d}{dx} \tag{6.63}$$

The double integration of (6.63) allows to reduce this equation to the basic linear relationship of the type

$$\begin{aligned}
&Y(x) = \sum_{s=0}^{3} C_sX_s(x),\\
&Y(x) = y_2(x) - \langle \ldots \rangle,\\
&X_0(x) = \int_0^x (\ln(x) - \ln(u))y_2(u)\frac{du}{u} - \langle \ldots \rangle, C_0 = -r_0,\\
&X_1(x) = \int_0^x y_2(u)\frac{du}{u} - \langle \ldots \rangle, C_1 = -r_1,\\
&X_2(x) = \ln^2(x) - \langle \ldots \rangle, X_3(x) = \ln(x) - \langle \ldots \rangle
\end{aligned} \tag{6.64}$$

The fitting parameters $C_{2,3}$ including the values of the derivatives of the first and the second order in the initial point x_0=0 are not essential for the further calculations and can be omitted. After evaluation of the parameters $r_{0,1}$ by means of the LLSM the desired power-law exponents $H_{1,2}$ can be calculated from the quadratic equation

$$H^2 + r_1H + r_0 = 0. \tag{6.65}$$

After calculation of the roots from (6.65), other three constants $B_{0,1,2}$ are calculated by the LLSM from (6.58). Therefore, in some general case five fitting parameters $H_{1,2}$ and $B_{0,1,2}$ pretend to describe the rather long segments of the chosen TLS and reduced the segment of data to three (hypothesis 1) or 5 (hypothesis 2) fitting parameters only. We should stress also that the power-law exponents evaluated from (6.65) can be also complex-conjugated. By a similar way, one can transform the third hypothesis to the basic linear relationship of a type (6.59). This function satisfied to the Euler's differential equation of the third order

$$D^3y_3(x)+r_2D^2y_3(x)+r_1Dy_3(x)+r_0y_3(x)=K_3, D=x\frac{d}{dx} \tag{6.66}$$

Transforming this differential equation to the corresponding integral equation we obtain the linear relationship of the type

$$\begin{aligned}
&Y(x)=\sum_{s=0}^{6}C_sX_s(x),\\
&Y(x)=y_3(x)-\langle\ldots\rangle,\\
&X_0(x)=\frac{1}{2}\int_0^x(\ln(x)-\ln(u))^2y_3(u)\frac{du}{u}-\langle\ldots\rangle, C_0=-r_0,\\
&X_1(x)=\int_0^x(\ln(x)-\ln(u))y_3(u)\frac{du}{u}-\langle\ldots\rangle, C_1=-r_1,\\
&X_2(x)=\int_0^x y_3(u)\frac{du}{u}-\langle\ldots\rangle, C_2=-r_2,\\
&X_s(x)=\ln^s(x)-\langle\ldots\rangle, s=1,2,3.
\end{aligned} \tag{6.67}$$

After calculation of the coefficients $C_{0,1,2}$ by the LLSM, one can calculate the desired Hurst exponents from the cubic equation

$$H^3+r_2H^2+r_1H+r_0=0 \tag{6.68}$$

The desired constants $A_{0,1,2,3}$ are calculated from the GH(3) by the LLSM, as well. These key formulas provides the desired fit of the calculated ratio $R(ka)/S(ka)$ up to K=[N/a] with a =L.

BIBLIOGRAPHY

[1] D. Baleanu, A. K. Golmankhaneh, and A. K. Golmankhaneh, "The dual action of fractional multi time hamilton equations," *International Journal of Theoretical Physics*, vol. 48, no. 9, pp. 2558–2569, 2009.

[2] A. Carpinteri and F. Mainardi, eds., *Fractals and Fractional Calculus in Continuum Mechanics*, vol. 378 of *CISM International Centre for Mechanical Sciences.* Vienna: Springer, 1997.

[3] R. Hilfer, ed., *Applications of Fractional Calculus in Physics.* Singapore: World Scientific, 2000.

[4] A. A. Kilbas, H. M. Srivastava, and J. J. Trujillo, *Theory and Applications of Fractional Differential Equations*, vol. 204 of *North-Holland Mathematics Studies.* Amsterdam: Elsevier, 2006.

[5] R. L. Magin, *Fractional Calculus in Bioengineering.* Redding, CT: Begell House, 2006.

[6] K. S. Miller and B. Ross, *An Introduction to the Fractional Calculus and Fractional Differential Equations.* New York: John Wiley & Sons, 1993.

[7] K. B. Oldham and J. Spanier, eds., *The Fractional Calculus: Theory and Applications of Differentiation and Integration to Arbitrary Order.* New York: Academic Press, 1974. Edited volume (Mathematics in Science and Engineering, Vol. 111).

[8] I. Podlubny, *Fractional Differential Equations*, vol. 198 of *Mathematics in Science and Engineering.* San Diego, CA: Academic Press, 1998.

[9] J. Sabatier, O. P. Agrawal, and J. A. Tenreiro Machado, eds., *Advances in Fractional Calculus.* Dordrecht: Springer, 2007.

[10] S. G. Samko, A. A. Kilbas, and O. I. Marichev, *Fractional Integrals and Derivatives.* Amsterdam: Gordon and Breach Science Publishers, 1993.

[11] G. M. Zaslavsky, *Hamiltonian Chaos and Fractional Dynamics.* Oxford: Oxford University Press, 2005.

[12] B. J. West, M. Bologna, and P. Grigolini, *Physics of Fractal Operators.* Institute for Nonlinear Science, New York, NY: Springer, 2003.

[13] V. V. Uchaikin, *The Method of Fractional Derivatives.* Art-i-Shock, 2008. Publisher given as Art-i-Shock in the original .bib; DOI not found.

[14] A. Le Méhauté, R. R. Nigmatullin, and L. Nivanen, *Flèches du temps et géométrie fractale.* Hermes Paris, 1998.

[15] R. R. Nigmatullin and A. Le Mehaute, "Is there geometrical/physical meaning of the fractional integral with complex exponent?," *Journal of Non-Crystalline Solids*, vol. 351, no. 33-36, pp. 2888–2899, 2005.

[16] D. Baleanu, A. K. Golmankhaneh, A. K. Golmankhaneh, and R. R. Nigmatullin, "Newtonian law with memory," *Nonlinear Dynamics*, vol. 60, no. 1, pp. 81–86, 2010.

[17] D. Baleanu and R. Agarwal, "Fractional calculus in the sky," *Advances in Difference Equations*, vol. 2021, 02 2021.

[18] G. S. Teodoro, J. T. Machado, and E. C. De Oliveira, "A review of definitions of fractional derivatives and other operators," *Journal of Computational Physics*, vol. 388, pp. 195–208, 2019.

[19] D. Baleanu and A. Fernandez, "On fractional operators and their classifications," *Mathematics*, vol. 7, no. 9, p. 830, 2019.

[20] A. Fernandez, M. A. Özarslan, and D. Baleanu, "On fractional calculus with general analytic kernels," *Applied Mathematics and Computation*, vol. 354, pp. 248–265, 2019. Duplicate of fernandez2019 (kept to preserve original key).

[21] Y. Luchko, "Fractional derivatives and the fundamental theorem of fractional calculus," *Fractional Calculus and Applied Analysis*, vol. 23, no. 4, pp. 939–966, 2020.

[22] R. R. Nigmatullin, W. Zhang, and I. Gubaidullin, "Accurate relationships between fractals and fractional integrals: new approaches and evaluations," *Fractional Calculus and Applied Analysis*, vol. 20, no. 5, pp. 1263–1280, 2017.

[23] J. Feder, *Fractals.* Physics of Solids and Liquids, New York, NY: Springer, 2013.

[24] S. Reid, G. E. Bell, and G. Edgemon, "The use of skewness, kurtosis and neural networks for determining corrosion mechanism from electrochemical noise data," tech. rep., NACE International, Houston, TX (United States), 1998.

[25] E. Y. Budnikov, A. Maksimychev, A. Kolyubin, and S. Timashev, "Space and time correlations in a dissipative structure emerging in an electrochemical system with a cation-exchange membrane," *Russian journal of electrochemistry*, vol. 37, no. 1, pp. 80–87, 2001.

[26] S. Martemianov, N. Adiutantov, Y. K. Evdokimov, L. Madier, F. Maillard, and A. Thomas, "New methodology of electrochemical noise analysis and applications for commercial li-ion batteries," *Journal of Solid State Electrochemistry*, vol. 19, no. 9, pp. 2803–2810, 2015.

[27] E. Astafev, "Electrochemical noise measurement methodologies of chemical power sources," *Instrumentation Science & Technology*, vol. 47, no. 3, pp. 233–247, 2019.

[28] E. Astafev, "Electrochemical noise of a li-ion battery: measurement and spectral analysis," *Journal of Solid State Electrochemistry*, vol. 23, no. 4, pp. 1145–1153, 2019.

[29] S. F. Timashev, *Flicker-Noise Spectroscopy.* Moscow: Fizmatlit, 2007. In Russian; bibliographic info in the original .bib indicated a book.

[30] S. F. Timashev and Y. S. Polyakov, "Analysis of discrete signals with stochastic components using flicker noise spectroscopy," *International Journal of Bifurcation and Chaos*, vol. 18, no. 09, pp. 2793–2797, 2008.

[31] R. Yulmetyev, P. Hänggi, and F. Gafarov, "Stochastic dynamics of time correlation in complex systems with discrete time," *Physical Review E*, vol. 62, no. 5, p. 6178, 2000.

[32] R. Yulmetyev, P. Hänggi, and F. Gafarov, "Quantification of heart rate variability by discrete nonstationary non-Markov stochastic processes," *Physical Review E*, vol. 65, no. 4, p. 046107, 2002.

[33] A. V. Mokshin, R. M. Yulmetyev, and P. Hänggi, "Simple measure of memory for dynamical processes described by a generalized Langevin equation," *Physical review letters*, vol. 95, no. 20, p. 200601, 2005.

[34] H. Sheng, Y. Chen, and T. Qiu, *Fractional Processes and Fractional-Order Signal Processing.* Signals and Communication Technology, London: Springer, 2012.

[35] R. R. Nigmatullin, R. A. Giniatullin, and A. I. Skorinkin, "Membrane current series monitoring: essential reduction of data points to finite number of stable parameters," *Frontiers in computational neuroscience*, vol. 8, p. 120, 2014.

[36] C.-K. Peng, S. Havlin, H. E. Stanley, and A. L. Goldberger, "Quantification of scaling exponents and crossover phenomena in nonstationary heartbeat time series," *Chaos: an interdisciplinary journal of nonlinear science*, vol. 5, no. 1, pp. 82–87, 1995.

[37] R. R. Nigmatullin, "Recognition of nonextensive statistical distributions by the eigencoordinates method," *Physica A: Statistical Mechanics and its Applications*, vol. 285, no. 3-4, pp. 547–565, 2000.

[38] R. R. Nigmatullin, P. Lino, and G. Maione, *New Digital Signal Processing Methods.* Cham: Springer, 2020.

CHAPTER 7

Conclusion and Future Opportunities

This chapter synthesizes the core message of *Reading Randomness (I)*: randomness is not mere noise but a carrier of information that can be read using a rigorous, measurement-aware workflow. We consolidate the roles of fractal geometry, fractional operators, sparse amplitude–frequency decompositions, and generalized correlation measures, then articulate open problems and roadmaps for the next decade of research and applications. The outlook spans algorithms, theory, validation and standards, datasets and software, and translational opportunities in engineering, geophysics, biomedicine, materials, and finance.

7.1 WHAT WE HAVE LEARNED

Reading vs. explaining. We distinguished "reading"—the disciplined extraction of robust, comparable *signatures*—from mechanistic explanation. Readings include fractional memory kernels, Prony-like sparse spectra, generalized correlations, and discrete geometrical invariants (DGIs). Explanations should be tested against such readings, not the reverse.

The three pillars. (i) *Fractals → fractional operators*: averaging over fractal-like sets induces nonlocal kernels and fractional derivatives/integrals [1, 2, 3, 4]. (ii) *Fluctuation metrology and sparse spectra*: pre-smoothing and non-orthogonal expansions (NOCFASS/NAFASS) reveal stable atoms for multi-periodic and long-memory data [5, 6, 7, 8]. (iii) *Generalized correlations*: internal vs. external correlations can be separated and estimated robustly, complementing spectral views [9].

A pragmatic workflow. Instrument-aware curation → exploratory reading → baseline integer-order models (and stress-tests) → fractional kernels and sparse atoms → validation and uncertainty (surrogate tests, cross-validation, identifiability checks).

DOI: 10.1201/9781003794691-7

7.2 SYNTHESIS OF METHODS

7.2.1 Non-orthogonal amplitude–frequency analysis

Let $x(t)$ be a pre-smoothed signal. NAFASS represents

$$x(t) \approx \sum_{k=1}^{K} a_k \cos(2\pi f_k t + \phi_k), \quad (f_k \text{ non-integer spaced}), \tag{7.1}$$

with band selection and recursion for $\{f_k\}$ governed by a local generalized geometric mean (GGGM) law [7]. The non-orthogonal atoms allow stable extraction of quasi-periodic components even under long memory.

7.2.2 Discrete geometrical invariants (DGI)

For a trendless sequence $\{y_n\}$, DGI builds a compact 3D surface S whose shape encodes invariants; classification proceeds by geometric proximity on S. DGIs can separate intrinsic randomness from apparatus-induced artifacts and compare sources as disparate as irrational-number sequences and earthquake records [9].

7.2.3 Generalized correlations

Given two records x, y, we consider correlation decompositions

$$\rho_{xy}(\tau) = \rho_{xy}^{\text{int}}(\tau) + \rho_{xy}^{\text{ext}}(\tau), \tag{7.2}$$

where ρ^{ext} accounts for the common measurement/actuation pathway. Robust estimators rely on blocked surrogates, odd/even splitting, and multi-resolution averages; they provide reliable complements to spectra when quasi-periodic atoms are weak.

7.2.4 Fractional operators and memory kernels

We used Caputo-type operators

$${}_0^C D_t^{\alpha} x(t) = \frac{1}{\Gamma(1-\alpha)} \int_0^t \frac{\dot{x}(\tau)}{(t-\tau)^{\alpha}}\, d\tau, \qquad 0 < \alpha < 1, \tag{7.3}$$

and generalized Prabhakar kernels with Mittag–Leffler functions for rich hereditary effects [3, 10, 11]. Variable-order and distributed-order variants capture evolving or heterogeneous memory.

7.3 CROSS-DOMAIN LESSONS

7.3.1 Electrochemistry and materials

Anomalous relaxations and broad spectra are naturally modeled by fractional kernels; sparse atoms isolate periodic drive or switching artifacts. Diagnostic pipelines benefit from NOCFASS + Prabhakar fits, with uncertainty reported via bootstrap surrogates.

7.3.2 Geophysics

Seismic precursors present mixed signatures (quasi-periodic bursts on long-memory background). DGIs offer compact descriptors; generalized correlations quantify coupling across stations while removing apparatus effects.

7.3.3 Biosignals

Heart-rate variability, EEG, and gait exhibit long-range dependence. Fractional operators encode memory; Prony-like atoms detect rhythms. Careful curation avoids confounding respiration/motion artifacts; clinical translation hinges on uncertainty and interpretability [12].

7.3.4 Infrastructure and finance

Structural health monitoring and financial ticks share nonstationarity and heavy tails. Fractional differencing baselines (ARFIMA) remain useful [13], but fractional kernels + sparse atoms frequently improve reading stability, especially under irregular sampling.

7.4 OPEN PROBLEMS AND FUTURE WORKS

7.4.1 Theory

1. **Kernel discovery.** Prove identifiability and consistency for data-driven discovery of fractional kernels (variable/distributed order); study equivalence classes under reparameterization.

2. **Sparse atoms with memory.** Establish recovery guarantees for non-orthogonal expansions under long-memory noise; characterize bias under spectral leakage.

3. **Correlation decompositions.** Formalize internal/external splits; build invariance results under sampling and instrument cascades.

4. **Bridging ARFIMA and fractional operators.** Map ARFIMA d parameters to operator kernels; unify inference under continuous/discrete-time limits [13].

7.4.2 Algorithms

1. **SINDy-F / Operator inference.** Extend sparse identification to include fractional operators and Prabhakar kernels; develop convex/greedy solvers with stability certificates.

2. **Band selection and NIHS.** Automate band selection in NOCFASS/NAFASS; study the generalized geometric-mean recursion for frequency dispersion; quantify identifiability of NIHS coefficients [7].

3. **Streaming/real-time reading.** Design sublinear-memory algorithms for on-device reading; develop drift detectors for kernel change-points.

4. **UQ at scale.** Fast bootstrap/surrogate engines exploiting FFT and low-rank updates; Laplace/VI approximations for kernel posteriors.

7.5 FINAL REMARKS

Reading randomness is a stance: treat fluctuations as structured carriers of information, readable via carefully engineered pipelines. We have argued for three pillars, a pragmatic workflow, and a cross-domain practice grounded in fractional operators, sparse atoms, and generalized correlations. The future work outlined here envisions a community effort—theory, algorithms, standards, and translation—to make readings stable, comparable, and trustworthy at scale.

BIBLIOGRAPHY

[1] I. Podlubny, *Fractional Differential Equations.* Academic Press, 1999.

[2] S. G. Samko, A. A. Kilbas, and O. I. Marichev, *Fractional Integrals and Derivatives: Theory and Applications.* Gordon and Breach, 1993.

[3] F. Mainardi, *Fractional Calculus and Waves in Linear Viscoelasticity.* Imperial College Press, 2010.

[4] B. J. West, M. Bologna, and P. Grigolini, *Physics of Fractal Operators.* Institute for Nonlinear Science, New York, NY: Springer, 2003.

[5] R. R. Nigmatullin and I. A. Gubaidullin, "Nafass: Fluctuation spectroscopy and the prony spectrum for description of multi-frequency signals in complex systems," *Communications in Nonlinear Science and Numerical Simulation,* vol. 56, pp. 252–269, 2018.

[6] R. R. Nigmatullin, V. S. Alexandrov, P. Agarwal, S. Jain, and N. Ozdemir, "Description of multi-periodic signals generated by complex systems: Nocfass—new possibilities of the fourier analysis," *Numerical Algebra, Control and Optimization,* 2022.

[7] R. R. Nigmatullin, A. A. Khamzin, and Y. Chen, "Approximation of aperiodic signals using non-integer harmonic series: The generalized Nafass approach," *Fractal and Fractional,* vol. 7, no. 7, p. 498, 2023.

[8] J. A. T. Machado and A. M. Lopes, "Approximation of data using non-integer harmonics series," *Nonlinear Dynamics,* vol. 89, no. 4, pp. 2845–2854, 2017.

[9] R. R. Nigmatullin, J. T. Machado, and R. Menezes, "Self-similarity principle: The reduced description of randomness," *Central European Journal of Physics,* vol. 11, pp. 724–739, 2013.

[10] T. R. Prabhakar, "A singular integral equation with a generalized Mittag-Leffler function in the kernel," *Yokohama Mathematical Journal*, vol. 19, pp. 7–15, 1971.

[11] A. A. Kilbas, H. M. Srivastava, and J. J. Trujillo, *Theory and Applications of Fractional Differential Equations.* Elsevier, 2006.

[12] R. L. Magin, *Fractional Calculus in Bioengineering.* Redding, CT: Begell House, 2006. Source file listed year=2010; Begell House hardcover edition commonly cited as 2006.

[13] J. R. M. Hosking, "Fractional differencing," *Biometrika*, vol. 68, no. 1, pp. 165–176, 1981.

Index

www.ingramcontent.com/pod-product-compliance
Ingram Content Group UK Ltd.
Pitfield, Milton Keynes, MK11 3LW, UK
UKHW052052230926
475702UK00025B/1132
* 9 7 8 1 0 4 1 3 4 8 1 4 6 *